Inverse Problems in Electric Circuits and Electromagnetics

Inverse Problems in Electric Circuits and Electromagnetics

V.L. Chechurin
Saint Petersburg State Polytechnical University
St. Petersburg, Russia

N.V. Korovkin
Saint Petersburg State Polytechnical University
St. Petersburg, Russia

M. Hayakawa
The University of Electro-Communications
Chofu City, Tokyo

N.V. Korovkin
Saint Petersburg State Polytechnical University
St. Petersburg, Russia

V.L. Chechurin
Saint Petersburg State Polytechnical University
St. Petersburg, Russia

M. Hayakawa
The University of Electro-Communications
Chofu City, Tokyo

Inverse Problems in Electric Circuits and Electromagnetics

e-ISBN 0-387-46047-0

ISBN 978-1-4419-4139-8 **e-ISBN 978-0-387-46047-5**

Printed on acid-free paper.

9 8 7 6 5 4 3 2 1

springer.com

Contents

Preface..ix

Chapter 1 Inverse Problems in Electrical Circuits and Electromagnetic Field Theory 1

1.1 Features of inverse problems in electrical engineering........................1
1.1.1 Properties of inverse problems..................................6
1.1.2 Solution methods..12
1.2 Inverse problems in electric circuits theory....................................18
1.2.1 Formulation of synthesis problems............................18
1.2.2 The problem of constructing macromodels (macromodeling) of devices..................................26
1.2.3 Identifying electrical circuit parameters.....................29
1.3 Inverse problems in electromagnetic field theory...........................33
1.3.1 Synthesis problems..35
1.3.2 Identification problems..43

References...45

Chapter 2 The Methods of Optimization of Problems and Their Solution 47

2.1 Multicriterion inverse problems...47
2.2 Search of local minima...59
2.3 Search of objective functional minimum in the presence of constraints..68
2.4 Application of neural networks...85
2.5Application of Volterra polynomials for macromodeling..................98
2.6 Search of global minima..105

2.6.1 The multistart method and cluster algorithm.............107
2.6.2 "Soft" methods...109

References 119

Chapter 3 The Methods of Solution of Stiff Inverse Problems 121

3.1 Stiff inverse problems...121
3.2 The principle of quasistationarity of derivatives and integrals..136
3.3 Using linear relationships for solving stiff inverse probems..151
3.4 The problems of diagnostics and the identification of inverse problems in circuit theory..............................156
3.4.1 Methods of identification of linear circuits....................159
3.4.2 Error of identification problem solution.......................161
3.5 The method of stiff diagnostics and identification problems solutions..168
3.5.1 Application of the principle of repeated measurements for solution of electric circuits' identification problem.............................168
3.5.2 Definition of linear connections between parameters of circuit mathematical models...............169
3.5.3 Algorithm and results of electric circuits' identification problem solution using repeated measurements...172
3.6 Inverse problems of localization of disturbance sources in electrical circuits by measurement of voltages in the circuit's nodes...181

References..191

Chapter 4 Solving Inverse Electromagnetic Problems by the Lagrange Method 193

4.1 Reduction of an optimization problem in a stationary field to boundary-value problems..193
4.2 Calculation of adjoint variable sources..................................202
4.3 Optimization of the shape and structure of bodies in various classes of media...213
4.4 Properties and numerical examples of the Lagrange method............220
4.4.1 Focusing of magnetic flux..................................221
4.4.2 Redistribution of magnetic flux 223

4.4.3 The extremum of electromagnetic force....229
4.4.4 Identification of substance distribution....231
4.4.5 Creation of a homogeneous magnetic field....234
4.5 Features of numerical optimization by the Lagrange method....239
4.6 Optimizing the medium and source distribution in non-stationary electromagnetic fields....242

References....249

Chapter 5 Solving Practical Inverse Problems 251

5.1 Search for lumped parameters of equivalent circuits in transmission lines....251
5.2 Optimization of forming lines....262
5.3 The problems of synthesis of equivalent electric parameters in the frequency domain....275
5.4 Optimization of current distribution over the conductors of 3-phase cables....285
5.5 Search of the shape of a deflecting magnet polar tip for producing homogeneous magnetic field....296
5.6 Search of the shape of magnetic quadrupole lens polar tip for accelerating a particle....301
5.7 Optimum distribution of specific electric resistance of a conductor in a magnetic field pulse....306

References....315

Appendices

Appendix A A Method of Reduction of an Eddy Magnetic Field to a Potential One....317
Appendix B The Variation of a Functional....323

Index....325

Preface

The design and development of electrical devices involves choosing from many possible variants that which is the best or optimum according to one or several criteria. These optimization criteria are usually already clear to the designer at the statement of the design problem. The methods of optimization considered in this book, allow us to sort out variants of the realization of a design on the basis of these criteria and to create the best device in the sense of the set criteria.

Optimization of devices is one of the major problems in electrical engineering that is related to an extensive class of inverse problems including synthesis, diagnostics, fault detection, identification, and some others with common mathematical properties. When designing a device, the engineer actually solves inverse problems by defining the device structure and its parameters, and then proceeds to deal with the technical specifications followed by the incorporation of his own notions of the best device. Frequently the solutions obtained are based on intuition and previous experience. New methods and approaches discussed in this book will add mathematical rigor to these intuitive notions.

By virtue of their urgency inverse problems have been investigated for more than a century. However, general methods for their solution have been developed only recently. An analysis of the scientific literature indicates a steadily growing interest among scientists and engineers in these problems. As a result, there has been an increase in the number of publications of new methods of solution of inverse problems as well as their active application in practice. It is essential that methods of solution of inverse problems find application not only for the development of new devices, but also for the modernization of existing equipment with the purpose of improving its characteristics or the extension of its operational life.

Inverse problems that are significant for practical purposes are, as a rule, solved numerically. The increase in efficiency of computers has allowed us to put into practice many new and effective methods of solution of inverse problems. Therefore we have focused the book on an account of methods oriented towards numerical solutions. We have not included analytical methods because they are not so effective for optimization of designs of real devices or physical properties of materials. Furthermore, an exposition of analytical methods would substantially expand the volume of this book and would result in excessive complication.

Inverse problems in the theory of electric circuits and in electromagnetic field theory have some particular features. We, however, notice numerous common features of these problems that allow their exposition to be combined within the limits of one book. In particular, the solution of inverse problems in the theory of circuits and in field theory is based on the same mathematical apparatus, namely, methods of solution of incorrect problems and methods of optimization. This material is presented throughout the book from a general point of view.

Together with well-known methods of solution of inverse problems that have been widely used in practice, we have described some methods based on ideas borrowed from various areas of science. Generic methods of minimization of functionals, or methods based on the application of neural networks can be treated among them. The method of Lagrange multipliers explicitly considered in the book was found to be very effective for the solution of optimization problems in electromagnetic field theory.

In the first chapter a classification of inverse problems is given with an analysis of their properties, and we describe the basic methods for the numerical solution of inverse problems in electrical engineering. In the second chapter methods of searching for local and global minima of functionals are discussed, as well as methods of searching for the minima of functionals in the presence of constraints on the parameters to be optimized. In the third chapter we discuss methods of solution of inverse problems in the theory of electric circuits. Special attention is given to the solution of stiff inverse problems, in particular problems of identification that are characterized by the presence of measurement errors. The fourth chapter is concerned with a systematic account of the Lagrange Method of continuous multipliers, which is applied to optimization in an electromagnetic field characterized by quite a large number of parameters. The fifth chapter presents examples that demonstrate the solution of practical inverse problems in the theory of electric circuits and in electromagnetic field theory, thus illustrating the effectiveness of methods considered in this book.

Alongside the classical methods of solution of inverse problems in electrical engineering, up-to-date methods have also been investigated by the authors themselves within the limits of their scientific activities, as well as by their colleagues: A.Adalev, A.Potienko, T.Minevich, A.Plaks, M.Eidmiller and others. The authors have also used the results of scientific studies of research workers of the faculty of Fundaments of Theoretical Electrical Engineering in the State Polytechnic University, St.-Petersburg (Russia) concerned with the solution of problems involving the analysis of electric circuits and electromagnetic fields. These include using the method of the scalar magnetic potential for the analysis of direct current magnetic fields (K.S.Demirchian). The authors are very grateful to professor K.S.Demirchian and professor P.A.Butyrin for useful discussions and valuable advice that helped improve the book and the understanding of distinctive features of the solution of inverse problems in electrical engineering.

The authors express their appreciation to Mr. R.Hogg, Mr. D.Bailey, and Mr. M.Repetto for helpful discussion of optimization problems in electromagnetics, as well as to Mr. Kh.Partamyan and Mr. B.DeCarlo for their help during the writing of the book.

The authors thank professors I.G.Chernorutski and E.B.Soloviova, whose scientific ideas have helped us with the preparation of this work.

The authors believe that this book will be useful for engineers, scientific researchers, postgraduate students and students major in electrotechnology, electrical power systems, and other specialties.

It is hoped that this book will promote further interest in inverse problems in electrical engineering among university students, lecturers and research workers.

Chapter 1. Inverse Problems in Electrical Circuits and Electromagnetic Field Theory

1.1 Features of inverse problems in electrical engineering

When analyzing electric devices, their parameters are assumed to be known. To analyze an electric circuit, its topology, element parameters, and characteristics of sources must be specified. Circuit currents and voltages, as well as other derived quantities such as real or reactive power, can then be determined. Similarly, the frequency-dependent or transient characteristics of the circuit can be determined.

In the analysis of electromagnetic fields, the sources (e.g. currents and charges), their distribution in space and their time dependences are typically specified. Furthermore, such functions as $\mu(x,y,z)$ (magnetic permeability), $\varepsilon(x,y,z)$ (dielectric permittivity) and $\sigma(x,y,z)$ (electric conductivity), which describe the spatial dependence of the properties of ferromagnetic, dielectric and conducting mediums, must be specified. Based upon the system geometry, the sources and material properties, the differential or integral characteristics of the field (such as field intensity, flux density) and integral properties (such as capacitance, inductance) as well as forces, moments of forces and pressures can be determined.

Electrical engineering problems where the system structure and the properties of its constituent elements must be determined in order to achieve a specified performance are referred to as inverse problems. For electric circuits, the solution of the inverse problem is the topology of the circuit, values of circuit elements parameters, and time-dependent variations of sources required to achieve a specified circuit behavior [1]. In the case of electromagnetic fields, an inverse problem would be to determine the spatial distribution of sources and materials to achieve a specified field distribution [2].

In inverse problems, the parameters of devices are determined from the conditions (criteria) contained in the formulation of the problem. These criteria express the desired optimal behavior and/or optimal characteristics of the system. Some examples of such criteria are listed below:

- the reactive power consumed by an electric circuit should have a preset value;

- the voltage between the nodes of an electric circuit during a transient should be less than a preset value;
- the distribution of the magnetic field strength within a given spatial region should be as close as possible to uniform; and
- the geometry of a conductor cross-section should produce minimum losses in the conductor.

The variables of an inverse problem, i.e. variables which describe the characteristics of the device for which the problem is specified, are interrelated according to the following operator equation:

$$\mathbf{A}(\mathbf{p})\mathbf{w} = \mathbf{v}, \tag{1.1}$$

where $\mathbf{A}$ is a matrix, a differential operator or any nonlinear operator, $\mathbf{p} = (p_1, p_2, \cdots, p_n)^T$ is an unknown vector subject to definition (e.g. the vector of the device optimized parameters or a vector $\mathbf{v}$ of sources), $\mathbf{w} = (w_1, w_2, \cdots, w_M)^T$ is the vector of variables, and $\mathbf{v} = (v_1, v_2, \cdots, v_N)^T$ is the vector of sources.

Equation (1.1) provides a means to calculate the variables $\mathbf{w}$ of the inverse problem for any $\mathbf{p}$. In most cases, a transformation from the variables $\mathbf{w}$ to some other variables $\mathbf{y}$ is necessary to simplify expressions for the criteria of the inverse problem. The vector $\mathbf{y}$ shall further be referred to as the vector of characteristics. The transformation to the vector of characteristics $\mathbf{y}$ can be presented in the following form:

$$\mathbf{y}(\mathbf{w},\mathbf{p}) = \mathbf{f}(\mathbf{w},\mathbf{p}),$$

where $\mathbf{f}$ is frequently a nonlinear vector function.

When formulating and solving inverse problems the concept of a functional may be used. A functional I is defined as a scalar function determined on some set of functions. The terms "functional" and "objective function" will be used interchangeably in this discussion [3].

As an example, a functional might be the definite integral of the function $\sum_k \left|\tilde{y}_k - y_k(\mathbf{p},\mathbf{w})\right|$ or its maximum value in a certain interval. In electrical engineering a typical functional specifies the proximity of the required variable $\tilde{\mathbf{y}}$ to the function $\mathbf{y}$:

$$I(\mathbf{p},\mathbf{w}) = \left\|\tilde{\mathbf{y}} - \mathbf{y}(\mathbf{p},\mathbf{w})\right\|.$$

Any of the following generally used norms may be applied to characterize the proximity of functions $\tilde{\mathbf{y}}$ and $\mathbf{y}$ [3]:

$$\|\cdot\|_2 \equiv \left(\frac{1}{T-t_0} \int_{t_0}^{T} h^2(t)dt \right)^{1/2}, \quad \|\cdot\|_q \equiv \left(\frac{1}{T-t_0} \int_{t_0}^{T} h^q(t)dt \right)^{1/q} \text{ or } \|\cdot\|_\infty \equiv \max_{t \in [t_0, T]} |h(t)|,$$

where $h(t)$ is an arbitrary function and $q = 2,4,\cdots$. Note that when using the norm $\|\cdot\|_\infty$, the functional $I(\mathbf{p},\mathbf{w})$ is non-differentiable since the function "max" is generally not continuous.

Let us assume that a vector $\tilde{\mathbf{y}}$ of an inverse problem defines the optimal characteristics of a device. Then the criterion of this inverse problem can be expressed as:

$$I(\mathbf{p},\mathbf{w}) \xrightarrow[\mathbf{p}]{} \min .$$

Let's consider examples. In the theory of linear DC circuits, the vectors of currents **I** and voltages **U** in circuit branches correspond to the vector of variables **w,** whereas the vectors of current sources **J** and voltage sources **V** in circuit branches correspond to the vector of sources **v**. Kirchhoff's equations and Ohm's law $\mathbf{DI} = -\mathbf{DJ}, \mathbf{CU} = \mathbf{CV}, \mathbf{RI} = \mathbf{U}$, can be written as:

$$\underbrace{\begin{bmatrix} \mathbf{D} & \mathbf{0} \\ \mathbf{0} & \mathbf{C} \\ \mathbf{R} & -\mathbf{1} \end{bmatrix}}_{\mathbf{A}} \cdot \underbrace{\begin{bmatrix} \mathbf{I} \\ \mathbf{U} \end{bmatrix}}_{\mathbf{w}} = \underbrace{\begin{bmatrix} -\mathbf{D} & \mathbf{0} \\ \mathbf{0} & \mathbf{C} \\ \mathbf{0} & \mathbf{0} \end{bmatrix} \cdot \begin{bmatrix} \mathbf{J} \\ \mathbf{V} \end{bmatrix}}_{\mathbf{v}},$$

where **C** and **D** are the matrixes of contours and cutsets of the circuit graph and **R** is a diagonal matrix of the circuit branches resistances. Let objective variables be the resistance of the first and second circuit branches; then $\mathbf{p}=(R_1, R_2)^T$. If the inverse problem involves finding of such a vector **p**, which makes the voltage transfer factor $K_U^{(k,i)}(\mathbf{p}) = U_k / U_i$ closest to a prescribed value $\tilde{\mathbf{y}} = \tilde{K}_U^{(k,i)}$, then the vector of characteristics of the inverse problem contains only one element $\mathbf{y}(\mathbf{p}) = K_U^{(k,i)}(\mathbf{p})$, i.e. it is a scalar. The criterion of the inverse problem can be written down, for example, as:

$$\left| \tilde{K}_U^{(k,i)} - K_U^{(k,i)}(\mathbf{p}) \right| \xrightarrow[\mathbf{p}]{} \min .$$

Let's consider an electrostatics problem. The vector **w** is formed by the potential φ of the electrostatic field created by electric charges in nodes of the grid. Then the vector **v** is formed by charges with density ρ located in the grid nodes. These quantities are connected by the equation:

$$\operatorname{div}(\varepsilon \operatorname{grad} \varphi) = -\rho ,$$

where $\varepsilon(x,y,z)$ characterizes the medium distribution. The operator $\mathbf{A}$ in this problem is represented by $\operatorname{div}(\varepsilon \operatorname{grad})$. Let the problem be finding of an $\varepsilon(x,y,z)$ value, which makes the electric field intensity vector flux, $y(\mathbf{p}) = \Psi_E = \int_S(-\operatorname{grad}\varphi)ds$ through a surface S, reach its maximum. Hence, the vector $\mathbf{p}$ is formed by ε values in each mesh of the grid. Then the criterion of this inverse problem can be expressed as:

$$-|y(\mathbf{p})| \xrightarrow[\mathbf{p}]{} \min .$$

In the circuit theory, and in the electromagnetic field theory, inverse problems of various types have to be solved [1,2]. Their definitions may differ essentially. Problems of each type have features that should be taken into account in their solving.

Let's examine in more detail different types of inverse problems that can be solved in electrotechnics. These problems can be divided into groups of synthesis and identification problems. Furthermore, the synthesis problems involve problems of structural and parametrical synthesis. Problems of diagnostics, macromodeling and defectoscopy are related to the group of identification problems.

Synthesis Problems

An electric circuit synthesis problem, i.e. creation of an electric circuit with given properties, includes two stages: searching of the circuit structure and searching of parameter values of its elements (the vector $\mathbf{p}$). A known problem of such type is the synthesis of electric filters when the structure and parameters of the filter are determined according to a given ideal gain-frequency characteristic $K(\omega)$. Since construction of a filter with ideal characteristic $K(\omega)$ is not possible, search for electric circuits that have characteristics close to the ideal by one or several criteria of proximity is necessary.

In the electromagnetic field theory, the synthesis problem involves finding a source characterized by the value $\mathbf{v}$, which creates a field closest to the sought field. For example, the required field should be uniform within some area or generally have a prescribed distribution in space and/or in time along a line, on a surface, or in a volume. In this case, the required currents or charges form the vector $\mathbf{p}$. Similarly, the search problems of media distributions generally determine structures and body shapes, which make the field or some of its integrated characteristics least of all differ from the prescribed quantities. In that case, the vector $\mathbf{p}$ is formed by the sought distribution of the media characteristics.

In electromagnetic field synthesis problems, sometimes the forms of domain and subdomain boundaries can be sought. Such problems are referred to as mobile boundary problems. Finding of a body form for a prescribed structure can also be considered as a problem with an unknown, or as a mobile boundary problem.

In many problems of electric circuits or electromagnetic field synthesis that can be solved in practice, usually a prototype of the required circuit or the electrical device is known. In the problem of improvement of prototype properties, the search of its optimal (best) parameters is referred to as a problem of optimization or optimal design.

The same problem can be considered either as a problem of optimization, or as a synthesis problem. Usually the term "synthesis" is used in electric circuits theory and the term "optimization" is more often applied in the electromagnetic field theory.

Identification Problems

An identification problem is in finding of mathematical models of electric devices. They are usually built on the basis of a known data set, connecting the input and output characteristics of the object. In most cases, the set of such data is limited, so the resulting mathematical model describes the device with an error.

Problems of diagnostics of electric circuits are related to the identification problems. In this case, the problem is in finding the kind of defect and its location in the circuit on the basis of the measured currents and voltages, as well as in finding of the deviations of parameters of circuit elements from their prescribed values.

In macromodeling problems, it is necessary to construct mathematical models of the electric devices on the basis of available sets of input and output signals of the devices. In macromodeling the device is considered as a multiterminal network. The macromodel handles only "external" variables, whereas the behavior of "internal" variables is not reproduced.

Solving an identification problem in the electromagnetic field theory requires distinguishing the unknown distribution of sources or media, on the basis of data of field measurements at some surface. In such problems, one must suggest about properties of media and the sources located in areas inaccessible to measurements, on the basis of field characteristics obtained by measurements on an accessible surface.

Such problems should be solved, for example, in geological prospecting when the distribution of electric potential measured on the ground surface allows identifying the media structure under the ground surface. In crack detection problems (nondestructive testing), the cracks in the metal are identified by the measured distribution of eddy currents and induced voltages.

1.1.1. Properties of inverse problems

Despite the wide variety of inverse problems in the electric theory, they can be characterized by some general properties, which we shall discuss further. Some of these properties are close to the properties of analysis problems. At the same time, inverse problems possess a number of inherent features [4,5,6].

Restrictions

One of the most important properties discussed below is related to the availability of conditions imposed on the problem's desired solution. For inverse problems, the presence of restrictions of technical and geometrical character is typical. The former are defined by the allowable values of parameters and characteristics of circuit elements. For example, resistance, inductance, and capacity of elements can be within certain limits. Parameters of real electric circuit elements cannot be lower or higher than some values. The loss power W in resistors and field strength E on surfaces of conducting bodies should not exceed allowable values. The dielectric permittivity of substances, as well as their other physical properties, lay within certain limits. It cannot be less than absolute dielectric permittivity of vacuum and practically cannot exceed a certain value. Due to restrictions of geometrical character, the sizes of bodies cannot go beyond set limits.

Non-uniqueness

Inverse problems can have non-unique solutions. For example, a voltage gain coefficient $K(s)=1/(Ts+1)$ can belong to electric circuits with series connected r, L, as well as with series connected r, C (here s is a Laplace operator and T is a constant).

As an illustration of properties of the inverse problems, let's consider the problem of finding a spatial current distribution which provides the desired magnetic field on the surface S of ideal ($\mu=\infty,\sigma=0$) ferromagnetic half-space (Fig. 1.1).

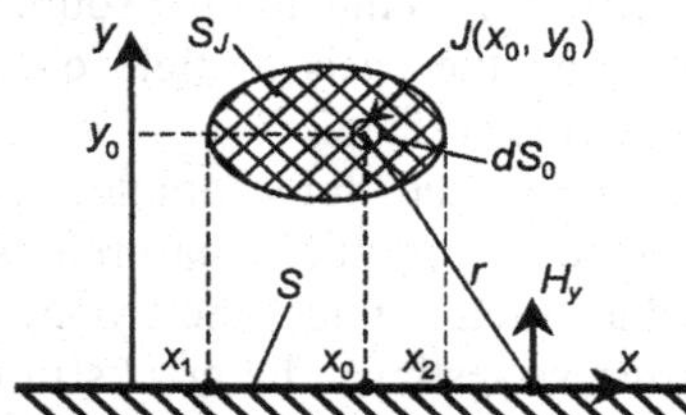

Fig. 1.1. Search of current distribution in the area S_J which provides the strength $H_y(x)$ of the magnetic field on the line y=0

It is required to allocate the electric current within the area S_J so that the normal to the line $y=0$ component H_y of the plane-parallel magnetic field strength has to vary under the law $H_w(x)$.

This problem has a non-unique decision. Let, for example, the desired field $H_w(x)$ be the field created by a linear current located at the point with coordinates x_0, y_0 inside the area S_J. Then electric currents regularly distributed on any conductor of circular section, with the center located at the point with coordinates x_0, y_0 inside the area S_J, can create on the line $y=0$ exactly the same field as the said linear current.

Let V_J be the allowable area of distribution of a current with density $J(x,y,z)$. We can prove that the problem of synthesis of a magnetic field in the area V_i, limited by the surface S_i, has a non-unique decision. Let's suppose that the areas V_J and V_i do not intersect and the current distributed within the area V_J, which creates the desired field in the area V_i, is known. It can be added up with such a current J_0 distributed in V_J, which does not create any fields in the area V_i. Then the sum of currents with densities J and J_0 is also a solution. Hence, there is an infinite number of ways of assigning sources which are not creating any fields in the area V_i. At any distribution of current J_0 within the area V_J, it is always possible to zero the tangent component of the magnetic field strength H_τ created by this current on the surface S_i, by placing a layer of current on a surface enclosing V_J.

Stability

Alongside with non-uniqueness, the inverse problems in electrical engineering are characterized by poor stability of solutions, just as in many analysis problems. Here, the stability is understood in the sense that for any solution of equation type $\mathbf{A(p)w} = \mathbf{v}$ there is a constant K, so that $\|\mathbf{p}\| \le K\|\mathbf{v}\|$.

Thus, if for the stable equation $\mathbf{A(p)w} = \mathbf{v}$ some approximate solution $\hat{\mathbf{p}}$ of the equation $\mathbf{A}(\hat{\mathbf{p}})\mathbf{w} = \hat{\mathbf{v}}$ is found (owing, for example, to the errors of the numerical solution), then $\|\mathbf{p} - \hat{\mathbf{p}}\| \le K\|\mathbf{v} - \hat{\mathbf{v}}\|$.

In case of poor stability, there are significant changes of the solution for small changes of inputs because of a large constant K compared to $\|\mathbf{v}\|$.

Let's consider the problem of solution stability for the problem of finding the current density $J(x,y)$ discussed above (see Fig. 1.1), to change the magnetic field strength $H_w(x)$. We can write down an equation connecting the density J of the sought current with the field $H_w(x)$ on the $y=0$ line.

The magnetic field strength at points $(x,0)$, created by the current $di = J(x_0, y_0)ds_J$, is equal (according to the Biot-Savart law and the method of images):

$$dH_y = \frac{(x-x_0)J(x_0,y_0)}{\pi r^2}ds_J,$$

where $r^2 = (x-x_0)^2 + y_0^2$. The equation for finding of the current density $J(x_0,y_0)$ becomes

$$\int_{S_J} \frac{(x-x_0)J(x_0,y_0)}{\pi r^2}ds_J = H_w(x).$$

The solution of this equation, which is a Fredholm equation of the first kind with a kernel $(x-x_0)/\pi r^2 = K(x,x_0)$, is unstable. Indeed, let required function $g(q)$ in the Fredholm equation of the first kind $\int_a^b K(x,q)g(q)dq = H_w(x)$ receive disturbance $\Delta g(q) = exp(j\omega q)$. We then have:

$$\Delta H_w(x) = \int_a^b K(x,q)\Delta g(q)dq = \int_a^b K(x,q)exp(j\omega q)dq =$$

$$= (j\omega)^{-1} exp(j\omega q)K(x,q)\Big|_{q=a}^{q=b} - (j\omega)^{-1}\int_a^b \frac{\partial K(x,q)}{\partial q} exp(j\omega q)dq \equiv O(\omega^{-1}).$$

For a large ω the quantity $\Delta H_w = O(\omega^{-1})$ is small and, hence, at small disturbances $\Delta H_w(x)$ we obtain large disturbances of the solution $\Delta g(x)$. Thus, the discussed inverse problem of magnetic field synthesis is characterized by instability of its solution.

Stiffness

It has been noted above that the solution of inverse problems is reduced to finding of extrema of functionals. In practice, for the majority of cases the search of extrema is carried out numerically. Frequently, any satisfactory solutions cannot be found for comprehensible periods of time even with the use of very powerful computers. Such difficulties may occur because of a specific property of functionals under minimization, referred to as "stiffness", which is discussed below.

Let's consider an inverse problem aiming to search a vector $\mathbf{p} = (p_1, p_2)^T \in \Pi$ representing the parameters of an electric circuit at which its full power $S(\mathbf{p}) = \sqrt{P^2(\mathbf{p}) + Q^2(\mathbf{p})}$ at the set sources is minimal. Here P

and Q are the active and reactive powers, and Π is the area of allowable values for the vector of parameters $\mathbf{p}$. We assume that for all $\mathbf{p}\in\Pi$ the inequality $|Q| >> P$ is feasible, that is typical of electric power devices. Then the criterion of the inverse problem can be written as:

$$I(\mathbf{p}) = P^2(p_1,p_2) + Q^2(p_1,p_2) \xrightarrow[p_1,p_2\in\Pi]{} \min .$$

Let the expression $Q(p_1,p_2)=0$ define some line $p_1 = q(p_2)$ (Fig. 1.2). As $|Q| >> P$, then the rate of change of the $I(\mathbf{p})$ functional along the line $p_1 = q(p_2)$ is much less than along the line perpendicular to it. Therefore, the functional value will grow quickly when moving away from this line. Functionals, which have equal level surfaces of similar form, are referred to as ravine or rigid surfaces. It can be shown that the matrixes of second derivatives (Hesse matrixes) of such functionals have eigenvalues with strongly differing absolute values.

The minimum $I(\mathbf{p})$ should be searched along the line $p_1 = q(p_2)$, which determines the so-called bottom of the ravine. In the case of using a gradient method for searching of the $I(\mathbf{p})$ minimum, the process of minimization quickly results to the bottom of the ravine. Due to the inequality $|Q| >> P$, the gradient near to the line $p_1 = q(p_2)$ is almost perpendicular to this line. Therefore the further motion to a minimum at the ravine bottom (along the line $p_1 = q(p_2)$) occurs very slowly.

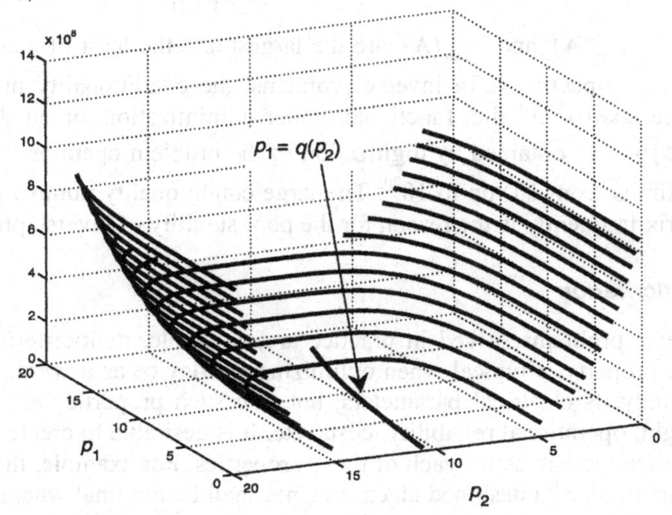

Fig. 1.2. Level lines of a ravine functional representing an electric circuit having reactive power considerably higher than its active power

In this example the equation of the bottom of the ravine is known and therefore can be used to accelerate the search of $I(\mathbf{p})$ minimum. However, generally this equation is not known. Therefore, there are serious obstacles to using gradient methods for minimization of ravine functionals.

Determination of the stiff functionals will be discussed further in Chapter 2. As previously noted, the Hesse matrix of a rigid functional has strongly differing eigenvalues. Let's consider the properties of these matrices in more detail.

The sensitivity of the equation $\mathbf{Aw} = \mathbf{v}$ solution to the change of $\Delta\mathbf{v}$ of its right-hand member, is determined by the relative change $\|\Delta\mathbf{w}\|/\|\mathbf{w}_w\|$ of the solution $\mathbf{w}$ norm at the relative change of the norm of the right-hand member $\|\Delta\mathbf{v}\|/\|\mathbf{v}\|$, where $\mathbf{w}_w$ is the exact solution of the equation $\mathbf{Aw} = \mathbf{v}$. By use of inequality $\|\Delta\mathbf{w}\| \le \|\mathbf{A}^{-1}\|\|\Delta\mathbf{v}\|$, one may come to the following relationship:

$$\frac{\|\Delta\mathbf{w}\|}{\|\mathbf{w}_w\|} \le \|\mathbf{A}\| \cdot \|\mathbf{A}^{-1}\| \frac{\|\Delta\mathbf{v}\|}{\|\mathbf{v}\|}.$$

This shows that the sensitivity of the equation solution to the change of its right-hand member depends on the quantity of $\mathrm{cond}(\mathbf{A}) = \|\Delta\mathbf{A}\|\|\mathbf{A}^{-1}\|$, referred to as the conditionality number of the matrix $\mathbf{A}$. For symmetric matrices:

$$\mathrm{cond}(\mathbf{A}) = \left|\frac{\lambda_{\max}(\mathbf{A})}{\lambda_{\min}(\mathbf{A})}\right|,$$

where $\lambda_{\max}(\mathbf{A})$ and $\lambda_{\min}(\mathbf{A})$ are the largest and the least of matrix $\mathbf{A}$ eigenvalues, respectively. In inverse problems, the conditionality number of the Hesse matrix of the functional under minimization or of the equation $\mathbf{A}(\mathbf{p})\mathbf{w} = \mathbf{v}$, obtained by digitization of the problem operator, can lie within the limits from 10^2 up to 10^{20}. The large conditionality number of the Hesse matrix frequently is the reason for the poor stability of inverse problems.

Multicriterion

Inverse problems solved in practice are, as a rule, multicriterion problems. This property is typical when optimizing the device as a whole, considering not only its electrical parameters, but also such properties as, for example, weight, operational reliability, costs, etc. It is desirable to create a device that has extreme values for each of these properties. For example, the power consumption y_1 of a designed electromagnet shall be minimal whereas the developed electromagnetic force y_2 simultaneously shall be close to a set value $\tilde{y}_2$. Formally, these criteria can be written down as:

$$\|y_1(\mathbf{w},\mathbf{p})\| \xrightarrow[\mathbf{p}]{} \min,$$
$$\|\tilde{y}_2 - y_2(\mathbf{w},\mathbf{p})\| \xrightarrow[\mathbf{p}]{} \min.$$

By virtue of contradictory of separate criteria, it is impossible to create devices for which each of them equals its extreme value.

Discreteness

One of the important properties of inverse problems in electrical engineering is that sought parameters of circuit elements, as well as desired distributions of the medium, are not described by, continuous functions. Indeed, a calculated electric circuit can be realized only using elements with nominal values of parameters. The solution of inverse problems in the electromagnetic field theory can result in a sectionally homogeneous medium, with non-continuous characteristics $\mu(x,y,z)$, $\varepsilon(x,y,z)$, $\sigma(x,y,z)$. Therefore, the search of solutions for conditions of discrete-type behavior for the vector **p** parameters demands taking proper account the non-differentiability of sought after functions.

For numerical solution of inverse problems, the continuous operators are replaced by their discrete analogues. At such transition, it is necessary to take into account properties of the discrete equations, which can differ from properties of the corresponding continuous equations. In some cases, the inverse problem which is described by the continuous equation $\mathbf{A}(\mathbf{p})\mathbf{w} = \mathbf{v}$ has a unique solution, however the solution of the corresponding discrete equation can be non-unique. For example, at a small number of measured values the discrete equation $\mathbf{A}(\mathbf{p})\mathbf{w} = \mathbf{v}$ can have some solutions. In other cases, it may have not any solutions at all.

Work content

Inverse problems are characterized by high work content during their solution by iterative methods. Each step of iterations includes a solution of analysis problems for some intermediate value of the vector **p** of parameters. The number of parameters, i.e. the number of vector **p** components can reach tens of thousands, which is typical for problems concerning the search of medium distributions in the electromagnetic field.

Incorrectness

Above-mentioned properties of the inverse problems in electrical engineering allow concluding, that in majority they belong to a class of so-called incorrect problems. The solution of the equation $\mathbf{A}(\mathbf{p})\mathbf{w} = \mathbf{v}$ is a correct problem if the following conditions are true: a) the solution of the equation exists for any right-hand member, b) the solution is unique, and c) the solution depends continuously of the initial data, i.e. the right-hand member $\mathbf{v}$ of the equation [6]. This implies that for the above-considered problems, these conditions can be violated. Therefore, their solution requires application of special methods.

1.1.2. Solution methods

In electrical engineering, inverse problems are solved by both analytical and numerical methods. The opportunity of an analytical solution is attractive because it does not require (or required only at the final stage of the solution) carrying out time-consuming numerical calculations. In some elementary cases, it is possible to establish analytic connections between the sought after parameters of the device and criteria of the inverse problem. Then the inverse problem becomes essentially simpler, as the analytical solution of the equation $\mathbf{A}(\mathbf{p})\mathbf{w} = \mathbf{v}$ is actually used. An example of such a problem is the search of field sources distributed in a homogeneous linear medium. If relations between density of the source and the strength of the field created by it are known, then the so-called direct methods can be applied.

Furthermore, we shall examine only numerical methods of solution of inverse problems in electrical engineering.

Narrowing of the solution searching area

Let's consider an approach which allows narrowing of the area of required solutions with the purpose of reducing the initial inverse problem to such one that has a unique solution. This approach uses *a priori* data concerning the properties of solutions. These data usually appear as restrictions for the solution search area and allow achieving its uniqueness in many cases.

For example, the problem of choice of the unique solution for the above considered synthesis problem of circuit voltage gain coefficient $K(s) = \dfrac{1}{Ts+1}$ could be solved by comparison of its possible solutions. Therefore, if from the technical point of view it is expedient to construct electric circuits without inductive elements, then the area of solutions in view of this criterion is narrowed. As a result, it is possible to find a circuit with resistors and capacitors. To obtain a unique solution it is expedient to also impose an a priori condition of minimality of the number of its elements.

The restrictions of integrated characteristics, such as the volume of used materials, e.g. the copper in wires or iron in magnetic cores, the losses power, weights of the devices, etc. can be attributed to *a priori* data for narrowing the area of possible solutions. Let, in the above considered problem (see Fig. 1.1), the required field $H_w(x)$ be a linear current field. It has been shown that this problem has a non-unique solution. However, it becomes a unique solution under the additional condition of minimum cross-sections S_J for current-carrying wires.

In some problems, when searching for a particular substance distribution in an electromagnetic field, the area of possible solutions can be narrowed by entering restrictions on the accuracy of processing of bodies' surfaces.

Thus, narrowing the search area of the solution by use of the information related to its properties in some cases allows obtaining unique solutions of inverse problems.

Methods of regularization

Let's examine inverse problems relating to the search of a distribution of sources forming the vector **v**. In electric circuits, they are represented by current and voltage sources, and in the electromagnetic field, by densities of electric charges and currents distributed in space and time. In this case, in the equation $\mathbf{Aw} = \mathbf{v}$ we accept $\mathbf{p} \equiv \mathbf{v}$, and the vector **w** is considered a preset value. For the equation $\mathbf{Aw} = \mathbf{v}$ we shall write it down as: $\mathbf{Bp} = \mathbf{w}$, where $\mathbf{B} = \mathbf{A}^{-1}$. The right-hand member **w** in the last equation is a preset value, whereas the quantities **p** included in its left-hand member should be found.

As it has been shown above in the study of stability properties of the inverse problems, a small error in calculation of **w** can result in large errors of the solution or even to its absence. In particular, for diagnostics problems the right-hand member of the equation $\mathbf{Bp} = \mathbf{w}$ is determined according to results of measurements with some error. In such cases, it is expedient to search not the exact, but the approximate solution.

Under the approximate solution of the equation $\mathbf{Bp} = \mathbf{w}$, referred to as generalized or quasi-solution, the solution is assumed at which the norm of discrepancy $\|\mathbf{Bp} - \tilde{\mathbf{w}}\|$ of the equation reaches its exact limit. The quasi-solution of the equation $\mathbf{Bp} = \mathbf{w}$, at the approximate right-hand member **w** always exists, as against its exact solution. It simultaneously delivers a minimum to the functional

$$I(\mathbf{p}) = \int_a^b [\mathbf{Bp}(q) - \tilde{\mathbf{w}}(q)]^2 dq \,,$$

where q – are time (for electric circuits) and/or spatial (for electromagnetic fields) coordinates.

Methods of regularization are applied to find stable quasi-solutions for inverse problems. It is well-known that the problem of the solution of the operational equation $\mathbf{Bp} = \mathbf{w}$ can be reduced to the problem of search of a minimum for the functional $I(\mathbf{p})$. Applying the method of weighed least squares results in the functional

$$I(\mathbf{p}) = \int_a^b [\mathbf{Bp}(q) - \tilde{\mathbf{w}}(q)]^2 \rho(q) dq, \qquad \rho(q) > 0,$$

where $\rho(q) > 0$ is positive on an interval a, and b is the weighting function.

During regularization of the problem, it is necessary to write down and minimize the so-called Tikhonov's functional [6]

$$T(\mathbf{p}, \alpha) = \int_a^b [\mathbf{Bp}(s) - \tilde{\mathbf{w}}(s)]^2 \rho(s) ds + \alpha \Omega(\mathbf{p}),$$

where $\Omega(\mathbf{p})$ is a specially selected positive functional, referred to as the stabilizer, $\alpha > 0$. In particular, it can be the functional $\Omega(\mathbf{p}) = |\mathbf{p}|^2 = \sum_k |p_k|^2$.

As it was noted above, at the numerical solution of inverse problems continuous operators shall be replaced by their discrete analogues. After this replacement, we come to a system of algebraic equations having the form $\mathbf{Bp} = \mathbf{w}$. Let's consider an application of the regularization procedure when solving this type of system of discrete equations in which the vector $\mathbf{p}$ is the sought quantity.

When using the regularization procedure for solution of the equation $\mathbf{Bp} = \mathbf{w}$, a search of a vector $\mathbf{p}$ from the condition $I(\mathbf{p}) = \|\mathbf{Bp} - \tilde{\mathbf{w}}\|^2 + \alpha \|\mathbf{p} - \mathbf{p}_0\|^2 \xrightarrow[\mathbf{p}]{} \min$ shall be carried out. Here, α is the regularization parameter; $\mathbf{p}_0$ is some given vector that, in particular, can be equal to zero if there are not any reasons for its choice. Under such stated conditions the so-called normal solution of the problem can be found, which is an approximate solution. At $\alpha \approx 0$ vector $\mathbf{p}$ will differ little from the sought quantity. In this case, the finding of $\mathbf{p}$ is reduced to the solution of the linear system of equations $(\mathbf{B} + \alpha \mathbf{1})\mathbf{p} = \tilde{\mathbf{w}} + \alpha \mathbf{p}_0$. Here, $\mathbf{1}$ is the unity matrix. The component $\alpha \mathbf{1}(\alpha > 0)$ makes the last system well conditional, so it can be solved, for example, by the Gauss method. However, in case of large values for α its solution can strongly differ from the solution of the initial system. Therefore, it is necessary to choose those least values for the parameter α, at which the conditionality of the regularized systems of equations becomes comprehensible.

There are various methods for finding the parameter α. In particular, it can be sought by the discrepancy of the equation; after calculation of the discrepancy $\mathbf{r}(\alpha)=\mathbf{Bp}(\alpha)-\mathbf{w}$, its norm shall be compared with the known error of the right-hand member $\mathbf{w}$ of the equation and with the margin error of matrixes $(\Delta\mathbf{B})\mathbf{p}$. If the discrepancy norm is larger than these errors, then the parameter α is large and needs reducing. Otherwise, it should be increased. Generally, the search procedure for parameter α is not so easy and its finding more resembles an art, which in order to master is only possible after solving numerous types of these problems.

Methods of solution of underdetermined and overdetermined problems

When solved numerically, the analysis problems reduce to systems of algebraic equations with a square matrix. For the solution of inverse problems, in some cases it is necessary to solve systems of equations with a rectangular matrix where the number of rows is not equal to the number of columns.

As an example let's consider the problem of finding of currents $i_1, i_2,\dots, i_m$ of the coaxial circular contours creating magnetic induction on their axis z, which must change by a set law $B(z)$.

Let's assume the values of a magnetic induction in m points on the z-axis are set or obtained by measurements. The system of equations $\mathbf{Bv}=\mathbf{w}$ connecting the magnetic induction ($\mathbf{w}$) in z-axis points and currents of contours ($\mathbf{v}$), is a nonsingular square matrix $\mathbf{B}$ of the size $m\times m$ and with elements

$$b_{p,k}=\frac{\mu_0 R_0^2}{2\sqrt{[R_0^2+(z_p-z_k)^2]^3}},$$

where $p,k=\overline{1,m}$ and z_p, z_k are the coordinates of a point on the z-axis with a set magnetic induction, and of the contour k having current i_k, accordingly, R_0 is the contour radius (Fig. 1.3).

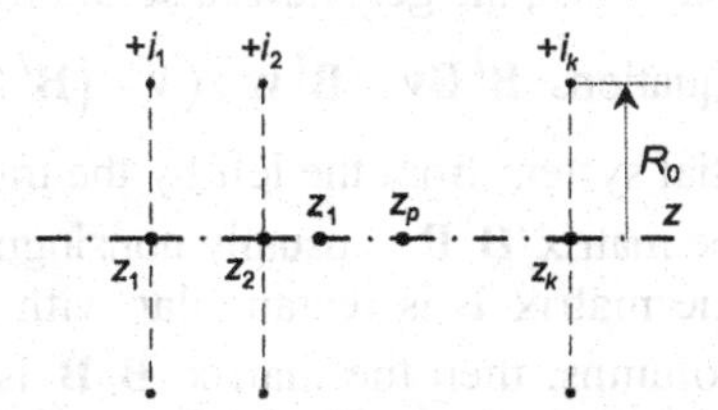

Fig. 1.3. Search of radii and coordinates of loops with currents which provide a given distribution of the magnetic induction along the z-axis

The solution of the system $\mathbf{Bv} = \mathbf{w}$ can be written down as $\mathbf{v} = \mathbf{Aw}$, (note, that $\mathbf{B} = \mathbf{A}^{-1}$).

The number of points on the axis (q – by designation) can be both more or less than the number m of contours. If the currents are found by results of measurements of the magnetic induction, then the proportion between numbers q and m is determined by availability of axis points for measurements. In such problems, usually $q > m$. In practice, sometimes it is required to obtain prescribed dependence $B(z)$ on the axis z, having a limited number of contours with currents. For those cases the criteria is the same $q > m$. The matrix $\mathbf{B}$ in the equation $\mathbf{Bv} = \mathbf{w}$ for $q > m$ is rectangular and vertically extended. If the measurement error of the magnetic induction can be neglected, the system of equations $\mathbf{Bv} = \mathbf{w}$ for $q > m$ has an exact solution. However, in the case of setting arbitrary values for magnetic induction $B(z)$ at q points on the axis z, the system of the algebraic equations at $q > m$ may lack any solutions. For example, at m=1, i.e. when we have only one contour with a current, at $q = 2$ it is not possible to find a satisfactorily accurate solution, if on two points of the axis the magnetic induction should have different signs.

In the cases where the number of contours m with the sought currents is more than the number of points q of axis, at which the induction should have preset values ($q < m$), the matrix $\mathbf{B}$ of the equations system is also rectangular, but is extended horizontally. The system $\mathbf{Bv} = \mathbf{w}$ will thus have infinitely many solutions, so it is possible to add them up with equations imposing additional conditions for the choice of sources (for example, on radii of contours or distances between them). Such conditions allow us to obtain a unique solution.

If the matrix $\mathbf{B}$ is square and nonsingular, i.e. its determinant is not equal to zero, then there is an inverse matrix to it and the sought vector can be found by the expression $\mathbf{v} = \mathbf{Aw}$. For the rectangular matrix $\mathbf{B}$, the inverse matrix does not exist. When solving equations with rectangular matrices the so-called generalized or pseudo-solutions must be sought. If the number of rows of the matrix $\mathbf{B}$ is more than the number of columns, the generalized solution can be found as the solution of the system of equations $\mathbf{B}^T\mathbf{Bv} = \mathbf{B}^T\mathbf{w}$, ($\mathbf{v} = \left(\mathbf{B}^T\mathbf{B}\right)^{-1}\mathbf{B}^T\mathbf{w}$), derived by multiplication of the initial system from the left by the transposed matrix $\mathbf{B}^T$, since in inverse problems the matrix $\mathbf{B}^T\mathbf{B}$ is usually nonsingular.

On the other hand, if the matrix $\mathbf{B}$ is rectangular with a number of rows less than the number of columns, then the matrix $\mathbf{B}^T\mathbf{B}$ is a singular matrix and the above-considered method of solution cannot be applied. In this case, obtaining a solution for the so-called pseudo-inverse matrix $\mathbf{B}^+$ should be used. It can be shown that the matrix $\mathbf{B}^+$ is uniquely determined by its following properties: $\mathbf{BB}^+\mathbf{B}=\mathbf{B}$, $\mathbf{B}^+\mathbf{BB}^+=\mathbf{B}^+$, $(\mathbf{BB}^+)^T=\mathbf{BB}^+$, $(\mathbf{B}^+\mathbf{B})^T=\mathbf{B}^+\mathbf{B}$.

With the use of a pseudo-inverse matrix, the solution of the system of equations $\mathbf{Bv} = \mathbf{w}$ can be represented as:

$$\mathbf{v} = \mathbf{B}^{+}\mathbf{w} + (\mathbf{1} - \mathbf{B}^{+}\mathbf{B})\xi, \tag{1.2}$$

where ξ is an arbitrary vector. The first term of the solution $\mathbf{v}_0 = \mathbf{B}^{+}\mathbf{w}$ corresponds to the solution $\mathbf{v}$ with the minimal Euclidean norm; therefore, this solution is frequently referred to as the solution by the least-squares method.

Note that the form of Eq. (1.2) for the solution is rather general. So, if $\mathbf{B}^T\mathbf{B}$ is a nonsingular matrix, then at $\xi = 0$ Eq. 1.2 gives the earlier obtained solution $\mathbf{v}=(\mathbf{B}^T\mathbf{B})^{-1}\mathbf{B}^T\mathbf{w}$. If $\mathbf{B}$ is a square nonsingular matrix, then the solution is $\mathbf{v} = \mathbf{Aw}$.

Methods of search for global and local minima

One of the features of inverse problems in electrical engineering is that the objective functions have, as a rule, a non-unique minimum. In most cases, it is impossible to assert a priori the existence of several minima. Therefore finding of a parameters vector delivering a minimum to the functional does not mean that there is no other parameters vector at which the objective function has a still smaller value.

Methods used for finding of the global minimum demand much more time than methods for finding of a local minimum. Indeed, at the worst it is necessary to find all local minima, and then by their comparison allocate the global minimum. To reduce the search time for the global minimum, methods of casual search or entering elements of randomness in the search algorithm can be applied. The most widely used methods of search of the global extremum, namely the method of simulated annealing and the evolutionary method, are discussed in Chapter 2.

Solution methods of multicriterion problems

Each of the criteria of an inverse problem should be complied with as a result of search for the optimum parameters, and can be considered as a component of some vector. As stated above, by virtue of discrepancy of some criteria finding of parameters at which all components of a vector simultaneously accept extreme values is impossible. Therefore, for a found solution one or several criteria can be complied with, while the others cannot.

To compare several solutions that comply with various criteria the following reasoning can be used. The solution is considered effective, if at improvement of any criterion there will be at least one other criterion, which thus will worsen. In others words, if for a given vector of parameters the improvement of any of criteria results to worsening of any other criterion, then this vector of parameters gives an effective solution. The area of all effective solutions is defined by the solution of the problem of multiobjective optimization. The finding of the area of all effective solutions demands a great deal of calculations. Therefore, in practice usually only some points belonging to it are found which then are regarded as the sought solution. To find these points the vector of criteria shall be reduced by means of various methods to a scalar for solving the subsequent unicriterion inverse problem by known methods. Further examination of the problems of multiobjective optimization shall be discussed in Section 2.5.

In conclusion, let's summarize the basic features of the inverse problems in electrical engineering.

The basic types of inverse problems are the problems of synthesis and identification. The synthesis problems include stages of structural and parametrical synthesis that are referred to as parametrical optimization. Problems of identification include macromodeling and diagnostics.

Inverse problems are characterized by the following properties: they are multicriterion problems with non-unique solutions, poor stability, discreteness of sought values and large work content. In mathematics, they are related to incorrect problems.

1.2 Inverse problems in electric circuits theory

Inverse problems in electric circuits theory are usually referred to as synthesis problems [7, 8, 9]. Standard synthesis problems involve creating a circuit possessing a set of specified properties. In this section basic types and features of inverse problems encountered in practical electric circuits theory are considered. Macromodeling and identification problems are also discussed.

Furthermore, the inverse problems of electric circuits theory are considered in the following sequence:

- structural synthesis;
- parametric synthesis (parametric optimization);
- macromodeling;
- parameters identification.

1.2.1. Formulation of synthesis problems

Let's consider an electric circuit with N inputs and M outputs. Let $\mathbf{v}=(v_1,v_2,\ldots v_N)^T$ be the vector of input signals (actions) and $\mathbf{w}=(w_1,w_2,\ldots w_M)^T$ the vector of output signals (circuit responses). We assume that the vector **v** belongs to the set of allowable input signals s_v, and the vector **w** to the set of output signals s_w. Signal transformation by the electric circuit will be presented by the operator f, carrying out mapping of the set s_v onto the set s_w: (f: $s_v \rightarrow s_w$).

Let us for example say that the input and output signals are sine waves of the same frequency. Then their sets represent spaces of vectors of complex amplitudes and dimensions N and M, respectively. The operator f in this case can be characterized by a complex matrix **A**, such that:

$$\forall \mathbf{v} \in s_v, \quad \mathbf{w} \in s_w \qquad \mathbf{A}\mathbf{w} = \mathbf{v}\,.$$

The last relationship does not impose any restrictions on whether the matrix **A** will be realized in a class of linear, nonlinear or passive electric circuits. It does not limit the number and types of constituent elements of the synthesized circuit. That is, it allows using bipolar, tripolar and/or two-port (e.g. transformers) elements, etc. at synthesis. Similarly, the amplitudes of the input and output signals are not limited.

To formulate synthesis problems, sets s_v, s_w and the operator f should be characterized more comprehensively. Thus, for the example presented above within the sets s_v and s_w, subsets S_v and S_w of sine wave signals with amplitudes $V_{\min} < V < V_{\max}$ and $W_{\min} < W < W_{\max}$, where $V_{\min}, V_{\max}, W_{\min}, W_{\max}$ are given quantities, can be allocated. Further on within the set Φ of operators f, carrying out the transformation f: $S_v \rightarrow S_w$, a subset F of operators that are realized, for example, by linear circuits, containing only bipolar elements and not containing inductance coils, can be allocated.

This process that involves narrowing s_v, s_w, Φ to S_v, S_w, F represents the "engineering" phase of the synthesis problem solution. At this stage any requirements concerning the designed circuit, including inconsistent ones, can be formulated.

In electrical engineering it is customary to group devices according to their functional purposes. Transformers, generators, amplifiers, rectifiers, inverters, filters, etc. can be treated as examples of such groups. Assignment of a device to one or another group allows simplifying the synthesis problems considerably. Specific features of devices in each group allow offering sufficiently narrow sets of input and output signals and reducing the element basis (i.e. describing the operator F properties in greater detail). So, for example, for the synthesis of filters, the set of elements used can consist only of resistors, condensers, plus current and voltage controlled sources. At synthesis of rectifiers only nonlinear elements with switch type voltage-current characteristics can be considered, etc. At the same time, the operator F simplifies limits to the range of circuits that can be obtained as a result of solving the synthesis problem.

Let's assume that the "engineering" phase of the synthesis problem solution is completed and sets S_v, S_w and the operator F have been determined. Then the problem of synthesis of the required electric circuit can be formulated as follows: to find the structure and parameters of an electric circuit, which implements the operator F mapping the set S_v onto the set S_w (F: $S_v \to S_w$).

Thus, the formulated synthesis problem has the properties of inverse problems indicated in Section 1.1. In particular, the operator F implementing the required mapping, as a rule, is not unique. Let's consider definitions of inverse problems in more detail.

Structural synthesis of electric circuits

The problem of structural synthesis involves finding of the form of operators F that allows carrying out the required mapping F: $S_v \to S_w$ for the sets of input and output signals. One of the important and intricate challenges arising at structural synthesis involves providing the realizability of the operator F for a class of devices that are available to the designer.

Let's consider an example. Let sets of input and output signals contain one element each: $S_v{:}v(t) = 1(t)$, and $S_w{:}\tilde{w}(t)$, where $\tilde{w}(t)$ is a given function of time. We shall assume that $v(t)$ and $\tilde{w}(t) = \tilde{u}_R(t)$ are input and output voltages of a circuit. It is required to synthesize an operator F that is carrying out the transformation $S_v \to S_w$ in a class of linear *RLC* circuits (Fig. 1.4). The specific type of circuit and its parameters are not stipulated at the definition of the problem and has been represented as the so-called black box.

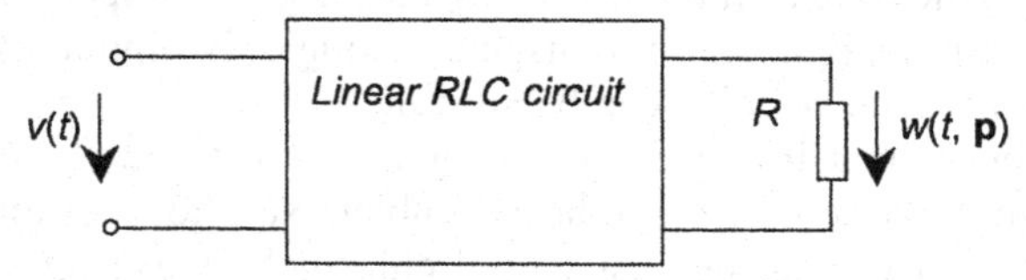

Fig. 1.4. *RLC*circuit, realizing linear transformation of signals

Due to the circuit linearity, the input $v(t) = 1(t)$ and the output $w(t,\mathbf{p}) = u_R(t,\mathbf{p})$ signals are connected by the following operator expression:

$$\mathbf{A}(\mathbf{p})u_R(t,\mathbf{p}) = q \cdot 1(t),$$

$$\mathbf{A} \equiv \frac{d^n}{dt^n} + p_{n-1}\frac{d^{n-1}}{dt^{n-1}} + \quad \cdots \quad + p_1\frac{d}{dt} + p_0,$$

$$\mathbf{p} = (q, p_0, p_1, \cdots, p_{n-1})^T \in \Pi, \quad p_k > 0, \quad k = \overline{0, n-1}, \quad p_0 > q,$$

where $\mathbf{p}$ is the vector of parameters of the operator $\mathbf{A}$ and Π is a space of vectors with positive components. Here it is supposed that in the synthesized circuit there will be no contours, which are passing only through condensers and voltage sources, as well as sections, which are passing only through coils and current sources.

The circuit output voltage $u_R(t)$ can be presented as $u_R(t,\mathbf{p}) = w(t,\mathbf{p}) = \mathbf{A}^{-1}(\mathbf{p})\cdot q$. The criterion at which the synthesized circuit carries out the transformation $S_v \to S_w$ can be presented as:

$$\int_0^\infty \left[\tilde{u}_R(t) - u_R(t,\mathbf{p})\right]^2 dt \xrightarrow[\mathbf{p}\in\Pi]{} \min. \tag{1.3}$$

Realization of the operator **A** within the class of linear *RLC* circuits also requires the fulfillment of Roth-Hurwitz conditions [10]:

$$p_{n-1} > 0,\quad \mathbf{det}\begin{bmatrix} p_{n-1} & p_{n-3} \\ 1 & p_{n-2} \end{bmatrix} > 0,\ \cdots,\ \mathbf{det}\begin{bmatrix} p_{n-1} & p_{n-3} & p_{n-5} & \cdots & 0 \\ 1 & p_{n-2} & p_{n-4} & \cdots & 0 \\ \vdots & \vdots & \vdots & \ddots & \vdots \\ 0 & 0 & 0 & \cdots & p_0 \end{bmatrix} > 0. \tag{1.4}$$

Thus, the first part of the structural synthesis problem is solved; the form of the operator that implements the required transformation and fulfills its realizability conditions is found.

On the following step of structural synthesis we search for the electric circuit topology and types of elements realizing this operator. Let $n=1$, then $\mathbf{p} = (q, p_0)^T$, and the operator **A** is given by:

$$\mathbf{A} = \frac{d}{dt} + p_0. \tag{1.5}$$

To realize this operator any of the circuits shown on Fig. 1.5 can be used. Obviously, other circuits can also be offered. In this example, the choice of the operator form is obvious enough because of the simplicity of the problem; its operator is linear, there are no restrictions on amplitudes of input and output signals ordinarily used in practice, and sets of signals contain only one element each. Nevertheless, even in this simplified example the solution of structural synthesis problem is ambiguous. Experience shows that it is ambiguous in most cases.

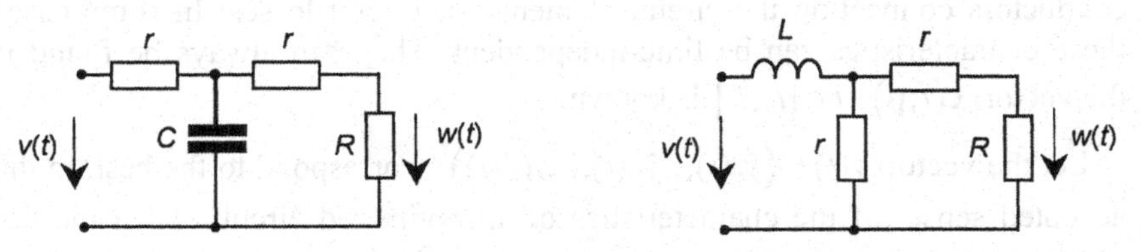

Fig. 1.5. Electric circuits showing the operator (1.5)

The result of the structural synthesis is a set of circuits or operators with unknown parameters. Therefore, the feature of structural synthesis problems is their incompleteness in the sense that the quality of a solution becomes clear only after the solution of the parametric synthesis problem.

Parametric synthesis of electric circuits (parametric optimization)

At parametric synthesis it is presumed that the topology of a circuit and types of its constituent elements are known. It is required to determine parameters of these elements on the basis of inverse problem criteria. As noted above, these criteria describe the best, in the accepted sense, of the characteristics of a synthesized circuit, and also establish restrictions on the ranges of change for its elements, parameters.

Parametric synthesis can be carried out in time, frequency or operator domains. Solutions of problems in various domains have specific features, however their formulations are close enough.

Let's consider a problem of parametric synthesis in the time domain. Let processes in a circuit be described by a system of state equations (the method of state variable analysis)

$$\begin{aligned} &\frac{d\mathbf{x}}{dt} = \mathbf{f}(t,\mathbf{x},\mathbf{p}), && t \in [t_0, T], \\ &\mathbf{y}(t) = \mathbf{g}(t,\mathbf{x},\mathbf{p}), && \mathbf{x}(t_0) = \mathbf{x}_0, \end{aligned} \tag{1.6}$$

where $\mathbf{x}(t)$ is the vector of state variables, $\mathbf{p} \in \Pi$ is the vector of device parameters subject to definition, $\mathbf{y}(t)$ is the vector of optimized characteristics of the circuit which can be calculated by state variables, and $\mathbf{f}$ and $\mathbf{g}$ are generally nonlinear vector functions. Numerical or analytical solutions of the system of Eq. (1.6) at a certain vector of parameters $\mathbf{p}$ allow determining $\mathbf{x}(t,\mathbf{p})$ in the whole interval $[t_0, T]$.

Vector $\mathbf{y}(t)$ defines the engineering characteristics of an optimized circuit accepted at the problem's formulation - e.g. maximal voltages between some nodes of the circuit in the whole interval $[t_0, T]$, the power of electromagnetic radiation created by the circuit in the surrounding space, sections of the conductors connecting the circuit elements, or circuit losses. In some cases these characteristics can be time-independent. They can always be found if the vector $\mathbf{x}(t,\mathbf{p})$, $t \in [t_0, T]$ is known.

Let the vector $\tilde{\mathbf{y}}(t) = \left(\tilde{y}_1(t),\ \tilde{y}_2(t), \ldots, \tilde{y}_m(t)\right)^T$ correspond to the best, in the accepted sense, of the characteristics of an optimized circuit. Then the dependences

$$\mathbf{y}(t,\mathbf{x}(t,\mathbf{p}))=(y_1(t,\mathbf{x}(t,\mathbf{p})),\quad y_2(t,\mathbf{x}(t,\mathbf{p})),\quad \cdots,\quad y_m(t,\mathbf{x}(t,\mathbf{p})))^T$$

can be found from Eq. (1.6) in the interval $[t_0,T]$ at an arbitrary vector $\mathbf{p}$ (similar to the statement in the last sentence of the previous paragraph). The problem of parametrical synthesis will consist in determining the vector $\mathbf{p}$ from the conditions:

$$\left\|\tilde{y}_k(t)-y_k(t,\mathbf{x}(t,\mathbf{p}))\right\| \xrightarrow[\mathbf{p}\in\Pi]{} \min, \quad k=\overline{1,m}, \quad t\in[t_0,T]. \tag{1.7}$$

It is apparent that the problem of parametrical synthesis is reduced to a problem of vector or multicriterion optimization. Criteria (see Eq. (1.7)) can be inconsistent. So, for example, at synthesis of an amplifier, $\tilde{y}_1(t)$ can define the allowable value of the dissipated power, and $\tilde{y}_2(t)$, the lower limit of the amplifier output power. Both conditions are justified from the engineering point of view, but are inconsistent, as the increase of output power results in a rise of dissipated power.

The parametrical synthesis problem formulation in the frequency domain, as a whole, is close to its formulation in the time domain. Parametrical synthesis in the frequency domain assumes the description of circuit properties by means of frequency characteristics; therefore it can only be carried out for linear circuits without internal sources of electromagnetic energy. This condition, on the one hand, considerably reduces the generality of the approach. However, on the other hand, it simplifies the parametrical synthesis problem.

Linear electric circuit's frequency characteristics are defined as follows:

$$X_k(j\omega,\mathbf{p})=\frac{G_{k,r}(j\omega,\mathbf{p})}{H_{k,q}(j\omega,\mathbf{p})}, \quad k=\overline{1,m}, \quad \omega\in[\omega_l,\ \omega_u],$$

$$\mathbf{X}(j\omega,\mathbf{p})=(X_1(j\omega,\mathbf{p}),X_2(j\omega,\mathbf{p}),\ \cdots,X_m(j\omega,\mathbf{p}))^T,$$

$$\mathbf{Y}(\omega,\mathbf{p})=(Y_1(\omega,\mathbf{X}(j\omega,\mathbf{p})),Y_2(\omega,\mathbf{X}(j\omega,\mathbf{p})),\ \cdots,Y_m(\omega,\mathbf{X}(j\omega,\mathbf{p})))^T.$$

Here ω is the circular frequency, $\mathbf{p}\in\Pi$ is the vector of electric circuit parameters subject to definition, G_k and H_k are polynomials of r and q degree from ω, with coefficients depending on elements of the vector $\mathbf{p}$; $\mathbf{Y}$ is the vector of circuit optimized characteristics which can be calculated from $X_k(j\omega,\mathbf{p}), k=\overline{1,m}$. The vector $\mathbf{Y}$ includes those of the circuit characteristics which are defined by technical or economic requirements to the synthesized circuit.

Let the vector $\tilde{\mathbf{Y}}(\omega)=(\tilde{Y}_1(\omega),\ \tilde{Y}_2(\omega),\ \cdots,\tilde{Y}_m(\omega))^T$ define the best, in the accepted sense, of the circuit characteristics in the frequency range $[\omega_l,\omega_u]$.

Parametrical synthesis problems in the frequency domain involve defining the vector **p** from the conditions:

$$\left\| \tilde{Y}_k(\omega)\text{-}Y_k(\omega,\mathbf{p}) \right\| \xrightarrow[\mathbf{p}\in\Pi]{} \min, \quad k=\overline{1,m}, \quad \omega \in [\omega_l, \omega_u]. \tag{1.8}$$

Apparently, the parametrical synthesis problem (see Eq. (1.7)) in the frequency domain is also reduced to a vector or multicriterion optimization problem.

Elements $X_k(j\omega,\mathbf{p})$ of the vector **X** are generally complex functions. Elements of vectors **Y** and $\tilde{\mathbf{Y}}$ in Eq. (1.8) and further are assumed real. Such assumption does not reduce the generality of the problem formulation. Indeed, for example, let the following $X_1(j\omega,\mathbf{p}) = g(\omega,\mathbf{p}) - jb(\omega,\mathbf{p})$ represent the circuit input conductivity. Let some optimal values $\tilde{y}_1(t) = \tilde{g}(\omega)$ and $\tilde{y}_2(t) = \tilde{b}(\omega)$ also be known. Then, in this case the condition of optimality (see Eq. (1.8)) can be expressed through real functions:

$$\begin{cases} \left\| \tilde{g}(\omega)\text{-}g(\omega,\mathbf{p}) \right\| \xrightarrow[\mathbf{p}\in\Pi]{} \min \\ \left\| \tilde{b}(\omega)\text{-}b(\omega,\mathbf{p}) \right\| \xrightarrow[\mathbf{p}\in\Pi]{} \min \end{cases}, \omega \in [\omega_l, \omega_u].$$

Due to the linearity of the synthesized circuit, analytical dependences of frequency characteristics $X_k(j\omega,p)$, $k=\overline{1,m}$, from the parameters of optimization **p** and frequencies ω, can be derived. Advanced programs of electric circuits symbolic analysis allow obtaining these expressions rapidly and in compact forms [11,12]. This feature of the optimization problem solution in the frequency domain allows us to speed up the calculation of Eq. (1.8) functionals considerably and to obtain for them, and their gradients rather simple analytical expressions.

Let us consider parametrical synthesis problems in operator domain. As is well-known, a formal replacement $s \leftrightarrow j\omega$ allows changing over from frequency characteristics to their images and back. Similarly, a Laplace transformation allows changing over from differential equations in time domain to their images. Therefore, the features of synthesis problems considered above for time and frequency domains are valid for synthesis problems in operator domain.

It is expedient to carry out parametrical synthesis in operator domain for obtaining analytical solutions of inverse problems for rather simple circuits and target functions. Such a simple T-shaped circuit has been considered above (see Fig.1.4). Let $n=1$. The vector $\mathbf{p}=(q,p_0)^T$ of parameters, included in the operator **A**, can be found from the problem is (see Eq. (1.3)) solution under the limiting conditions (see Eq. (1.4)). Let the solution be vector $\mathbf{p}=(2,4)^T$, then the equation $\mathbf{A}(\mathbf{p})u_R(t,\mathbf{p})=q$ becomes:

$$\frac{d}{dt}u_R(t)+4u_R(t)=2,$$

which results in,

$$\tilde{U}_R(s)=\frac{2}{s(s+4)}.$$

Figure 1.5 shows diagrams of electric circuits that realize the operator **A**. Let's determine the parameters r, R, C of the circuit elements (shown on Fig. 1.5a) by solving the parametrical synthesis problem for this circuit. The voltage on the resistor R is given by

$$U_R(s)=\frac{R}{s\left[sC(r+R)+2r+R\right]}.$$

Comparing $\tilde{U}_R(s)$ and $U_R(s)$, we obtain a system of equations

$$\begin{cases} R=2, \\ C(r+R)=1, \\ 2r+R=4, \end{cases}$$

one of the solutions being R=2, r=1, C=1/3. Other solutions are also possible. As stated above, a typical parametrical synthesis problem has non-unique solutions in most cases.

It was assumed above that required parameters change continuously. It is not always true, because in many practical cases they can have only discrete values. Rigidly defined series of possible values are typical, for example, for electronic circuits parameters.

The parametrical synthesis problems on a discrete set of parameters can be solved beginning with the assumption that all parameters are continuous. Then the obtained values can be replaced by the nearest in the series of nominal values. Such replacement will worsen the quality of the solution. Perhaps, accomplishing a search of extremums in the domain of parameters' discrete values, it would be possible to find another, better solution than the one obtained by replacement of the optimum parameters by the nearest parameters in the series of nominal values.

Another possible way of finding a solution of the parametrical synthesis problem with discrete parameters is the use of methods of minimization that do not demand differentiation of the functional. Particularly, methods based on genetic algorithm [13], related to such methods, are discussed further [14,15].

1.2.2. The problem of construction macromodels (macromodeling) of devices

Macromodeling or, equivalently, construction of integrated models of electric devices, is close to the above-considered synthesis problems, but it has a number of specific features. Integrated models (macromodels) are a convenient and commonly used means for describing functions of complex devices and systems. When the equivalent circuit of a device or the system of equations describing its behavior are rather complex or not sufficiently accurate, solving of the analysis problem is impossible. To analyze systems including such devices, integrated models are used which allow approximate representation of signal transformation in the "input - output" mode without detailed elaboration of the internal processes. All this, while assuming that macromodels will be substantially simpler than full models, which are taking into account processes running inside the devices.

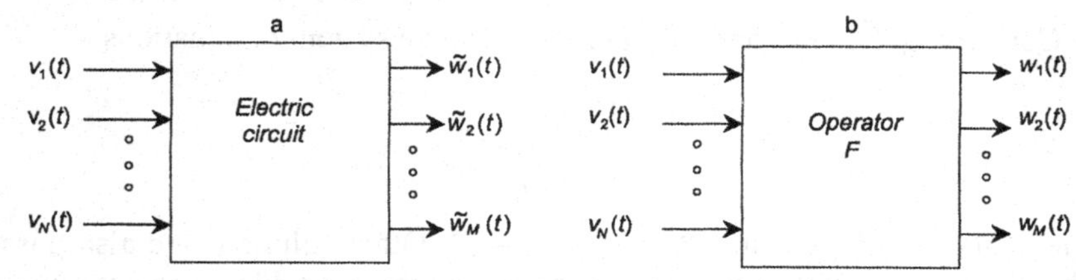

Fig. 1.6. Modeled subject (electric circuit) **a**) and its macromodel **b**)

Macromodeling is an actual problem, especially for objects lacking full descriptions. It can be, for example, a new device without sufficiently investigated properties. Macromodeling of such devices is sometimes the sole opportunity to describe their performance.

Let the electric circuit for which we intend to construct a macromodel in time domain be a multiport with N inputs and M outputs (Fig. 1.6). Each element $\mathbf{v}(t)=(v_1(t),v_2(t),\ldots,v_N(t))^T$ of the set S_v of input signals, and each element $\tilde{\mathbf{w}}(t)=(\tilde{w}_1(t),\tilde{w}_2(t),\ldots,\tilde{w}_M(t))^T$ of the set S_w of time-dependent output signals, represents a N- and M-dimensional vector, respectively.

In general, a circuit macromodel is presented by an operator, mapping the set S_v of admissible input (test) signals $\mathbf{v}$ on the set S_w of output signals $\mathbf{w}$, F: $S_v \rightarrow S_w$. We shall emphasize again the importance of the fact that numerous internal variables of the initial circuit do not participate in the macromodel synthesis. Therefore, dimensions of vectors $\mathbf{v}$ and $\mathbf{w}$ used for the solution of the macromodeling problem can be much less than the necessary number of variables for a full circuit description.

The problem of creating a macromodel involves synthesizing an operator F and defining its parameters $\mathbf{p}$, which delivers the solution of the system of inequalities:

$$\begin{gathered}\left\|\tilde{w}_k(t) - w_k(t,\mathbf{p})\right\| < \delta_k, \quad k = \overline{1,M}, \\ \mathbf{w} = \left(w_1(t), w_2(t), \ \cdots, \ w_M(t)\right)^T, \ \mathbf{v} = \left(v_1(t), v_2(t), \ \cdots, \ v_N(t)\right)^T, \\ \mathbf{w} = F(\mathbf{v}), \ \ \mathbf{v} \in S_v, \ \ \mathbf{w} \in S_w, \ \ \tilde{\mathbf{w}} \in \tilde{S}_w,\end{gathered} \tag{1.9}$$

where $\tilde{\mathbf{w}}$ and $\tilde{S}_w$ are the output signal and the set of output signals generated by the electric circuit subject to macromodeling, and $\delta_k, k = \overline{1,M}$ is the error by which the macromodel reproduces the real object properties.

To reveal the macromodeling object properties one can feed test input signals $\mathbf{v}(t)$ and simultaneously measure the output signals $\mathbf{w}(t)$. This is referred to as testing of the macromodeling object. To solve a macromodeling problem the following steps shall be taken:

- setting the form of the mathematical description of the macromodeling object (defining the type of the operator F);
- developing the method of operator F parameters definition by testing results; and
- finding the minimal set S_v of test signals, sufficient for definition of operator parameters with specified accuracy.

The distinction of macromodeling from synthesis is that the operator F can be physically unrealizable. A macromodel should reflect with adequate accuracy only the link between input and output signals. Any requirements for its realization as a physical device may not be imposed. The freedom of choice of the operator F allows offering various general description forms establishing analytical dependence of output signals from the inputs when creating macromodels of nonlinear circuits. The most investigated ways of description of these operators are Volterra functional series [16,17,18,19,20], Volterra - Picard series [21] or polynomials of split signals [8]. The relationship between input and output signals is searched in the form of functional series. They can be considered as multivariate polynomial expansion of output signals.

Let us consider a simplified case when input $v(t)$ and output $w(t)$ signals are scalar functions of time. Then the Volterra functional series gives an unequivocal link between input and output signals in nonlinear stationary circuits with zero initial conditions (on the assumption that the series converge) is given by:

$$w(t)=\sum_{k=1}^{\infty}\underbrace{\int_{-\infty}^{\infty}\cdots\int_{-\infty}^{\infty}}_{k} h_k(\tau_1,\ \tau_2,\ \ldots,\ \tau_k)\prod_{r=1}^{k}v(t-\tau_r)d\tau_1 d\tau_2 \ldots d\tau_k\,, \qquad (1.10)$$

where $h_k(\tau_1,\tau_2,\ldots,\ \tau_k)$ is the Volterra kernel of k order.

The first summand (k=1) of a Volterra series is the linear convolution integral corresponding to a linear circuit with a kernel (or pulse response) $h_1(\tau)$. The subsequent summands (k>1) represent nonlinear (with respect to input signals) convolutions with kernels $h_k(\tau_1,\tau_2,\ldots,\tau_k)$, which are the nonlinear circuit multivariate pulse responses of k order.

When macromodeling nonlinear circuits representations of a Volterra series kernel in the domain of Fourier-images are frequently used:

$$H_k(j\omega_1,j\omega_2,\ \ldots j\omega_k)=$$
$$=\underbrace{\int_{-\infty}^{\infty}\cdots\int_{-\infty}^{\infty}}_{k} h_k(\tau_1,\ \tau_2,\ \ldots,\ \tau_k)\ e^{-j(\omega_1\tau_1+\omega_2\tau_2+\ \ldots\ +\omega_k\tau_k)}d\tau_1 d\tau_2\ \ldots\ d\tau_k,$$

where $H_k(j\omega_1,j\omega_2,\ \ldots j\omega_k)$ is the Volterra series kernel in the frequency domain.

Links between input and output variables can also be described by means of polynomials of split signals [9]:

$$w(t)=\sum_{d_1=0}^{K_1}\sum_{d_2=0}^{K_2}\cdots\sum_{d_m=0}^{K_m}C_{d_1d_2\ldots d_m}\left[v_1(t)\right]^{d_1}\left[v_2(t)\right]^{d_2}\cdots\left[v_m(t)\right]^{d_m}, \qquad (1.11)$$

where $v_1(t),v_2(t),\ \ldots,\ v_m(t)$ are the so-called split signals. The main property of these signals is that they are not vanishing, intersecting, self-intersecting or tangent in any point within their definitional domain. Usually values $K_i,\ \ i=\overline{1,m}$ and m are not very large and do not exceed 10. Higher values of these parameters signify a fast increase of model dimensionality and complexity of the approximation problem.

Eqs. (1.10) and (1.11) define the type of macromodel operator F. Coefficients $h_k(\tau_1,\tau_2,\ldots,\ \tau_k)$ and $C_{d_1d_2\ldots d_m}$ are required macromodel parameters forming the vector **p**. Then macromodel creating will include determining these parameters by means of measuring (or calculating) circuit responses to test signals $v_\ell(t)\in S_v,\ \ \ell=1,2,\ \cdots$ (i.e. functions $w_\ell(t)\in S_w,\ell=1,2,\cdots$).

Equation (1.10) gives a more general form of a macromodel in comparison with Eq. (1.11) since its use allows definition of kernels $h_k(\tau_1, \tau_2, ..., \tau_k)$ at any test signals in the frequency or time domain. When using polynomials of split signals, it would be assumed that a set of input signals close to the set of signals at which the modeled device works shall be considered.

One of the remarkable properties of the approach considered above is to create macromodels where the required parameters of macromodels are linearly included into Eqs. (1.10, 1.11). Indeed, let's assume that the integral of the squared difference of the device and model output signals

$$\left\| \tilde{w}_k(t) - w_k(t) \right\| = \int_0^\infty \left(\tilde{w}_k(t) - w_k(t) \right)^2 dt$$

is used as a norm in the macromodeling problem (see Eq. (1.9)). Then, due to the above-mentioned linearity, this functional is quadratic. Since any quadratic functional is non-negative everywhere and has a unique minimum, then the problem (see Eq. (1.9)) always has a unique solution.

This approach to create macromodels has much in common with construction of so-called neural networks that will be discussed later, in Section 2.4. Here, we shall note that the neural network, which approximates a nonlinear operator, consists of nonlinear parameters.

Thus, the process of solving the problem (see Eq. (1.9)) becomes essentially complicated, since it may have (and frequently has) local minima.

A macromodel must reflect with specified accuracy the object properties on a wider class of signals than testing signals. Usually this requirement is difficult to make feasible, since there is little *a priori* information about the object. Therefore, it is necessary to be content with the hypothesis that the chosen model is adequate. Then an *a posteriori* check of this hypothesis shall be carried out. After acceptance, the model is tested by means of signals that are different from those used for its construction. The same signals are used for testing the object. Then the congruence degree between model and object responses are estimated. This is the same approach as used in neural networks.

1.2.3. Identifying electrical circuit parameters

Electrical circuit identification is the means of definition of its equivalent circuit parameters according to measurements of circuit responses to set actions. Let's assume that the structure of the circuit (or the type of operator *F*) is known. Then the identification problem turns to a problem of finding of sets of input (test) S_v and output (measured) S_w signals that allow unique identification of the circuit parameters.

If there are no restrictions imposed upon the sets S_v and S_w, the identification problem is solvable and has a unique solution. Indeed, if it were possible to apply any test signals to all circuit elements and carry out all necessary measurements then the finding of electric circuit parameters would be a trivial problem. To formulate nontrivial identification problems we shall first analyze its features.

Let's consider properties and features of identification for problems regarding passive DC linear circuits. It is assumed that the structure of the circuit, that is the circuit graph, is also known. For experimental determination of the circuit's elemental parameters, the following is required:

- to choose a tree in the circuit graph;
- to connect a voltage or current source to the circuit in an arbitrary point;
- to measure the voltages (by means of voltmeters) of all branches, corresponding to the circuit graph tree and to calculate voltages of all remaining branches;
- to measure the currents (by connecting ammeters) of all branches, corresponding to circuit graph links and to calculate currents in all remaining branches; and
- to determine the resistance of each of the branches as the ratio of the branch voltage to its current.

In most practical cases using this algorithm for a solution is unacceptable. The main difficulty is the necessity to connect ammeters into circuit branches. For the majority of devices (for example, printed-circuit boards) the connection of ammeters is difficult or even impossible. In most cases only the circuit nodes are accessible for measurement. Therefore only the voltages between circuit nodes can be measured. So for solving an identification problem, several voltage measurements must be performed and circuit parameters shall be calculated by the results of the measurements.

Let's consider an example of the identification problem solution for the circuit shown in Fig. 1.7a. Circuit nodes accessible for connection of measuring instruments are indicated as numerated points. The following experiments shall be performed:

- by connecting a voltage source V to the node 1 (Fig. 1.7a.), the current $I_1^{(1)}$ and the voltage $V_{20}^{(1)}$ can be measured (Fig. 1.7b.). In this case the problem of ammeter connection does not arise, as it is connected in a branch, external with respect to the identified circuit;
- by connecting a known current source J to the node 2, the voltages $V_{10}^{(2)}$ and $V_{20}^{(2)}$ can be measured (Fig. 1.7c.).

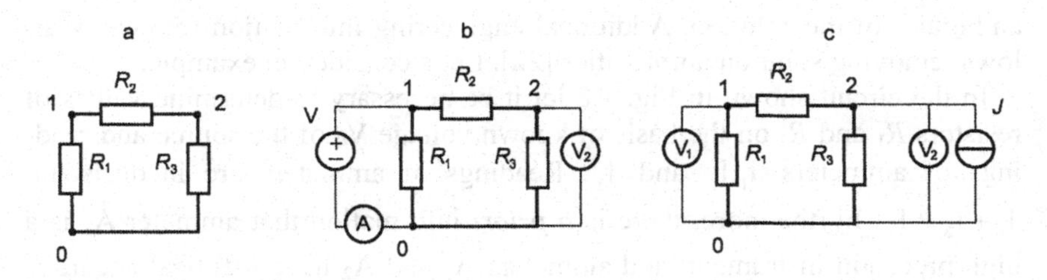

Fig. 1.7. Measured electric circuit **a)** and circuit diagrams for connection of measuring instruments and sources **b)**, **c)** to determine parameters of its elements

Values R_1 and R_3 are determined from the following system of equations (assuming that ammeters and voltmeters are ideal):

$$I_1^{(1)} = \frac{1}{R_1}V + \frac{1}{R_3}V_{20}^{(1)},$$

$$J = \frac{1}{R_1}V_{10}^{(2)} + \frac{1}{R_3}V_{20}^{(2)}.$$

Hence, R_1 and R_3 are found and then R_2 is determined from the relation:

$$J = \frac{1}{R_3}V_{20}^{(2)} + \frac{1}{R_2}\left(V_{20}^{(2)} - V_{10}^{(2)}\right).$$

To solve this problem, other circuits for connection of sources and measuring instruments can be used that are different from the ones considered above. That will change the calculation algorithm of circuit parameters accordingly. This example shows that identification of parameters, even in such elementary cases, requires statement and performance of a series of experiments, as well as finding an algorithm to calculate circuit parameters by experimental results.

In practice various restrictions are imposed upon the choice of sources, measuring instruments and their connection circuits. Here are some examples of the most general rules and restrictions:

- sources should be such that at their connection, currents and voltages could not damage the circuit;
- the calculation algorithm of circuit parameters should be numerically stable for the use of experimental data; and
- experiments and the number of used instruments should be as few as possible.

Properties of measuring instruments also highly affect the results. Therefore, one of the features of these problems is the use of instrument mathematical models. Specific problems at identification arise when measuring instruments are nonideal and their readouts are inconsistent, which results in ambiguity of the solution. Additional engineering information frequently allows removing solution ambiguities [22]. Let's consider an example.

In the circuit shown in Fig. 1.8 let it be necessary to determine values of resistors R_1 and R_2 on the basis of known voltage V_0 of the source and readings of ammeters I_1, I_2 and I_3. Readings of ammeters are inconsistent: $I_1 + I_2 \neq I_3$. Furthermore, there is *a priori* information that ammeter A_3 is a high-precision instrument, and ammeters A_1 and A_2 have identical accuracy ratings and their readings are less valid.

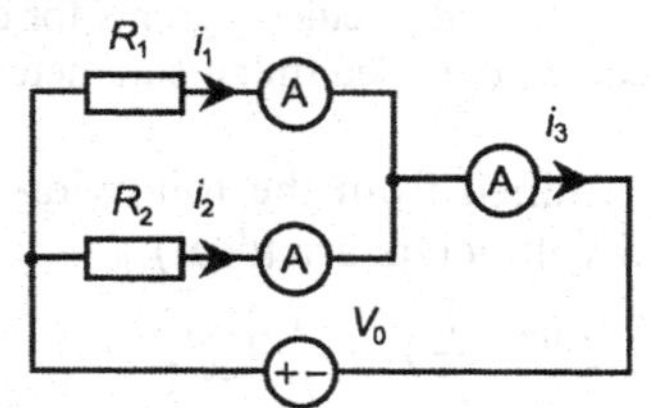

Fig. 1.8. Circuit diagram illustrating the connection of ammeters for different classes of accuracy

Taking into account that ammeter A_3 is a high-precision instrument, for the currents i_1, i_2 in the circuit we can write down:

$$i_1 + i_2 = I_3 . \tag{1.12}$$

Eq. (1.12) can be supplemented by the condition of minimum mean deviation of required currents i_1, i_2 from readings of ammeters I_1, I_2:

$$\left(1 - \frac{i_1}{I_1}\right)^2 + \left(1 - \frac{i_2}{I_2}\right)^2 \xrightarrow[i_1,\, i_2]{} \min . \tag{1.13}$$

Combined solution of Eqs. (1.12) and (1.13) allows us to find currents i_1, i_2:

$$i_1 = I_1 \frac{I_2^2 + I_1\left(I_3 - I_2\right)}{I_1^2 + I_2^2}, \qquad i_2 = I_2 \frac{I_1^2 + I_2\left(I_3 - I_1\right)}{I_1^2 + I_2^2}. \tag{1.14}$$

Let's estimate the effect of using Eq. (1.14). Let $\tilde{i}_1 = 5$ A, $\tilde{i}_2 = 1$ A be the true values of currents. Measured values by ammeters are I_1=5.3 A, I_2=1.05 A, and I_3=5.95 A ($I_1 + I_2 \neq I_3$). Currents $i_1 = 4.92$ A, $i_2 = 1.03$ A are calculated from Eq. (1.14) and they are essentially closer to the true values than originally measured currents. Using thus found i_1, i_2 values, resistances R_1 and R_2 can be calculated:

$$R_1 = \frac{V_0 - r_{A_1} i_1 - r_{A_3}(i_1 + i_2)}{i_1}, \quad R_2 = \frac{V_0 - r_{A_2} i_2 - r_{A_3}(i_1 + i_2)}{i_2},$$

where $r_{A_1}, r_{A_2}, r_{A_3}$ are the resistances of ammeters.

Use of *a priori* information allows us to remove ambiguous interpretations of experimental results and to obtain more accurate solutions.

When solving problems of this type the way of using available additional information is important. Its correct use and good mathematical description allow us to obtain correct solutions. Examples of such additional information can be:

- nameplate data of the electric circuit elements, such as maximum deviations of the elements' parameters from their nominal values, aging rate of elements, temperature-dependence of parameters, etc.;
- for periodical measurements - results of previous measurements; or
- understanding of the devices' operation principle. If, for example, there is *a priori* information that the device is a properly functioning filter, then the condition $U_{in} >> U_{out}$ is valid within its stopband.

In summary, let's note the basic features of inverse problems in electric circuits theory. In electric circuits synthesis problems it is possible to allocate phases of structure synthesis of a circuit and synthesis of its elements' parameters. The structural synthesis phase is reduced to the search of circuit topology. The characteristic feature of this phase is the solution ambiguity. The interrelation of synthesis phases is apparent by the fact that the structural synthesis can be estimated only after finding of the circuit elements' parameters, i.e. after the conclusion of the parametrical synthesis phase.

1.3 Inverse problems in electromagnetic field theory

Field synthesis and identification problems in electromagnetic field theory are regarded as inverse problems. As noted above in Section 1.1, synthesis problems are frequently referred to as problems of optimization.

Statement and solution of inverse problems in connection with calculation of electromagnetic fields are of interest for designing, manufacturing and modernization of electrical devices such as electrophysical facilities, electrical machines and equipment, high-voltage installations, induction heating apparatus, etc. [2].

Inverse problems to be solved in these areas can be divided into two groups: search of spatial and time-dependent distributions of field sources, and search of shapes and structures of bodies affecting electromagnetic field distribution. Examples of problems of both groups are discussed below.

Designing active shields intended for reducing field strength is a typical problem for finding spatial distributions of sources. Another similar type of problem is the search for the forms of current coils in order to obtain a specified field within a certain area.

Problems of finding the bodies' shapes and structures arise when defining the shapes for pole faces in electric machines and devices, or when searching for optimal shapes for high-voltage electrodes, electromagnetic shields, and waveguides. So-called moving boundary problems when determining only the shape of interface between mediums with various properties, without taking interest in their inner structure are also related to problems of this type.

In this section, the features of inverse problems in electromagnetic field theory will be examined on the basis of general statements given in Section 1.1. Particularly, characteristic types of minimized functionals and operators associating vectors of input and output variables will be discussed as well as features arising from the solution of inverse problems for static and quasi-static electromagnetic fields.

For inverse problems in electromagnetic field theory, similar to electric circuits theory, input and output influences (or quantities) can be defined. The vector $\mathbf{v} = (v_1, v_2, ..., v_N)^T$ of input actions defines densities of extraneous sources of currents $J(x,y,z,t)$ and charges $\rho(x,y,z,t)$. The vector $\mathbf{w} = (w_1, w_2, ..., w_M)^T$ of output quantities defines field potentials or strength. In the equation $\mathbf{Aw} = \mathbf{v}$ connecting these quantities, the operator $\mathbf{A}$ can be both differential and integral. This operator's type depends on the selected way the electromagnetic field analysis problem is described.

Output quantities depend on both the input vector $\mathbf{v}$, and on the parameters vector $\mathbf{p}$ that defines medium properties and its spatial distribution. In particular, this vector of parameters $\mathbf{p}$ can define the spatial distribution of dielectric permittivity $\varepsilon(x,y,z)$, magnetic permeability $\mu(x,y,z)$, or specific electric conductivity $\sigma(x,y,z)$ of media. These quantities are designated below as $\xi(x,y,z)$. In the equation $\mathbf{Aw} = \mathbf{v}$, vectors of parameters $\mathbf{p}$ and input actions $\mathbf{v}$ appear in different ways; vector $\mathbf{p}$ is a part of operator $\mathbf{A}$ whereas vector $\mathbf{v}$ is not. Furthermore, finding $\mathbf{p}$ and $\mathbf{v}$ vectors are problems which have different degrees of complexity. As a rule, searching for the spatial distributions of media is more difficult than finding the input actions vector, i.e. densities of currents $J(x,y,z,t)$ and charges $\rho(x,y,z,t)$ of extraneous sources.

In practice, it is necessary to solve both types of problems, i.e. searching for spatial distributions of media characteristics $\xi(x,y,z)$ and distributions of extraneous sources (currents and charges). In some cases searching for only media distributions cannot satisfy inverse problems criteria. These problems require simultaneous finding of the media characteristics ε,μ,σ and densities of currents $J(x,y,z,t)$ and charges $\rho(x,y,z,t)$ of extraneous sources.

1.3.1 Synthesis problems

As noted in the previous section, synthesis problems in electric circuit theory can be divided into two groups of structural and parametrical synthesis. Similar divisions can be applied for synthesis problems in electromagnetic field theory. Problems of structural synthesis involve finding the general form of the parameters vector **p**.

Structural synthesis

Let's consider the stage of structural synthesis at solution of the inverse problem of finding vector **p** which describes the media spatial distribution. Search of this solution can be performed for various classes of media. If the purpose of the inverse problem is only to find the bodies' shapes, then solutions can be searched in the class of homogeneous media. When looking not only for shapes, but also for structures of bodies, it may be necessary to search for solutions in piecewise homogeneous or non-homogeneous media classes [23]. Then both linear and nonlinear isotropic or anisotropic media can be used. In the latter cases the parameters vector **p** includes the quantity $\xi(x,y,z)$, which describes the media properties. It can be a scalar or tensor constant, depending on field strength. Selection of medium type to be used at synthesis depends on the requirements of the device, as well as on practical realizability of the solution. Defining a media class subsequently used for finding a solution can be considered as a problem of structural (topology) synthesis.

The structural synthesis stage also should be realized when searching the input actions vector **v**.

For example, when searching for currents to obtain a specified distribution of magnetic induction on the axis of a magnetic system (see Fig. 1.3) the type of magnetic field sources should first be defined. They can be separate current-carrying coils, coaxial solenoids having finite lengths with various radiuses, or sets of permanent magnets.

At the first stage of an active screen design, the winding type (for example, concentrated or distributed coils) and its spatial layout (single-layer or multilayer) should be selected.

As distinct from problems of structural synthesis in electric circuits theory, similar problems in electromagnetic field theory, anticipating the search for parameters vector **p** and actions vector **v**, are frequently solved on the basis of available experience.

At the second stage of synthesis which, just as in the case of electric circuits theory can be referred to as parametrical synthesis, optimum parameters, namely distributions of media characteristics, as well as densities and coordinates of field sources should be determined.

Parametrical synthesis

The process of searching for the parameters vector **p**, as well as actions vector **v** generally is a multistep problem. At each step, it is necessary to solve a field analysis problem for some intermediate i.e. non-optimum values of vectors **p** and **v**. The solution of analysis problems is usually extensively time-consuming. This stage is very important since the effectiveness of field analysis substantially determines the efficiency of the initial inverse problem solution.

Field analysis problems can be formulated in the form of differential or integral equations. Differential equations are written down with respect to scalar or vector potentials or to the field intensity. For example, at static field analysis the equation $\operatorname{div}(\varepsilon \operatorname{grad}\varphi) = -\rho$ should be solved for the scalar potential φ. At direct current magnetic field analysis, the equation $\operatorname{rot}\left(\mu^{-1}\operatorname{rot}\boldsymbol{A}\right) = \boldsymbol{J}$ should be solved for the magnetic field vector potential $\boldsymbol{A}$ (here ρ and $\boldsymbol{J}$ are cubic densities of charges and electric currents, respectively).

Integral equations are written down with respect to densities ρ and $\boldsymbol{J}$ of sources located on the surfaces or within volumes of bodies of required shape and structure. Since at introduction of these sources the medium is reduced to be homogeneous, then field potentials and strength can be expressed by analytic relationships with respect to extraneous sources, and sources found as a result of integral equations solution. Therefore, the criteria of inverse problems can also be expressed through the densities ρ and $\boldsymbol{J}$ of the sources.

Let's further consider the character of the restrictions imposed on the required solutions of inverse problems in electromagnetic field theory, as well as types of the characteristics vector $\mathbf{y}(\mathbf{w},\mathbf{p})$.

Restrictions

For the solving of an inverse problem, restrictions of both geometrical and physical nature should be taken into account. Geometrical restrictions are defined by the admissible sizes l of areas in which the required sources and bodies can be located. They are caused by constructive reasons, reasons of mechanical strength and serviceability of the device, and can be expressed by a set of inequalities $l_{k,\min} < l < l_{k,\max}$, where $l_{k,\min}, l_{k,\max}$, $k = \overline{1,m}$ are accordingly minimal and maximal admissible sizes. High-low bias restrictions can be imposed on the materials physical properties $\xi_{\min} < \xi < \xi_{\max}$. One-sided restrictions can be imposed on quantities describing the electromagnetic field, i.e. field strengths $E < E_{\mathrm{adm}}$, $H < H_{\mathrm{adm}}$, power losses $Q < Q_{\mathrm{adm}}$, magnetic induction $B < B_{\mathrm{adm}}$, current density $J < J_{\mathrm{adm}}$, electromagnetic force $F < F_{\mathrm{adm}}$, inductance $L < L_{\mathrm{adm}}$, etc. One-sided restrictions can also be used when maximum admissible weight $P < P_{\mathrm{adm}}$, cost $K < K_{\mathrm{adm}}$, mechanical stress $\sigma < \sigma_{\mathrm{adm}}$, temperature $T < T_{\mathrm{adm}}$, etc. are known. These restrictions can be written down as systems of inequalities $g_k(\mathbf{p}) < 0$, $k = \overline{1,m}$ or systems of equations $h_i(\mathbf{p}) = 0$, $i = \overline{1,q}$. Note that in many cases these restrictions show nonlinear dependence on field vectors.

In multicriterion problems, there is freedom of choice for the objective function and its restrictions. Therefore, one of the restrictions can be considered as a criterion function. For example, let's assume that two requirements are imposed to search for the optimum shape for a ferromagnetic screen: a) the field in the shielded area should not exceed a specified value $(B_{sh} \le B_w)$ and b) the weight of the screen should be less than an admissible value $P < P_{\mathrm{adm}}$. If particularly the screen weight $I(\mathbf{p}) = P$ is assumed to be the criterion function, then the restriction condition will become $B(\mathbf{p}) \le B_w$.

Such choices usually occur when solving multicriterion problems of optimization that will be discussed in Section 2.5.

Characteristics of inverse problems

Inverse problem characteristics $\mathbf{y}(\mathbf{w},\mathbf{p})$, intended for writing down its criteria, are expressed through field potentials and strength obtained as a result of solution of electromagnetic field equations $\mathbf{A}\mathbf{w} = \mathbf{v}$. There are two types of characteristics: local and integral. Inverse problem local characteristics are expressed through the field potential or its derivative at a point. Examples of such characteristics are the module of electric field strength $y = E = |\mathrm{grad}\,\varphi_e|$, and electric field energy cubic density $y = W_e' = 0.5\varepsilon|\mathrm{grad}\,\varphi_e|^2$. Magnetomotive force between points a and b $y = F_{ab} = \varphi_{ma} - \varphi_{mb}$, electric voltage

$y = U_{ab} = \varphi_{ea} - \varphi_{eb}$, and the local electromagnetic force component along the ort $\boldsymbol{k}$ $y = f_k = [\boldsymbol{J}, \mathrm{rot}\boldsymbol{A}]_k$ are also local characteristics.

Inverse problem integral characteristics are usually connected to the field potentials or strength through linear, surface or volume integrals. For example, field strength vector flux, resistance, inductance, capacitance, field energy, power losses, etc. can be considered as integral characteristics. Some examples of such characteristics are shown below.

Magnetic flux through surface S can be expressed through scalar or vector magnetic potential: $y = \Phi = -\int_S \mu \frac{\partial \varphi_m}{\partial n} ds$, $y = \Phi = \oint_l \boldsymbol{A} d\boldsymbol{l}$, where n is normal to the surface S and l is the bounding contour of this surface. A body's electric capacitance is: $C = \frac{q}{\varphi_e} = \frac{1}{\varphi_e} \oint_S \rho_s ds$, where q and φ_e, respectively, are the body charge and potential, and ρ_s is the surface charge density. Therefore, $y = C = -\frac{1}{\varphi_e} \oint_S \varepsilon \frac{\partial \varphi_e}{\partial n} ds$. Magnetic field energy can be expressed as $W_m = 0.5 \int_V \boldsymbol{BH} dV$, so the inverse problem characteristic becomes $y = 0.5 \int_V \mu (\mathrm{grad}\varphi_m)^2 dV$ or $y = 0.5 \int_V \frac{1}{\mu} (\mathrm{rot}\boldsymbol{A})^2 dV$. Similar relations can express other integral characteristics of inverse problems.

Definition of inverse problem characteristics $\mathbf{y(w,p)}$ allows us to write down the problem as

$$\|\tilde{\mathbf{y}} - \mathbf{y(w,p)}\| \xrightarrow[\mathbf{p} \in \Pi]{} \min,$$

which requires preliminary definition of Π – the parameter $\mathbf{p}$ domain, and selection of the norm of difference between required $\tilde{\mathbf{y}}$ and obtained $\mathbf{y(w,p)}$ characteristics of the problem.

Functionals

Let's consider typical functionals for inverse problems in electromagnetic field theory. Further on, when discussing examples of static and quasistatic fields, concrete types of included inverse problems characteristics $\mathbf{y(w,p)}$ will be defined.

In some problems the functional determines maximal values of local characteristics $\mathbf{y(w,p)}$: $I(\mathbf{p}) = \max \mathbf{y}(\mathbf{w},\mathbf{p})$. For their solution it is required to find shapes of bodies at which the maximal value of a parameter, for example, the field strength, obtains its minimum value.

In many problems the functional determines the deviation of the problem's characteristic $\mathbf{y(w,p)}$ from its required characteristic $\tilde{\mathbf{y}}$, specified in the domain V, which can be a line segment, a surface or a volume. Usually a functional is given by

$$I(\mathbf{p}) = \int_V \left[\tilde{\mathbf{y}} - \mathbf{y}(\mathbf{w},\mathbf{p})\right]^2 \, dV .$$

If the characteristic $\mathbf{y(w,p)}$ is integral and the problem involves finding the parameters vector $\mathbf{p}$ that minimizes the difference between $\tilde{\mathbf{y}}$ and $\mathbf{y}(\mathbf{w},\mathbf{p})$, then the following functional is used:

$$I(\mathbf{p}) = \left[\tilde{\mathbf{y}} - \mathbf{y}(\mathbf{w},\mathbf{p})\right]^2 .$$

Furthermore, some typical inverse problems in various kinds of electromagnetic fields and their features will be discussed.

Inverse problems of electrostatics

For the solution of inverse problems in electrostatic fields the medium is usually regarded as linear homogeneous or piecewise homogeneous. The electrostatic field potential satisfies to the equation $\operatorname{div}(\varepsilon \operatorname{grad}\varphi_e) = -\rho$ or to the equation $\operatorname{div}(\varepsilon \operatorname{grad}\varphi_e) = 0$.

In typical electrostatics problems, shapes of electrodes are searched at which the electric field in some area is homogeneous or is distributed according to a specified law [24]. Homogeneous fields are required for testing insulation materials, as well as for obtaining air gaps with maximal electric strength. In problems of electro-optics finding specified distributions of electric field, strength is required within crystals with dielectric permittivity $\varepsilon(x,y,z)$ dependent on the field strength. In such problems the inverse problem characteristic $\mathbf{y(w,p)}$ defines some component $E_k(\mathbf{p})$ of electric field strength. Taking into account that $\tilde{\mathbf{y}} = \tilde{E}_0 = \text{const}$, the functional becomes

$$I(\mathbf{p}) = \int_V \left[\tilde{E}_0 - E_k(\mathbf{p})\right]^2 \, dV .$$

The included vector $\mathbf{p}$ defines the shape of the electrodes' surface.

One of the typical problems in electrostatics is finding of shapes of electrodes (conductors) that provide specified electric strength, i.e. a specified level of electric field strength on their surfaces. In the case of optimal shaped electrodes, the voltage of an electric discharge is maximal.

Electric field strength on the surfaces of high-voltage elements electrodes is usually a function of surface points coordinates. As a rule it increases at the edges of electrodes. Violations of the condition $E \le E_{\text{adm}}$ lead to dielectric breakdown and loss of the device serviceability. Therefore, to find the optimal shape for electrodes the following condition must be met: $E_{\max} \leq E_{\text{adm}}$. Then electrodes' shapes should be found which provide minimal values for the fields' maximal strength on their surface, i.e. the condition $\min(\max E)$ must be fulfilled. In this case, criterion function can be written down as $I(\mathbf{p}) = \max\left|\text{grad}\varphi(\mathbf{p})\right|$. Electrodes of shapes satisfying this condition allow increasing voltages at set device sizes or reduced sizes at constant voltages.

Solution of such problems is usually very difficult. Their simplification is possible if the position of the point (or points) in which $E = E_{\max}$ can be specified *a priori*.

When searching of high-voltage electrodes shapes it is sometimes convenient to change from a min-max problem to a problem of maintaining a constant $\tilde{E}_0$ in points on the electrode surface *S*. In these cases the following objective function is used:

$$I(\mathbf{p}) = \int_S \left[\tilde{E}_0 - \left|\text{grad}\varphi(\mathbf{p})\right|\right]^2 ds .$$

Inverse problems of magnetostatics

When designing electric machines, devices, and other electric equipment, several parameters of interest are searched for, such as windings geometry, current distribution, and in some cases the spatial arrangements of permanent magnets. Problems involving the search for shapes and structures of ferromagnetic bodies usually have solutions that are more complex.

Problems of finding of windings, currents, and configurations, as well as layouts of contours with currents, are common in practice [25, 26]. They should be solved when designing active screens providing reduction of magnetic fields to a required level, or for obtaining change of magnetic induction by a specified law within a domain.

Devices of materials magnetization require specified fields within a certain volume. If a high accuracy solution is required, it is reached by relating distribution not only of conductors with currents, but also of ferromagnetic bodies.

There are problems of search of poles shapes for obtaining maximal mutual attraction force, or maximal magnetic flux (its concentration) within a certain area.

The magnetic fields of direct currents and permanent magnets can be described by means of scalar magnetic φ_m or vector magnetic potentials $\boldsymbol{A}$.

These potentials satisfy the equations $\operatorname{div}(\mu \operatorname{grad}\varphi_m) = -0$ and $\operatorname{rot}(\mu^{-1}\operatorname{rot}\boldsymbol{A}) = \boldsymbol{J}^e$, respectively.

The basic feature of an optimization problem when searching for structures of bodies in a magnetic field, is related to the nonlinearity of the dependence $\mu(B)$. As experience demonstrates, medium nonlinearity can substantially affect the results of inverse problems solution. For example, when finding optimum shape for a pole made of ferromagnetic material and providing uniform magnetic induction or its change by sine wave law along a given line, the solution result depends on the degree of saturation of the pole material.

The characteristic problem when searching of a pole optimum shape, usually is the magnetic induction $B_n(l)$ component normal to the line *ab,* where l is the coordinate of a point on this line. The objective functional can be given as

$$I(\mathbf{p}) = \int_a^b \left[\tilde{B}_{nw}(l)\text{-}B(\mathbf{p},l)\right]^2 dl\,.$$

Here, the parameters vector **p** defines the spatial distribution of medium magnetic permeability, i.e. the function $\mu(x,y,z)$, $\tilde{B}_{nw}(l)$ is the required magnetic induction along *ab*.

Problems involving the search of permanent magnets layouts to obtain required distribution of magnetic induction vector in some area are also nonlinear. This fact is due to nonlinearity of $\mu = \mu(H)$ not only for ferromagnetic media in magnets field, but also to nonlinearity of permanent magnets properties.

Similar to electrostatics problems, when searching for optimum shape of ferromagnetic shield, criterion function $I(\mathbf{p}) = \max\left|\operatorname{grad}\varphi_m(\mathbf{p})\right|$, as well as the function

$$I(\mathbf{p}) = \iint_S \left[\operatorname{grad}\varphi_m(\mathbf{p})\right] ds\,,$$

are used, i.e. a min-max problem is solved.

Here, φ_m is the magnetic field scalar potential and S is a surface enclosing the screen. The magnetic field's potential φ_m and its strength $\boldsymbol{H}\text{=-}\operatorname{grad}\varphi_m$ are calculated in the points of surface S. As noted above, the parameters vector $\mathbf{p}(x,y,z)$ defines the spatial distribution of the ferromagnetic material, i.e. the function $\mu(x,y,z)$.

Inverse problems of quasistatics

The problem of finding of optimal forms of coils to obtain required distribution of eddy currents in conducting bodies with the purpose of finding their uniform heating is typical for optimization in quasistatic electromagnetic fields changing by harmonic law. In other problems of this kind either currents or media distributions should be found, at which, for example, electromagnetic forces or moments of forces, exerted on constructional elements, take on their extreme values [27].

To calculate electromagnetic fields, differential equations should be solved with respect to complex quantities, namely the vector magnetic potential $\boldsymbol{A}$ and scalar electric φ_e potentials:

$$\text{rot}\left(\mu^{-1}\text{rot}\boldsymbol{A}\right) = j\omega\sigma\boldsymbol{A} - j\omega\sigma\,\text{grad}\varphi_e + \boldsymbol{J}^e,\ \text{div}\left(\text{grad}\varphi_e\right) = \text{div}\boldsymbol{A},$$

or with respect to complex function of current $\boldsymbol{T}$ (vector potential of electric current) and scalar magnetic potential φ_m:

$$\text{rot}\left(\mu^{-1}\text{rot}\boldsymbol{T}\right) = j\omega\sigma\boldsymbol{T} - \text{grad}\varphi_m,\ \text{div}\left(\mu\,\text{grad}\varphi_m\right) = \text{div}\boldsymbol{T}.$$

The feature of solution of problems of this type involves the necessity of calculation of complex quantities that results in doubling of the number of unknowns in comparison with the problems of static magnetic fields calculation.

In such problems restrictions on medium specific electric conductivity ($0 \le \sigma \le \sigma_{\max}$), on emitted power and temperature ($T \le T_{\text{adm}}$) are usually imposed.

There are very few solved optimization problems of quasistatic electromagnetic fields varying in time by an arbitrary law. There are, for example, problems of induction heating of conducting bodies by means of eddy currents flowing within their volume or on their surface [28]. The shapes and layouts of coils with currents as well as current time-dependence are determined, which provide uniform heating of the conducting body during a specified time interval, i.e. the dependence $J(x,y,z,t)$ is found.

The calculation of current density $J(x,y,z,t)$ to achieve the required purpose of optimization is similar to the problem of search of optimum control of a system with distributed parameters. Though methods of solution of these problems are well-known, finding these solutions is complicated further by the necessity to look towards a ladder-field problem. If windings are fed from a voltage source, then the required law of current variation can be obtained only by joint solution of the electric circuit and electromagnetic field equations. Electric circuit equation $i(t)R + d\Phi/dt = u(t)$, $\Phi = \int_S \boldsymbol{B}d\boldsymbol{s}$ (Φ is the magnetic flux linked to a winding, R is the circuit resistance) is supplemented with electromagnetic field equations:

$$\mathrm{rot}\left(\mu^{-1}\mathrm{rot}\boldsymbol{A}\right) = -\sigma\frac{\partial \boldsymbol{A}}{\partial t} - \sigma\,\mathrm{grad}\left(\frac{\partial \varphi_e}{\partial t}\right),$$

$$\mathrm{div}\left(\mathrm{grad}\varphi_e\right) = \mathrm{div}\boldsymbol{A}\,,\ \ \boldsymbol{B} = \mathrm{rot}\boldsymbol{A}\,.$$

At numerical calculations the number of optimized parameters, or, in other words, the number of components of the vector **p**, is a factor affecting the solution efficiency. It depends on the required solution accuracy and can vary over a wide range - from units up to hundreds of thousands. In the case for low desired accuracy, when the maximal relative deviation of design value from the required value is of order $10^{-2} \div 10^{-3}$, the number of optimization parameters can be $10 \div 15$. At deviations of the order of $10^{-4} \div 10^{-5}$ this number can increase significantly. Especially large numbers of parameters occur when searching medium structure and its distribution within a volume.

1.3.2 Identification problems

Alongside with synthesis problems, there are well-known identification problems concerning definition of shapes (in some cases, also structures) of bodies that disturb the uniformity of media, as well as problems of diagnostics (defectoscopy). In diagnostics problems, shapes and structure of defects (especially cracks) are determined by means of analyzing the distribution of eddy currents in conducting bodies.

Presence of minerals in the ground breaks its electric or magnetic uniformity that results in fields being distorted in the earth's stratum and to redistribution of potentials of electrodes used to feed current into the ground. Analysis of the potential distribution on the ground surface allows one, as a result of the inverse problem solution, to determine the character of infringement of the ground medium uniformity and judging which minerals are present. This, in turn, enables us to draw a conclusion on expediency of investments in the mining of certain deposits that can considerably cut down expenses for costly well boring. In a similar fashion, by using the inverse problem solution, one may find the position of a hidden, e.g. underwater, object by measuring electric potentials on the water's surface produced by field sources distributed in the water.

Let's write down the identification problem's equation, its characteristics and the minimized functional. The electric field in a conducting medium (for example, in the ground) with specific electric conductivity $\sigma(x,y,z)$ satisfies the equation $\mathrm{div}(\sigma\,\mathrm{grad}\varphi) = -\rho$. The quantity $\rho(x,y,z)$ defines the density of the electric current fed into the medium, and the potential $\varphi(x,y)$ is the potential available for measurement on the surface. Vectors of inputs **v** and outputs **w** in the inverse problem operator equation $\mathbf{A}(\mathbf{p})\mathbf{w} = \mathbf{v}$ are the quantities $\rho(x,y,z)$ and $\varphi(x,y)$, accordingly. The parameters vector **p** defines the required distribution of $\sigma(x,y,z)$ and the expression $\operatorname{div}\sigma\operatorname{grad}$ is the differential operator $\mathbf{A}(\mathbf{p})$.

The inverse problem characteristic **y(w,p)** is the potential $\varphi(x,y)$ in the points of the grounds' surface. In view of $\tilde{y} = \tilde{\varphi}(x,y)$, the minimized functional becomes

$$I(\mathbf{p}) = \iint_S \left[\tilde{\varphi}(x,y) - \varphi(\mathbf{p},x,y)\right]^2 dS .$$

Devices of so-called eddy current defectoscopy are commonly used in practice. They use electromagnetic signals as a source of a time-dependent test (probe) field. They induce eddy currents in conducting bodies (for example, pipes or metal bands) searching for defects. The character of eddy currents distribution depends on presence of cracks and on their shapes. Detecting a crack and determining its shape is possible by measuring the emf received by sensors located on the surface of the conducting body and comparing it with the emf of sensors located on bodies without defects [29,30,31].

Identification problems in the electromagnetic field theory, similar to electric circuits theory, become complicated in connection with the necessity for the search of a source, which is optimally arranged, and on account of measurement errors.

In summary, let's note the basic features of inverse problems in electromagnetic field theory.

Two basic types of problems can be emphasized: search of spatial distribution and time-dependence of densities of field sources and search of spatial distribution of media. Problems of search of media distribution are characterized by large numbers of parameters and demand much more time to find a solution.

Inverse problems characteristics can be divided into local and integral. The latter are rather various and define such quantities as conductivity, capacitance, inductance and mutual inductance, electromagnetic force, magnetic flux, power losses, electric and magnetic fields energy, etc.

For inverse problems in electrostatics, a min-max statement is typical when searching shapes of conductors' surfaces that provide minimal value for the maximal field strength.

Typical problems in magnetostatics, that involve finding of media distributions providing magnetic field uniformity in a certain area or its change according to a specified law, demand taking into account nonlinear dependence of media properties from the magnetic induction.

Inverse problems in quasistatic electromagnetic fields require joint solving of circuit and field equations.

Inverse problems in electromagnetic field theory are rather time-consuming, as the number of parameters, when determining the media properties by numerical optimization, can reach tens to hundreds of thousands.

Practical inverse problems in electromagnetic field theory can demand finding solutions with high accuracy when the local relative error of deviation between the inverse problem characteristic and the required one is of order 10^{-4}-10^{-5}.

For successful solution of inverse problems in electromagnetic field theory the use of effective methods of electromagnetic field numerical calculation is necessary.

References

1. Curtis, E.B., and J.A. Morrow (2000). *Inverse Problems for Electrical Networks*. Singapore, NY: World Scientific Publishing Co.
2. Neittaanmaki, P., M. Rudnicki, and A. Savini (1996). *Inverse Problems and Optimal Design in Electricity and Magnetism*. Oxford:Claredon Press.
3. Korn, G.A., T.M. Korn (2000). *Mathematical Handbook for Scientists and Engineers: Definitions, Theorems, and Formulas for Reference and Review*. New York:Dover Publication.
4. Hill, P.E., W. Murray, and M.H. Wright (1995). *Practical Optimization*. London: Academic Press.
5. Reklaitis, G.V., A. Ravindran, and K.M. Ragsdell (1983). *Engineering Optimization*. New York: John Wiley and Sons.
6. Tikhonov, A., and V. Arsenin (1977). *Solutions of ill-posed problems*. Washington D.C.:Winston.
7. Benenson, Z. et al. (1981). *Simulating and computing optimization of electronic devices* (in Russian). Moscow: Radio and Sviaz.
8. Lanne, A.A. (1969). *Optimal synthesis of linear electric circuits* (in Russian). Moscow: Sviaz.
9. Lanne, A.A. (1985). *Nonlinear dynamic systems: synthesis, optimization, identification* (in Russian). Leningrad: VAS.
10. Gantmacher, F.R. (1998). *The theory of matrices*. AMS, vol 1-2.
11. Gielen, G.G.E., P. Wambacq, and W. Sansen (1994). Symbolic Analysis Methods and Applications for Analog Circuits. *Proc IEEE*, vol 82, no2:287-303.
12. Wambacq, P., G.G.E. Gielen, and W. Sansen (1998). Symbolic Network Analysis Methods for Practical Analog Integrated Circuits: A Survey. *IEEE Trans. Circuits syst*, pt II, vol 45, no10:1331-1341.
13. Norenkov, I.P. (2000). *Basics on automatic design*. Moscow: MGTU.
14. Riss, F., and B. Sekefal'vi-Nad' (1979). *Lectures on Functional Analysis* (in Russian). Moscow:Mir.

15. Roy, B. (1972). Problems and Methods for Multiple Objective Functions. *Mathematical Programming*. vol 1, no2: 239-266.
16. Baesler, I., and I.K. Daugavet (1993). Approximation of Nonlinear Operators by Volterra Polynomials. *Amer Math Soc Transl* (2), vol 155: 47-57.
17. Volterra, V. (1959). *Theory of Functionals and of Integral and Integro-diffential Equations*. NY: Dover Publ.
18. Wiener, N. (1966). *Nonlinear problems in random theory.* Cambridge: Ma MIT.
19. Eykhoff, P. ed. (1981). *Trends and progress in system identification*. England: Pergamon, Oxford.
20. Eykhoff, P. (1974). *System Identification*. London:John Wiley and Sons.
21. Danilov, L.V. (1987). *Volterra-Picard series in nonlinear electric circuits theory* (in Russian). Moscow: Radio and Sviaz.
22. Demirchian, K.S., and P.A. Butyrin (1988). *Simulating and computings of electric circuits* (in Russian). Moscow: High School.
23. Dyck, D.N., and D.A. Lowther (1996). Automated design of magnetic devices by optimizing material distribution. *IEEE Trans Magn*, vol 32, no3:1188-1193.
24. Kim, D., et al. (2003). Generalized continuum sensitivity formula for optimum design of electrode and dielectric contours. *IEEE Trans on Magn*, vol 39, no3: 1281-1284.
25. Byun, J., et al. (2002). Topology Optimization for Superconducting Coil Distribution with Critical Current Constraint. *IEEE Conference on Electromagnetic Field Computation*, Perugia, Italy: 194.
26. Kwak, I., et al. (1998). Design Sensitivity of Electro-Thermal Systems for Exciting-Coil Positioning. *International Journal of Applied Electromagnetics and Mechanics*, vol 9, no3: 249-261.
27. Kim, D., et al. (2002). 3D optimal shape of ferromagnetic pole in MRI magnet of open permanent-magnet type. *IEEE Trans on Appl Superconductivity*, vol 12, no 1: 1467-1470.
28. Byun, J., and S. Hahn (1997). Optimal Design of Induction Heating Devices Using Physical and Geometrical Sensitivity. *Proceedings of the Fourth Japan-Korea Symposium on Electrical Engineering*, Seoul, Korea: 55-58.
29. Lee, H., et.al. (1994). An inverse analysis for crack identification in eddy current NDT of tubes. *IEEE Transactions on Magnetics*, vol 30, no5: 3403-3406.
30. Koh, C.S., et al. (1997). The application of artificial neural network to defect characterization in eddy current NDT. *NDT and E International*, vol 30,no5:328.
31. Pávó, J., and K. Miya (1994). Reconstruction of crack shape by optimization using eddy current field measurement. *IEEE Trans on Magnetics*, vol MAG-30:3407-3410.

Chapter 2. The Methods of Optimization Problems Solution

2.1 Multicriterion inverse problems

Any design electric device should satisfy several criteria, and each individual criteria k is associated with an objective functional $I_k(\mathbf{p})$. Therefore in most practical cases when simultaneous fulfilment of conditions $\min_{\mathbf{p}} I_k(\mathbf{p}), \quad k=\overline{1,N}$ is desirable, solving of multicriterion (multiobjective) inverse problems may be required [1]. Hence various criteria may have inconsistent character; individual objective functionals $I_k(\mathbf{p})$ cannot simultaneously take on their minimal values at the same vector $\mathbf{p}$.

Let's consider some examples. When designing an electromagnet, it is desirable that its winding dissipate minimal power, whereby maximizing the force developed by the electromagnet. From the engineering point of view both conditions are justified, but on the other hand they are inconsistent, as increasing the magnetic force demands an increase of current density and, hence, gives rise to losses.

At magnetic field synthesis, desire to provide maximum field density within a certain area conflicts with attempts to reduce the current and power consumption in the circuit that produces this field.

Condition to provide maximum rate of pulse rise on the output of a circuit usually contradicts the desire to have pulses with as flat a top as possible.

At synthesis of electric filters several inconsistent requirements can be produced, such as minimal deviation of its gain-frequency characteristic $K(\omega)$ from a specified one, minimal number of elements, minimal dissipation power, minimal delay at signal transmission, etc.

Let's consider an example of an inverse problem with two objective functionals. The problem includes finding of the shape of a polar tip at which its weight is minimal and the magnetic flux $\Phi_0 = -\mu_0 \int_a^b \frac{\partial \varphi_m}{\partial n} dl$ (here φ_m is the scalar magnetic potential), entering the area *abcde* (Fig.2.1) through the line *ab*, it is focused on the line *de* at the opposite side of this area. For the scalar magnetic potential, homogeneous boundary conditions of the first type ($\varphi_m = 0$) at the line *cde* and homogeneous boundary conditions of the second type $\left(\frac{\partial \varphi_m}{\partial n} = 0\right)$ at lines *ega* and *cfb* are assumed.

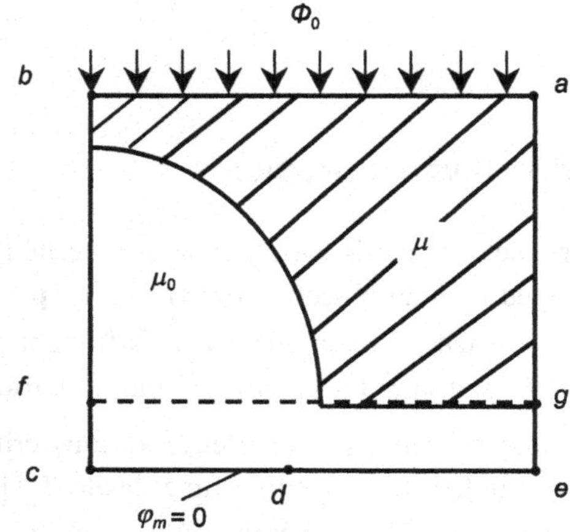

Fig. 2.1. Search of media distribution for focusing the magnetic flux on the segment *de*

A static magnetic field is assumed to be plane-parallel. Scalar magnetic potential satisfies the equation $\operatorname{div}\mu\operatorname{grad}\varphi_m = 0$ over all area *abce*. The area of the pole forming material location is confined within the contour *abfg*.

To solve this problem for the most elementary case it is necessary to find the function $\mu(x,y)$ in the specified area occupied by the pole material with typical constraints on material magnetic permeability $\mu_{\min} = \mu_0$ and $\mu_{\max} = \mu$. Then the pole shape will be defined as the boundary of media having minimum and maximum magnetic permeability. Components of parameters vector **p** will be magnetic permeability values within portions, into which *abfg* area is divided.

Here, we have two partial objective functionals. One of these is equal to $I_1(\mu) = \mu_0 \int\limits_e^d \frac{\partial \varphi}{\partial n} dl$ and is proportional to the magnetic flux through the line *de*. The second objective functional $I_2 = P$ is the pole weight which depends on its sectional area shown in Fig. 2.1 as the shaded portion.

Objective functionals I_1 and I_2 are interconnected. Let's assume, for example, that the pole's shape provides maximum magnetic flux through the line *de*. Reducing the pole weight *P*, the magnetic flux through the line *de* decreases and at the limit $I_2 = P = 0$ when there is no pole, we have $I_1 = -\Phi_0/2$ (at *de*=*cd*), that is field focusing does not occur. Therefore objective functionals I_1 and I_2 are inconsistent and cannot reach their minimal values at the same vector **p**, i.e. at a certain pole shape. Solutions can satisfy one or both criteria but not in full measure.

Situations similar to the above are considered typical for parametrical synthesis problems with purpose of designing maximum lightweight, power-saving and reliable devices - requirements that are formulated in partial criteria of inverse problems. Similar to the example discussed above, there are usually no solutions which simultaneously satisfy all criteria.

The area Π, where search of solutions parameters vector **p** is carried out, can be divided into three parts: the area of effective solutions A_1, of weakly effective solutions A_2 and the remaining part $A_3 = \Pi \setminus A_1 \setminus A_2$, which can be named as the area of ineffective solutions.

The main feature of effective solutions area (set) A_1 is that none of the inverse problem criteria can be improved in this area without the worsening of another criterion.

The main feature of weakly effective solutions area (set) A_2 is that in this area simultaneous improvement of all criteria of the inverse problem is impossible. That is, one or several criteria can be improved simultaneously without worsening any of the others. Correspondingly, in area A_3 simultaneous improvement of all criteria is possible.

Solution of a multicriterion inverse problem with inconsistent criteria involves finding the effective solutions area. Obviously, weakly effective solutions represent considerably less interest for practice in comparison with the effective ones. Application of weakly effective solutions is justified, because they frequently turn out as a result of solution of multiextremal problems.

A solution **p** is regarded as optimal by Pareto (W. Pareto, 1897) if a vector $\mathbf{p}_1$ can be specified, for which $I_k(\mathbf{p}_1) \leq I_k(\mathbf{p})$, $k = \overline{1,N}$, where at least one of the inequalities is strict. That is, a solution belongs to the area of effective or Pareto-optimal solutions if it is impossible to improve any of the individual criteria by its change, not having worsened another criterion at the same time.

The concept of Pareto-optimal set of vectors **p** does not provide the algorithm of its construction. Yet finding of boundaries for this set of solutions is highly interesting for practical purposes. Knowledge of a device Pareto set allows discarding all other solutions at further designing and thus reducing expenses for search of the best on the basis of any additional, frequently intuitive and badly formalizable criteria. In other words, it can be asserted that any solution $\mathbf{p}_{opt}$ belongs to the area A_1 of effective solutions.

Let's examine the disposition of areas A_1, A_2 and A_3 in the space of two inconsistent objective functionals I_1 and I_2. By lying off I_1 and I_2 values along plane coordinate axes, the areas A_1, A_2 and A_3 of solutions can be graphically represented (Fig. 2.2). Here the line *bc* defines the area A_1 of effective solutions. Indeed, when moving from a point on the line *bc* to another point on the same line, i.e. when reducing one of the objective functionals, the other objective functional apparently grows. Lines *ab* and *cd* define the area A_2 of weakly effective solutions, as the solution at the point *b* is preferable to the solution at the point *a*, and the solution at the point *c* is preferable, to the solution at the point *d*.

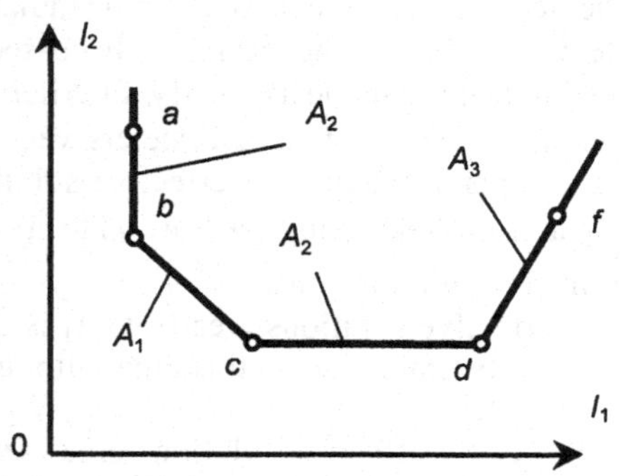

Fig. 2.2. Areas of effective, weakly effective and non-effective solutions of the optimization problem

The solution at the point *f* can be simultaneously improved for both objective functionals and, therefore, the line *df* defines the area A_3 of solutions. Adjacent points can be connected by curves - either concave or convex. Generally, in case of a large number of criteria, the set of Pareto-optimal solutions may consist of several disconnected subsets and finding of their boundaries may be a difficult task.

Hence, in many practical cases precise computations of effective solutions area demand too much time; simplified approaches must be applied that allow finding of approximate sets of effective and weakly effective solutions.

Let's consider some approaches of estimation of solutions sets A_1, A_2 and A_3. These approaches involve reducing the vector of partial objective functionals $I_k(\mathbf{p})$, $k=\overline{1,N}$ to a scalar and searching for a vector **p**, at which this scalar functional reaches its minimum.

In linear convolution method, partial objective functionals are united in a scalar by means of weighting coefficients $\alpha_k > 0$, $k=\overline{1,N}$. Thus, the problem is reduced to a search for the minimum of the scalar functional:

$$\min_{\mathbf{p}\in\Pi} \sum_{k=1}^{k=N} \alpha_k \left\| \tilde{\mathbf{y}}_k - \mathbf{y}_k(\mathbf{w},\mathbf{p}) \right\|. \tag{2.1}$$

Relationships between the coefficients α_k can be defined by our conceptions with regard to importance of either one or another criterion.

This approach transforms the problem with a vector objective functional to a problem of selection of coefficients α_k, which essentially determine the derived solution. Note that criteria included in Eq. (2.1) need preliminary normalizing to provide their comparison and, in particular, their summation.

In the above discussed problem of magnetic flux focusing, let's assume that the coefficient α_1 at the objective function $I_1 = -\Phi$ is equal to 0.7, and the coefficient α_2 at the objective function $I_2 = P$ is equal to 0.3. Then the scalar functional, formed by the method of linear convolution, becomes $I = 0.7I_1 + 0.3I_2$. The pole shape found by minimization of this functional certainly will differ from the pole shape found at use, for example, of the scalar functional $I = 0.8I_1 + 0.2I_2$. Thus, it is necessary to solve the problem of choice of coefficients α_k values.

This approach for construction of scalar objective functionals has a built-in contradiction that is shown below by an example. Assume that a problem of designing an electromagnet should be solved by inconsistent partial criteria having minimum dissipated power $P(\mathbf{p})$ and maximum developed magnetic force $F(\mathbf{p})$. Putative (desired) values of these quantities P_0 and F_0 may expediently be used as normalizing values. Then the scalar functional (see Eq. (2.1)) for this problem is given by

$$\min_{\mathbf{p}\in\Pi}\{\alpha_1 P(\mathbf{p})/P_0 + \alpha_2 F(\mathbf{p})/F_0\} = \min_{\mathbf{p}\in\Pi}\{\alpha_1 I_1(\mathbf{p}) + \alpha_2 I_2(\mathbf{p})\}. \tag{2.2}$$

Assume that both criteria are equally important. Then $\alpha_1 = \alpha_2 = 0.5$. After solving Eq. (2.2) and evaluating the results, we find $I_1(\mathbf{p}_{opt}) >> I_2(\mathbf{p}_{opt})$. This means that criterion I_2 did not make any essential impact upon the solution of Eq. (2.2) that does not correspond to our notion about the importance of this criterion. The reason of this discrepancy is in the wrong choice of normalizing values. However, to know their "correct" values it is necessary to know the solution of Eq. (2.2) which, in turn, depends on these values.

Though the choice of coefficients α_k, $k = \overline{1,N}$ presents certain difficulties, the method considered above of reducing of vector criterion to scalar is widely used. This fact in many respects relates to the work by prototypes in engineering practice, when the new device is close to a previously developed or well investigated one that allows setting good approximations for normalizing values.

Another approach to introduce the scalar objective functional is based on the principle that some objective functionals' values are regarded as optimum if they belong to some range, acceptable to the designer, that frequently occurs in technical problems. Assuming, for example, that the weight of a magnet pole cannot exceed a certain value P_{up}, the problem of magnetic flux focusing can be formulated as follows: to find a vector **p**, delivering minimum to the functional $I_1 = -\Phi$, provided that the pole weight $P \leq P_{up}$.

If in a multicriterion problem one of criteria, for example I_1, can be assigned as the main criterion, then other criteria can be considered as constraints of the form $I_k \leq \beta_k$. Then criteria $I_k, k = \overline{2,N}$ will be fulfilled approximately, since exact, i.e. minimal values of β_k are unknown. Lower estimations $\beta_{k,\min}$ for β_k can be derived, solving, by turns, problems of minimization of partial functionals

$$\mathbf{p}_{k,\min} = \arg \min_{\mathbf{p} \in \Pi} I_k(\mathbf{p}), \quad \beta_{k,\min} = I_k(\mathbf{p}_{k,\min}), \quad k = \overline{2,N}. \tag{2.3}$$

Here at minimization of the k-th functional remaining, $N-1$ criteria are not taken into account. It needs to be emphasized that as a result of the solution of "single-criterion" problems (see Eq. (2.3)), only lower estimations for β_k are found, whereas their values remain undetermined.

Let the first criterion be the main criterion. Then the multiextremal problem can be written down as follows:

$$\begin{aligned} &\min_{\mathbf{p} \in \Pi} \left\| \tilde{\mathbf{y}}_1 \text{-} \mathbf{y}_1(\mathbf{w},\mathbf{p}) \right\|, \\ &\left\| \tilde{\mathbf{y}}_k \text{-} \mathbf{y}_k(\mathbf{w},\mathbf{p})) \right\| \leq \beta_k, \qquad k = \overline{2,N}. \end{aligned} \tag{2.4}$$

This method of reducing the vector criterion to a scalar one, referred to as the method of the main criterion, has several advantages in comparison with the method considered above. However, the choice of coefficients β_k, $k = \overline{2,N}$ is as intricate as the choice of normalizing values. Generally, various sets of constraints β_k, $k = \overline{2,N}$ result in various solutions of Eq. (2.4).

Choice of the most "rigid" constraints $\beta_k = \beta_{k,\min}$, $k = \overline{2,N}$ will most probably lead to the absence of any solutions for the problem (see Eq. (2.4)). That is, no solution will be found which simultaneously satisfies all constraints. Easing of constraints $\beta_k = \alpha_k \beta_{k,\min}$, $\alpha_k > 1$, $k = \overline{2,N}$ by means of introduction of coefficients α_k reverts to the problem of their choice discussed above in connection with the method of linear convolution.

The condition of best choice of β_k, $k = \overline{2,N}$ requires that all constraints become equalities at the point of solution. Indeed, if one of the constraints (for example the k-th) in the solution of Eq. (2.4) holds as an inequality, then the value β_k can be reduced and thereby an optimum solution of Eq. (2.4) will be derived. On the whole, use of this approach is effective if there is additional information on the optimal values of β_k, $k = \overline{2,N}$, i.e. when a prototype of the design device is available. Therefore, methods ensuring choice of normalizing coefficients or coefficients β_k, $k = \overline{2,N}$ during the problem solution are of more interest.

Let's consider the following maximization problem:

$$
\begin{gathered}
F = \max_{\beta_k} \sum_{k=1}^{k=N} \lg(\beta_k), \\
\left\| \tilde{\mathbf{y}}_k - \mathbf{y}_k(\mathbf{w},\mathbf{p})) \right\| / y_{k0} \le \beta_k, \qquad k = \overline{1,N},
\end{gathered}
\tag{2.5}
$$

where all partial criteria (see Eq. (2.1)) are regarded as constraints. Here an increase of the maximized functional F is accompanied by tightening the constraints. Quantities $y_{k0} > 0$, $k = \overline{1,N}$ are normalizing coefficients. Uncertainty of choice of normalizing coefficients y_{k0}, $k = \overline{1,N}$ in Eq. (2.5) does not have a strong impact on the solution, and errors of some 1-2 order of magnitude at their choice are not of any crucial importance. It should be noted that this approach gives good results at other forms of maximized functional, in particular at $F = \max_{\beta_k} \sum_{k=1}^{k=N} \beta_k^2$.

To reduce a vector objective functional to a scalar one the so-called minimax method or method of minimax convolution can be applied. In this method, similar to the first of the methods discussed above (of linear convolution), coefficients α_k are used but all functionals $\alpha_k \left\| \tilde{\mathbf{y}}_k - \mathbf{y}_k(\mathbf{w},\mathbf{p}_j) \right\|$, $k = \overline{1,N}$ are considered equivalent without assigning any main criterion. Desired parameters vector $\mathbf{p}_{opt}$ is found at the condition of minimum for one of the functionals $\alpha_k \left\| \tilde{\mathbf{y}}_k \text{-} \mathbf{y}_k(\mathbf{w},\mathbf{p}_j) \right\|$, $k = \overline{1,N}$ having the maximal value, i.e. from the condition

$$I(\mathbf{p}) = \min_{\mathbf{p}\in\Pi} \max_{\underline{k=1,N}} \left(\alpha_k \left\| \tilde{\mathbf{y}}_k \text{-} \mathbf{y}_k(\mathbf{w},\mathbf{p}) \right\| \right).$$

Thus, in the minimax method the maximal objective functional takes on its least value at the desired parameters vector $\mathbf{p}_{opt}$.

Apparently, application of the minimax method also requires introduction of coefficients α_k, determining the degree of importance for each criterion. Therefore, when choosing either one or another set of coefficients α_k, various parameters vectors $\mathbf{p}$ will be found which are considered the best within the limits of the minimax method.

Each of the above considered methods of search of multicriterion problems solution, based on reducing a vector objective functional to a scalar, allows finding, at least, a weakly effective solution. In the case for two-criterion problems these methods permit graphic interpretation. Such interpretation allows clarification of what problems all areas A_1, A_2 and A_3 of solutions can be found, by means of these methods.

In the linear convolution method lines of constant values of scalar objective functional $I = \alpha_1 I_1 + \alpha_2 I_2$ are straight lines (Fig. 2.3) with angular coefficients $k = -\alpha_1 / \alpha_2$.

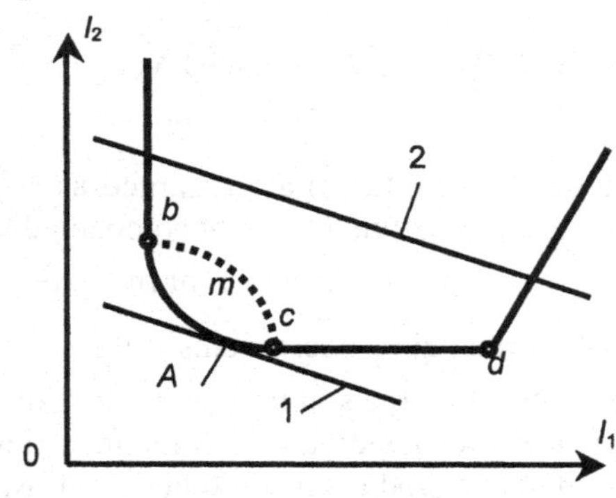

Fig. 2.3. Geometrical interpretation of the linear convolution method

In Fig. 2.3 lines 1 and 2 are shown, corresponding to constant values of I. Minimal values of I correspond to the points of line 1. For a pair of assumed values α_1 and α_2 the problem solution is determined by the point A, at which the straight line 1 and the solid curve *bmc*, defining the area of effective solutions, are tangent. Evidently, choice of other coefficients α_1 and α_2

will lead to change of I constant values lines slope and, correspondingly, to repositioning of the point of contact. Therefore, in this case the finding of all points of effective solutions area is possible by means of coefficients α_1 and α_2 variation.

However, in some cases the linear convolution method does not allow finding all effective solutions. Let, for example, effective solutions area be defined by the concave dashed line *bmc* (Fig. 2.3). At any relation of coefficients α_1 and α_2 determining the slope of scalar functional constant values line, only angular points belonging to the effective solutions area can be obtained. Thus, there are problems for which the linear convolution method does not allow finding of the whole area of effective solutions. Therefore, in the general case this method allows evaluating only the area of weakly effective solutions.

The main criterion method allows finding all effective solutions. Let, in the above example of magnetic flux focusing, the criterion $I_1 = -\Phi$ be chosen as the main criterion. As can be seen from Fig. 2.4, setting various $P \le P_{up}$, which determine the upper limit of the objective functional I_2, any of effective or weakly effective solutions can be obtained.

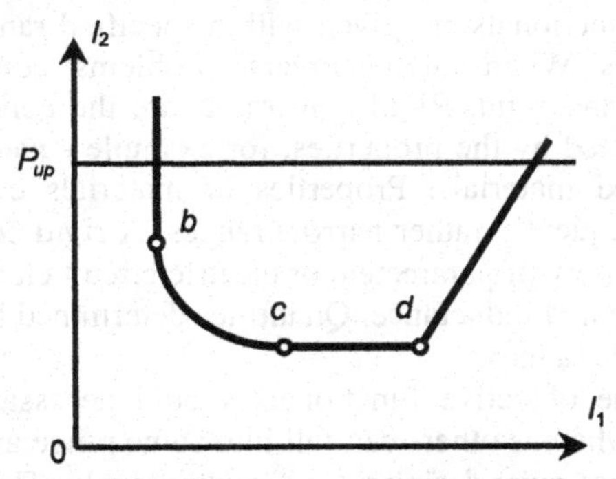

Fig. 2.4. Search of effective solutions using the main criterion method

The minimax method, as well as the main criterion method, enables finding all effective solutions of a multicriterion problem by searching of all possible values of coefficients α_k. This can also be confirmed by the following example of a two-criterion problem with scalar criterion $I = \min\ \max(\alpha_1 I_1, \alpha_2 I_2)$. Let's plot the line 1 according to equation $I_2 = (\alpha_1/\alpha_2) I_1$ (Fig. 2.5). At each point of this line, $\alpha_1 I_1 = \alpha_2 I_2$ and the function $\max(\alpha_1 I_1, \alpha_2 I_2)$ is decreasing when approaching to the coordinate origin. For any specified coefficients α_1 and α_2, this function takes minimum value at the intersection of the straight line 1 with the curve *bc* of effective solutions area, i.e. at the point *A*. Evidently, all effective solutions of this problem can be found by changing coefficients α_1 and α_2.

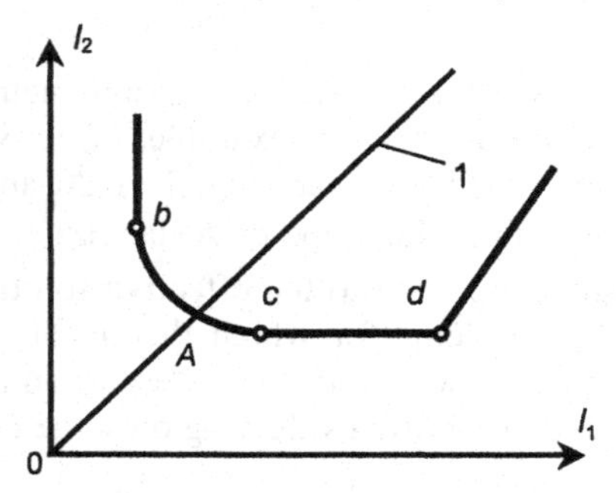

Fig. 2.5. Search of effective solutions using the minimax method

Finding effective solutions areas demands a great amount of calculations, in particular, for large numbers of criteria. Such calculations are required when designing expensive equipment.

Let's further consider solution methods of multicriterion problems where some objective functionals are given within specified ranges instead of having precise values. When solving inverse problems, constraints can be divided into weak and rigid. Rigid constraints are the constraints of physical character determined by the properties, for example - mechanical or electromagnetic, of used materials. Properties of materials can be chosen only within certain, frequently rather narrow ranges. A rigid constraint is also the condition of positivity of parameters of electric circuit elements such as resistance, capacitance and inductance. Quantities determined by weak constraints can vary over wide ranges.

Similarly, some objective functionals should necessarily reach minimal possible values, whereas others can fall into some range and, hence, the latter can be specified not quite definitely, i.e. ambiguously. Then the desired solution is not necessarily classed among feasible or infeasible solutions, but by virtue of indistinctness of constraints - among desirable or undesirable ones. In some cases indistinct descriptions of criteria and constraints can even lead to solutions more adequate to reality than definitely specified ones.

Not quite distinctly defined sets of criteria and constraints relate to so-called fuzzy sets [2,3]. For their analysis, suitable mathematical tools should be applied and rules for processing of fuzzy sets and logic rules for indistinct propositions (fuzzy logic), etc. should be established. Fuzzy sets mathematical tools can also be applied for calculating parameters that optimally satisfy inconsistent criteria.

There are means of quantitative evaluation for each indistinct proposition. The so-called membership function μ, defining the closeness of a proposition to the truth, can serve as an evaluation of this proposition truth degree.

In ordinary logic, 0 is assigned as value of a false proposition and 1 is assigned as value of a true proposition, so the membership function can be only 0 or 1. At fuzzy propositions the membership function can have intermediate values between 0 and 1. Function μ value can be used to estimate either objective functional I_K minimization degree, or constraint R_k fulfilment degree. Let, for example, losses W be acceptable if they are not to exceed 50 kW and are not acceptable if they are more than 60 kW. The membership function $\mu(W)$, shown in Fig. 2.6*a*, indicates the measure of various W values' acceptability within the range $0 \le W \le 60$ kW. The condition of closeness of current density J to the value 2A/mm^2 can be expressed, for example, by the membership function $\mu(J)$ shown in Fig. 2.6b.

In this way, membership functions can be constructed not only for objective functionals I_k, but also for constraints R_k. It should be noted that membership functions can be not only piecewise linear, as shown in above figures, but also nonlinear.

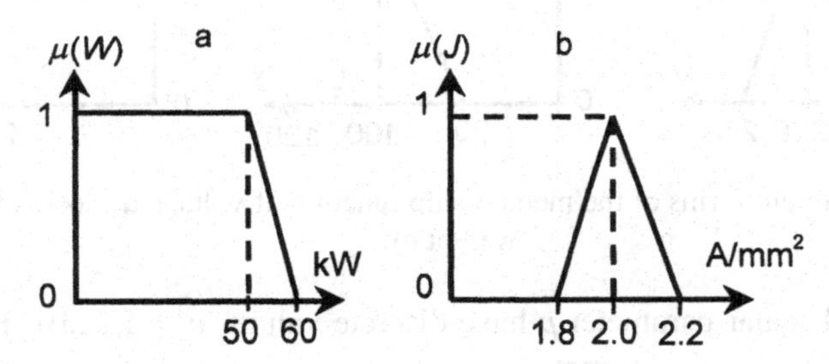

Fig. 2.6. Examples of losses membership function **a)** and current density **b)**

Sets of objective functionals $I_1(\mathbf{p}), I_2(\mathbf{p}),...,$ as well as constraints $R_1(\mathbf{p}), R_2(\mathbf{p}),...$ values can be considered as equivalent fuzzy sets characterized by membership functions $\mu(I_k(\mathbf{p}))$ and $\mu(R_k(\mathbf{p}))$.

If at $\mathbf{p}=\mathbf{p}_j$ one or more membership functions μ become zero, it means that either some optimization objectives have not been achieved or some constraints have not been fulfilled. Therefore, such a vector $\mathbf{p}_j$ of parameters cannot belong to the set of desired solutions. At the same time existence of a set of vectors $\mathbf{p}$ is possible, for which membership functions μ of objectives and constraints are positive, and which can be considered as a desired one. The vector $\mathbf{p}_{opt}$ that is preferable among the others should be chosen from this set. To search for the optimum parameters vector, various strategies can be used.

Let's consider an example of membership functions assigning and solution search strategy. Let, for example, there be partial criteria concerning device voltage and losses power, expressed by the following fuzzy propositions: voltage U should be close to 220 V, and losses power W should be equal approximately to 100 W. In addition, there is a rigid constraint on device weight g, which should not exceed 200 G. Permissible spreads of $I_1 = U$ and $I_2 = W$ can be specified as $\Delta U = \Delta I_1 = \pm 10\,\text{V}$, $\Delta W = \Delta I_2 = \pm 20\,\text{W}$. Let membership functions values $\mu_1 = \mu_2 = \mu_3 = 1$ correspond to values $U = 220\,\text{V}$, $W = 100\,\text{W}$ and $g = 200\,\text{G}$. By use of piecewise linear representation, the membership functions $\mu_1(I_1)$, $\mu_2(I_2)$ and $\mu_3(R)$ can be of the type shown in Fig. 2.7.

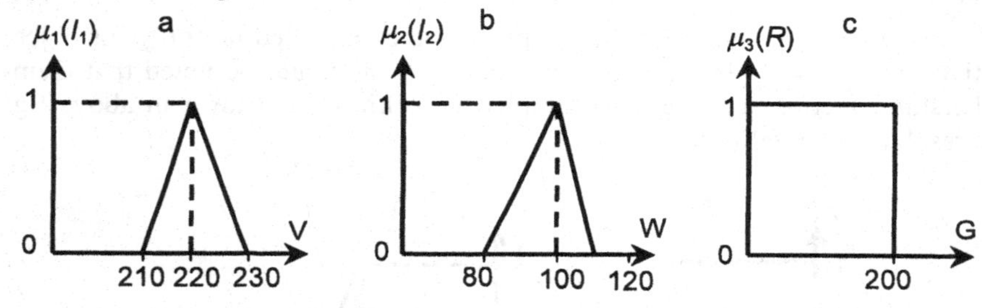

Fig. 2.7. Different forms of the membership function of voltage **a**), losses **b**), and weight **c**)

Let desired scalar parameter p have discrete values $p_j = 1,2...10$. For each value of parameter p_j, $j = \overline{1,10}$ values of voltage $U = I_1(p_j)$, losses power $W = I_2(p_j)$ and weight $G = R(p_j)$ can be found. Assume that membership functions $\mu_1(I_1(p_j))$, $\mu_2(I_2(p_j))$ and $\mu_3(R\,(p_j))$ correspond to some value p_j. Grade of membership of this set to the desired one can be estimated in various ways. For example, either by the minimal value of $\mu_{\min}(p_j) = \min\{\mu_1(I_1(p_j)), \mu_2(I_2(p_j)), \mu_3(R(p_j))\}$, or on some average from these values of the membership function. By means of choosing the minimum of μ_1, μ_2, μ_3 values for each of the possible values of parameters, a piecewise linear dependence $\mu_{glob} = \mu_{\min}(p_j)$ can be constructed shown, in Fig. 2.8. It is referred to as the global membership function.

The vector of parameters that provides maximal value of the global membership function μ_{glob} is considered as the vector $\mathbf{p}_{opt}$ of optimal parameters.

Similarly, by calculation, e.g., of root-mean-square values $\mu_{av}(\mathbf{p}_j)=\sqrt{\sum_{i=1}^{i=N}\mu_i^2(\mathbf{p}_j)}$ (N is the total number of criteria and constraints), $\mu_{av}(\mathbf{p}_j)$ dependence can be constructed and the vector $\mathbf{p}_{opt}$, at which the function $\mu_{av}(\mathbf{p}_j)$ reaches its maximum, can be determined.

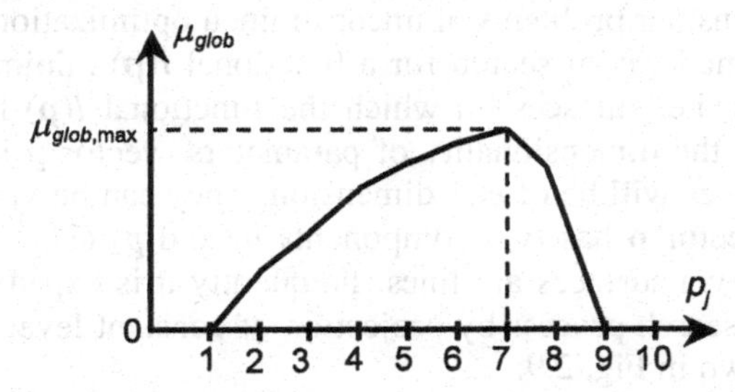

Fig. 2.8. Global membership function as a function of parameter p_j

Mathematical methods used for the solution of multicriterion problems generally allow narrowing of solutions area and only in special cases they lead to a unique solution. The fuzzy sets method also allows narrowing of optimal solutions area. If, as a result of application of known mathematical methods, there remains although narrow, but nevertheless a finite region of acceptable solutions in the range of parameters **p**, then so-called expert estimations are used. In many cases estimations by comparison of several solutions to one another cannot be formalized. In some cases experts give only intuitive solutions. However, this circumstance does not make mathematical methods less valuable. Their application allows narrowing the area of possible solutions by rejecting the worse ones. Then remaining solutions turn to be approximately equivalent.

2.2 Search of local minima

The majority of inverse problems discussed in Chapter 1 can be reduced to search of functional $I(\mathbf{p})$ lower boundary with constraints $g_i(\mathbf{p})\le 0, i=\overline{1,m}$, $h_i(\mathbf{p})=0,\ \ i=\overline{1,q}$, i.e. the so-called constrained minimization (optimization) problems. Actually finding a solution for unconstrained optimization problems is much simpler in comparison with constrained optimization problems. Therefore, problems of constrained minimization are frequently reduced to unconstrained problems. For this equality type, constraints are multiplied by scalar factors and added to the minimized functional. As for the inequality type, conditions such as $g_i(\mathbf{p}) \le 0,\ i=\overline{1,m}$, should be transformed to equalities by introduction of additional variables, and also included in the minimized functional. Thus, we arrive at problems of unconstrained optimization of a different type: the so-called expanded functional. Methods of transition to the expanded functional will be considered in Section 2.3 in more detail. In this section, we assume that such transitions have already been executed and consequently we consider problems of unconstrained optimization [4].

When studying methods of search for a functional $I(\mathbf{p})$ minimum its constant level surfaces, i.e. surfaces on which the functional $I(\mathbf{p})$ has constant values, are used. If the dimensionality of parameters' vector $\mathbf{p}$ is n, then its constant level surfaces will be of n–1 dimension. They can be visually represented when the vector $\mathbf{p}$ has two components p_1 and p_2 (Fig. 2.9). In that case the constant level surfaces are lines. Frequently it is expedient to illustrate the minimum search process by projection of constant level lines on the plane p_1, p_2, as shown in Fig. 2.9.

The configuration of level lines, as a rule, is unknown before optimization. They can be of various kinds, for example, circles (if $I(p_1,p_2)=p_1^2+p_2^2$) or ellipses (if $I(p_1,p_2)=ap_1^2+bp_2^2,\ a \ne b,\ a,b>0$). In the latter case, the functional will change more slowly along the larger main axis of the ellipse than along the smaller one. If lengths of the main axes are substantially different, we have a ravine type functional, which changes rather slowly when moving along its axis.

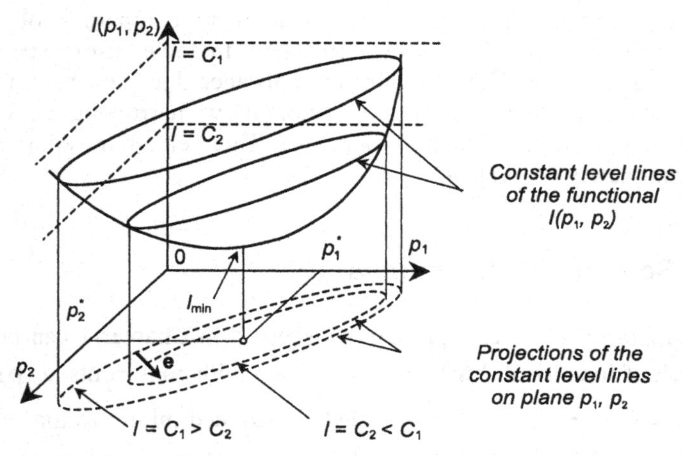

Fig. 2.9. Equal level lines of the objective function and their projection onto the plane p_1, p_2

Search of the parameters' vector $\mathbf{p}=(p_1,p_2,\ldots,p_n)^T$, which provides a minimal value of the functional, is carried out by means of the following iterative procedure:

$$\mathbf{p}^{k+1}=\mathbf{p}^k+s^k\mathbf{e}^k,$$

where k is the number of iteration, $\mathbf{e}^k$ is a vector determining the direction of change of optimization parameters, and s^k is the step of parameters change in the direction set by the vector $\mathbf{e}^k$. The step s^k can be both constant, and changing from iteration to iteration.

Methods of minimum search can be divided into two groups depending on the means of choice of the vector $\mathbf{e}^k$: those using derivatives of the functional in the point $\mathbf{p}^k$ for finding $\mathbf{e}^k$ (so-called gradient methods) and those not using them.

Let's consider methods of functional minimum search that do not require calculating its derivatives. In the simplest method, movement to a minimum is carried out by successive changes of optimization parameters' vector $\mathbf{p}$ components p_i. In this case vector $\mathbf{e}$ components are basis vectors (orts) of coordinates, therefore change of vector $\mathbf{p}$ i-th component is equivalent to change of the i-th coordinate. Each coordinate is changed, while keeping all other coordinates constant until the functional decreases. Then one must proceed to change the next coordinate. Change step dimension in the direction of each ort is determined by the method of one-parameter optimization.

This approach is realized in the Hook and Jeeves method. This method consists of two stages - the first stage to carry out so-called investigating search, and the second stage to search by a sample. Let $\mathbf{p}^0$ be the initial value of the variables' (coordinates) vector. On the i-th step of investigating search, $p_i^1=p_i^0+\Delta p_i, i=\overline{1,n}$. If the functional decreases for the coordinate i at $\Delta p_i>0$ then p_i^1 is selected as the new value with subsequent proceeding to change the next coordinate. In case of increase of the functional, the value $p_i^1=p_i^0-\Delta p_i$ should be accepted. The stage of investigating search is completed when changes of the functional at consecutive change of all coordinates are determined. At the stage of search by a sample the direction of perspective minimization vector $\Delta\mathbf{p}=(\Delta p_1,\Delta p_2,\ldots,\Delta p_n)^T$ shall be found, which then becomes the direction of change of the $\Delta\mathbf{p}$ value as long as the functional decreases. Further on, stages of investigating search and search by a sample shall be iterated.

Random search methods are also considered among methods of search of functional minimum without calculating its derivatives. These methods are simple in realization and they have low sensitivity to occurrence of errors during calculations. Random search iterative procedure is given by

$\mathbf{p}^{k+1} = \mathbf{p}^k + \boldsymbol{\alpha}^k$, where $\boldsymbol{\alpha}^k$ is a vector, having random variables as its components, with a given law of probability density distribution.

Random search methods differ by assumed (or corrected on each step of search) law of distribution of $\boldsymbol{\alpha}$. The extent of its change should be limited as points $\mathbf{p}^{k+1}$ should not be far from the points $\mathbf{p}^k$. In one modification, named the best probe method, m vectors $\boldsymbol{\alpha}$ are generated for each condition $\mathbf{p}^k$, and m values of the functional near the point $\mathbf{p}^k$ are calculated. Further, a vector $\boldsymbol{\alpha}$ is allocated, which provides maximum decrease of the functional to make a step in the direction of this vector $\boldsymbol{\alpha}$. Substantially, the functional gradient direction is statistically determined in this approach (interestingly, the number of tests m can be less than the number of coordinates n).

Algorithms of search methods, not using derivatives of the functional, are simple to understand and to realize. However, for the search for a minimum they require greater amounts of calculations of the functional in comparison with methods using functional derivatives, which shall be discussed below.

In the method of quickest descent the direction of the vector $\mathbf{e}$ is opposite to the functional gradient direction (or, equivalently, it is in the line of the functional antigradient). In this direction the functional $I(\mathbf{p})$ decreases with the maximal speed. The well-known, antigradient vector

$$\mathbf{e} = -\nabla I(\mathbf{p}) = -\left(\frac{\partial I}{\partial p_1}, \frac{\partial I}{\partial p_2}, \dots, \frac{\partial I}{\partial p_n}\right)^T$$

is normal to the level surface $I(\mathbf{p}) = \text{const}$, in the point of its calculation. Correspondingly, its projection is perpendicular to projections of constant level lines, as shown in Fig. 2.1.

Using a normalized vector of gradients, so that it has a unit length, we then assume that $\mathbf{e}^k = -\nabla I(\mathbf{p}^k) / \|\nabla I(\mathbf{p}^k)\|$. In this case the norm of increment $\Delta\mathbf{p}^k = s^k \mathbf{e}^k$ is defined by the size of the step s^k, which is assumed to be constant or chosen by a definite algorithm. In the method of quickest descent, steps s^k are found by the solution of the one-parametric problem:

$$I(\mathbf{p}^k + s^k \mathbf{e}^k) \xrightarrow[s^k]{} \min. \tag{2.6}$$

It is desirable to calculate the gradient vector $\Delta I(\mathbf{p})$ components $\partial I/\partial p_i$ analytically, but at the absence for such opportunity the following expression can be used:

$$\frac{\partial I}{\partial p_i} \cong \frac{I(p_1, p_2, \cdots p_i + \varepsilon_i, \cdots p_n) - I(p_1, p_2, \cdots p_i - \varepsilon_i, \cdots p_n)}{2\varepsilon_i}, i = \overline{1,n}.$$

This seemingly effective method has not been used expansively in practice because of its slow convergence at the minimization of ravine functionals. Figure 2.10 shows a fragment of ravine functional constant level lines projections (Rosenbroke) with a typical trajectory of descent to the minimum, obtained by the method of quickest descent. The antigradient vector **e** direction does not coincide with the direction of the ravine floor. Speed of change of the functional on ravine walls is much higher than the speed of its change along the ravine floor. Therefore, movement along the ravine floor in the method of quickest descent occurs slowly as it requires carrying out a large number of rather small steps.

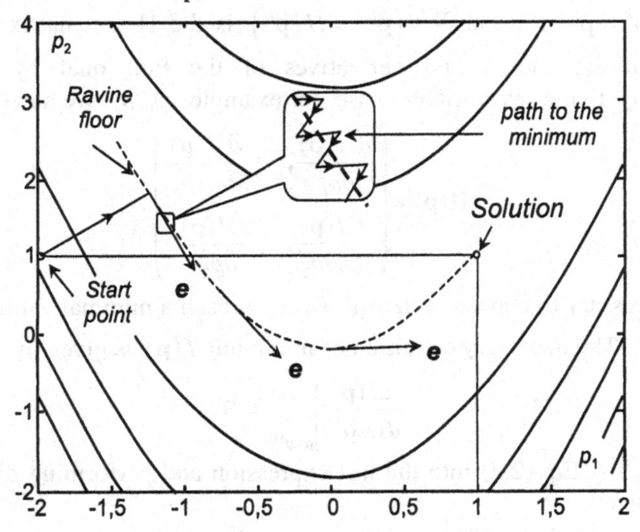

Fig. 2.10. Projections of ravine functional level lines and path of descent to its minimum, according to the gradient method

The choice of variables scales included in the functional significantly affects the speed of minimum search. Indeed, by means of changing the scales of variables it is possible to change the relief of functional constant level surfaces. If, for example, in the functional $I(\mathbf{p}) = p_1^2 + b^2 p_2^2$, which has ellipses as level lines, by changing the scale of the variable p_2 we can obtain a functional $I_1(\mathbf{p}) = p_1^2 + p_2^2$. Its minimum is reached at the same values of variables $p_1 = p_2 = 0$, however level lines represent circles instead of ellipses. A good choice of variables scales provides reduction of length of functional ravines, speeding up the search of minimum significantly.

Taking into account that the functional dependence from variable **p** is nonlinear, when searching of a minimum it is expedient to take into consideration not only its first $(\nabla I(\mathbf{p}))$, but also second derivative. Expanding $I(\mathbf{p})$ in Taylor's series about the point $\mathbf{p}=\mathbf{p}^k$ and keeping only the expansion components containing the first and second derivatives yields:

$$I(\mathbf{p})\cong I(\mathbf{p}^k)+\nabla I(\mathbf{p}^k)^T\,\Delta\mathbf{p}^k+0.5(\Delta\mathbf{p}^k)^T\nabla^2 I(\mathbf{p}^k)\,\Delta\mathbf{p}^k, \tag{2.7}$$

where $\Delta\mathbf{p}^k=\mathbf{p}-\mathbf{p}^k$, and $\nabla^2 I(\mathbf{p}^k)=H(\mathbf{p}^k)$ is the Hesse matrix with elements that are the second derivatives of the functional by variables $p_1,p_2,\dots,p_n$. In the two variables case, for example, p_1,p_2, we have

$$\mathbf{H}(\mathbf{p})=\begin{pmatrix}\dfrac{\partial^2 I(\mathbf{p})}{\partial p_1^2}, & \dfrac{\partial^2 I(\mathbf{p})}{\partial p_1\partial p_2}\\ \dfrac{\partial^2 I(\mathbf{p})}{\partial p_1\partial p_2}, & \dfrac{\partial^2 I(\mathbf{p})}{\partial p_2^2}\end{pmatrix}.$$

It is necessary to choose such $\Delta\mathbf{p}^k$ so as to reach a minimal value of $I(\mathbf{p})$ at $\mathbf{p}=\mathbf{p}^{k+1}$. The necessary condition of minimum $I(\mathbf{p})$ is given by

$$\left.\frac{dI(\mathbf{p})}{d(\Delta\mathbf{p}^k)}\right|_{\mathbf{p}=\mathbf{p}^{k+1}}=0.$$

Substituting Eq. (2.7) into the last expression and performing differentiation, we have:

$$\nabla I(\mathbf{p}^k)+\mathbf{H}^{-1}(\mathbf{p}^k)\cdot\Delta\mathbf{p}^k\Big|_{\mathbf{p}=\mathbf{p}^{k+1}}=0.$$

Then, taking into account that $\Delta\mathbf{p}^k=\mathbf{p}-\mathbf{p}^k$, we find:

$$\mathbf{p}^{k+1}=\mathbf{p}^k-\mathbf{H}^{-1}(\mathbf{p}^k)\cdot\nabla I(\mathbf{p}^k). \tag{2.8}$$

The last expression determines both search direction and step dimension. If the functional dependence of variable **p** is by square-law, then at such direction and a step of search the point of functional $I(\mathbf{p})$ minimum is reached by one step at any initial approximation $\mathbf{p}^0$. Generally, $I(\mathbf{p})$ dependence of the variable **p** is not by square-law and hence the number of steps to reach a minimum increases. Then parameter s^k (the step length) may be expediently introduced to rewrite the iterative Eq. (2.8) as

$$\mathbf{p}^{k+1} = \mathbf{p}^k + s^k \mathbf{H}^{-1}(\mathbf{p}^k) \cdot \nabla I(\mathbf{p}^k) = \mathbf{p}^k + s^k \mathbf{e}^k .$$

Analysis shows that the stated method, named the Newton method, converges to the solution $\mathbf{p}^*$ if the matrix $\mathbf{H}^{-1}(\mathbf{p})$ is positive-definite (i.e. has positive eigenvalues) over the whole domain of minimum search. The Hesse matrix $\mathbf{H}(\mathbf{p})$ is positive-definite if $I(\mathbf{p})$ is a strictly convex function. Generally, this condition does not always hold. Therefore the Newton method iterative procedure sometimes can diverge or lead not towards the point of $I(\mathbf{p})$ minimum, but away from it.

Distinctions in definition of search direction vector $\mathbf{e}^k$ by the method of quickest descent and Newton method can be seen in the following example regarding search of minimum of a square-law functional $I(\mathbf{p}) = p_1^2 + b^2 p_2^2$. On the condition that $b >> 1$, this functional has a ravine stretched along the variable p_1 (Fig. 2.11). The Hesse matrix of the functional is $\mathbf{H} = \begin{pmatrix} 2, & 0 \\ 0, & 2b^2 \end{pmatrix}$.

It has strongly differing eigenvalues $\lambda_1 = 2$ and $\lambda_2 = 2b^2$. When moving from a point A (Fig. 2.11), the search direction in the method of quickest descent is characterized by the vector $-\nabla I(\mathbf{p}_A)$, directed along the normal to a level line (not necessary to a minimum point), whereas at use of the Newton method the vector,

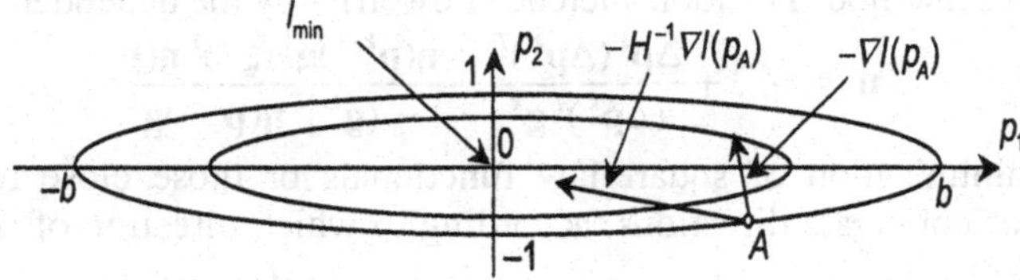

Fig. 2.11. Directions of movement by the method of quickest descent and by the Newton method

$$-\mathbf{H}^{-1}\nabla I(\mathbf{p}_A) = -\begin{pmatrix} \frac{1}{2}, & 0 \\ 0, & \frac{1}{2b^2} \end{pmatrix} \cdot \begin{pmatrix} \frac{\partial I}{\partial p_1} \\ \frac{\partial I}{\partial p_2} \end{pmatrix} = -\begin{pmatrix} \frac{1}{2}, & 0 \\ 0, & \frac{1}{2b^2} \end{pmatrix} \cdot \begin{pmatrix} 2p_1 \\ 2b^2 p_2 \end{pmatrix} = -\begin{pmatrix} p_1 \\ p_2 \end{pmatrix},$$

obviously, is directed to the point 0. Thus, during minimization the Newton method, as distinct from the method of quickest descent, allows faster movement along the ravine.

Difficulties when using the Newton method involve, firstly, the necessity of calculation and inversion of the Hesse matrix and, secondly, possible loss of its positive definiteness resulting in failure of convergence to a minimum by this method.

There are so-called quasi-Newtonian methods, in which the time-consuming calculation of the Hesse inverse matrix is replaced by a simpler calculation of its approximate value. Calculation of the matrix $\boldsymbol{\eta}(\mathbf{p}^k) \cong \mathbf{H}^{-1}(\mathbf{p}^k)$, approximating $\mathbf{H}^{-1}(\mathbf{p}^k)$, is possible without calculating second derivatives of the functional, but only finding variables $\mathbf{p}^k, \mathbf{p}^{k-1}$ and gradients $\nabla I(\mathbf{p}^k), \nabla I(\mathbf{p}^{k-1})$.

Matrix $\boldsymbol{\eta}(\mathbf{p}^k)$ is written down as $\boldsymbol{\eta}^k \cong \boldsymbol{\eta}(\mathbf{p}^{k-1}) + \Delta\boldsymbol{\eta}(\mathbf{p}^{k-1}) = \boldsymbol{\eta}^{k-1} + \Delta\boldsymbol{\eta}^{k-1}$, with a given initial matrix $\boldsymbol{\eta}(\mathbf{p}^0)$ (usually a unitary one) and calculating $\Delta\boldsymbol{\eta}(\mathbf{p})$ on each iteration by increments $\Delta\mathbf{p}^k = \mathbf{p}^k - \mathbf{p}^{k-1}$, $\mathbf{g}^k = \nabla I(\mathbf{p}^k) - \nabla I(\mathbf{p}^{k-1})$ according to a selected algorithm.

In BFGS method (Broyden-Fletcher-Goldfarb-Shanno), a more precise description of the matrix $\boldsymbol{\eta}^k$ can be given by the following recurrent dependence:

$$\boldsymbol{\eta}^k = \boldsymbol{\eta}^{k-1} + \left(\frac{1 + (\mathbf{g}^k)^T \boldsymbol{\eta}^{k-1} \mathbf{g}^k}{(\mathbf{g}^k)^T \Delta\mathbf{p}^k} \right) \cdot \frac{\Delta\mathbf{p}^k (\Delta\mathbf{p}^k)^T}{(\Delta\mathbf{p}^k)^T \mathbf{g}^k} - \frac{\Delta\mathbf{p}^k (\mathbf{g}^k)^T \boldsymbol{\eta}^{k-1} + \boldsymbol{\eta}^{k-1} \mathbf{g}^k (\Delta\mathbf{p}^k)^T}{(\mathbf{g}^k)^T \Delta\mathbf{p}^k},$$

and in DFP method (Davidon-Fletcher-Powell) – by the dependence

$$\boldsymbol{\eta}^k = \boldsymbol{\eta}^{k-1} + \frac{\Delta\mathbf{p}^k (\Delta\mathbf{p}^k)^T}{(\Delta\mathbf{p}^k)^T \mathbf{g}^k} - \frac{\boldsymbol{\eta}(\mathbf{p}^{k-1}) \mathbf{g}^k (\mathbf{g}^k)^T \boldsymbol{\eta}(\mathbf{p}^{k-1})}{(\mathbf{g}^k)^T \boldsymbol{\eta}(\mathbf{p}^{k-1}) \mathbf{g}^k}.$$

For minimization of square-law functionals or those close to them the method of conjugate directions, according to which direction of search on k-th iteration is determined by the expression $(\mathbf{e}^k)^T \mathbf{Q} \mathbf{e}^{k-1} = 0$, where $\mathbf{Q}$ is a positive-definite square matrix (for example, the Hesse matrix $\mathbf{Q} = \mathbf{H}$), can be effectively applied. The step s^k in the method of conjugate directions is selected the same way as in other gradient methods, for example, from the Eq. (2.1).

If $\mathbf{Q} = \mathbf{1}$, i.e. it is a unitary matrix, the condition $(\mathbf{e}^k)^T \mathbf{1} \mathbf{e}^{k-1} = 0$ shows that vectors $\mathbf{e}^k$ and $\mathbf{e}^{k-1}$ are mutually orthogonal. For an arbitrary matrix $\mathbf{Q}$ under the condition $(\mathbf{e}^k)^T \mathbf{Q} \mathbf{e}^{k-1} = 0$, the vector $\mathbf{e}^k$ is always orthogonal to the vector $\mathbf{Q}\mathbf{e}^{k-1}$. Therefore, $\mathbf{e}^k$ and $\mathbf{e}^{k-1}$ are called conjugate vectors. Thus, in this method the descent to the point of a minimum is carried out on conjugate, instead of orthogonal directions, as it is done in the method of quickest descent.

The number of linearly independent directions conjugated in relation to positive-defined matrix $\mathbf{Q}$ of order n, is equal to n. By execution of single descents along each of the conjugate directions the minimum of square-law functional of n variables will be found after n descents. As vectors of conjugate directions, eigenvectors of the matrix $\mathbf{H}=\mathbf{Q}$, which form a linearly independent system of n vectors, can be selected. At that, the order of choice of directions (eigenvectors) is of no importance.

This method can also be applied for search of the minimum for non-square-law functionals, however in that case the number of iterations will be infinite.

In the method of conjugate gradients the search direction $\mathbf{e}^k$ is determined by the expression $\mathbf{e}^k=-\nabla I\left(\mathbf{p}^k\right)+\omega^k\mathbf{e}^{k-1}$, where ω is a scalar. It should be selected so that directions $\mathbf{e}^k$ and $\mathbf{e}^{k-1}$ were conjugated i.e. to satisfy the condition $\left(\mathbf{e}^{k-1}\right)^T\mathbf{H}\mathbf{e}^k=0$. This requirement results in the following expression:

$$\mathbf{e}^{k+1}=-\nabla I\left(\mathbf{p}^{k+1}\right)+\mathbf{e}^k\frac{\left(\nabla I\left(\mathbf{p}^{k+1}\right)\right)^T\nabla I\left(\mathbf{p}^{k+1}\right)}{\left(\nabla I\left(\mathbf{p}^k\right)\right)^T\nabla I\left(\mathbf{p}^k\right)}.$$

Note that both methods of direct search (Hook and Jeeves method), and - to a greater extent - gradient, particularly, quasi-Newtonian methods, which have the property of fast convergence to a minimum, have found application for solution of optimization problems in electrical engineering.

Above discussed methods of numerical integration are poorly effective when searching of minima of rigid functionals. At the same time, as it will be shown further, rigidity is a characteristic property of inverse problems in electrical engineering. To search minima of such functionals special methods should be used. These methods will be discussed in Chapter 3.

As shown in Section 1.1, the Hesse matrix $\mathbf{H}(\mathbf{p})=\nabla^2 I(\mathbf{p})$ of a rigid functional $I(\mathbf{p})$ is weakly conditioned and, because of this, its equal level surfaces have specific ravine structure. Graphic representation of equal level lines allows showing only one-dimensional ravines (Fig. 2.3) for two-variable functionals. Generally, a ravine may be multidimensional. By definition [5] a doubly continuously differentiable functional $I(\mathbf{p}), \mathbf{p}\in\Pi$ is called r-ravine, if eigenvalues of the matrix $\mathbf{H}=I''(\mathbf{p})$ satisfy the following inequalities:

$$\lambda_1\geq\cdots\geq\lambda_{n-r}>>\left|\lambda_{n-r+1}\right|\geq\cdots\geq\left|\lambda_n\right|,$$

for all $\mathbf{p}\in\Pi$. The number r defines the dimensionality of the functional ravine in the domain Π. The quantity $S=\lambda_{\max}(\mathbf{H})/\left|\lambda_{\min}(\mathbf{H})\right|$, $\lambda_{\min}(\mathbf{H})\neq 0$ can serve as a degree of the functional "ravinity".

2.3 Search of objective functional minimum in the presence of constraints

When solving inverse problems it is necessary to take into consideration the conditions imposed on desired parameters, i.e. constraints of geometrical and physical character. Alongside with such "natural" constraints, introducing additional constraints is required in some problems, as a rule, inequalities having the form $I_k(\mathbf{p}) \geq 0$ or $I_k(\mathbf{p}) \leq 0$; for example, when one criterion is assumed to be the main one (i.e. the objective functional), and others are considered as conditions (see Section 2.1).

Sometimes these additional conditions (constraints) are introduced to improve inverse problem characteristics. So, as discussed in Chapter 1, the problem of field synthesis in a domain V_i has non-unique solutions because a zero field within V_i can be created in an infinite number of ways by the introduction of sources on the limiting surface of this domain. However, the solution becomes unique at the condition of minimality of the volume with the sources that provide the required field.

To find a minimum functional in problems with constraints on parameters' vector $\mathbf{p}$ is much more complex than in problems without constraints. A functional at the presence of constraints may not become its minimal value on the optimum vector $\mathbf{p}$. At the presence of constraints one must proceed to more complex conditions from the necessary condition $dI/d\mathbf{p} = 0$ of extremum of functionals without constraints. Moreover, in some cases with constraints the functional extremum may become non-unique though in their absence it was unique.

To simplify their solution, problems with constraints should be converted to problems without constraints, i.e. to problems of unconstrained optimization. The most widely used methods of reduction of functionals minimization problems with constraints to problems of unconstrained optimization are methods of penalty functions and of Lagrange multipliers [4,6]. These methods are discussed below.

Penalty functions method

Let the parameters' vector $\mathbf{p}$ satisfy the condition $h(\mathbf{p}) = 0$ at the minimization of the functional I. In the elementary variant of penalty functions method, an expanded functional $F(\mathbf{p}) = I(\mathbf{p}) + \sigma h^2(\mathbf{p})$ is created where the positive factor $\sigma > 0$ is a parameter. Apparently, searching of functional F minimum is a problem of unconstrained optimization. The component $\sigma h^2(\mathbf{p})$ to be added to the functional $I(\mathbf{p})$, and increasing its value, is considered a "penalty" which gets smaller as the constraint $h(\mathbf{p})=0$ holds more accuracy. Sometimes for more exact consideration of the constraint $h(\mathbf{p})=0$, the problem should be solved several times by increasing values of the factor σ. This will certainly complicate the problems solution.

If a constraint is given in the form of inequality $g(\mathbf{p})\geq 0$, a function $\varphi(\mathbf{p})$, which becomes zero in the case of its validity, can be expediently used. Function $\varphi(\mathbf{p})=g^2(\mathbf{p})[1-\operatorname{sign} g(\mathbf{p})]$ is an example of $\varphi(\mathbf{p})$. As described above, the expanded functional $F(\mathbf{p})$ can be written down, adding a component $\sigma\varphi(\mathbf{p}),\sigma>0$ to the functional $I(\mathbf{p})$.

At set constraints of the type $g(\mathbf{p})\geq 0$, $h(\mathbf{p})=0$, the following can be accepted as the expanded functional:

$$F(\mathbf{p})=I(\mathbf{p})+\sigma\left\{h^2(\mathbf{p})+g^2(\mathbf{p})\left[1\text{-sign}\, g(\mathbf{p})\right]\right\}.$$

The factor $\psi(\mathbf{p})=h^2(\mathbf{p})+g^2(\mathbf{p})\left[1\text{-sign}\, g(\mathbf{p})\right]$ at the coefficient σ becomes zero if constraints are valid. Indeed, expressions in square and curly brackets become zero only if $h(\mathbf{p})=0$ and $g(\mathbf{p})\geq 0$. At violation of constraints, i.e. at $g(\mathbf{p})<0$ (then sign $g(\mathbf{p})=-1$) and $h(\mathbf{p})\neq 0$, we have $\varphi(\mathbf{p})>0$. Then larger violations of constraints cause more increase of the functional $F(\mathbf{p})$.

If there are several constraints $g_i(\mathbf{p})\geq 0$ ($i=\overline{1,N_g}$), $h_i(\mathbf{p})=0$ ($i=\overline{1,N_h}$), the expanded functional in penalty functions method becomes

$$F(\mathbf{p},\sigma)=I(\mathbf{p})+\sigma\left\{\sum_{i=1}^{i=N_h} h_i^2(\mathbf{p})+\sum_{i=1}^{i=N_g} g_i^2(\mathbf{p})\,\left[1-\operatorname{sign} g_i(\mathbf{p})\right]\right\}.$$

Function $\varphi(\mathbf{p})$ can be assumed to be

$$\varphi(\mathbf{p})=1-\tanh(ag(\mathbf{p}))\ \text{ or }\ \varphi(\mathbf{p})=\frac{1}{1+\exp(ag(\mathbf{p}))},\quad a>0.$$

As opposite to function $\varphi(\mathbf{p})=g^2(\mathbf{p})[1-\operatorname{sign} g(\mathbf{p})]$, these functions are differentiate on $\mathbf{p}$.

The minimum of the expanded functional F can be located both inside the domain Π of allowed values of parameters $\mathbf{p}$ (where the min$F$ and min$I$ coincide as in this case $\sigma\psi(\mathbf{p})=0$), and outside the domain of Π. In the latter case the lower limit of functional F is reached on the boundary of domain Π. If the minimum of F is inside the domain Π, then the solution is found by a one step search of min$F$. Otherwise, repetitive steps of search are required, increasing σ at each step until the function $\psi(\mathbf{p})=0$ is close to zero within the specified error.

Outside the domain Π where constraints are not valid, and inside the domain Π where they are valid, the functional $F(\mathbf{p},\sigma)$ derivatives are essentially different. This difference is especially strong for large values of σ. Therefore, computing generally becomes more complex by increasing σ. As a result, finding the minimum $F(\mathbf{p},\sigma)$ located near the surface $\psi(\mathbf{p})=0$ is inconvenient using this method. The method of Lagrange multipliers lacks this disadvantage.

Lagrange multipliers method

This method is extensively used in solving optimization problems with the presence of constraints. Therefore, we shall discuss it in more detail.

In this method, constraints on optimization parameters are written down in the form of constraints-equalities. Therefore, a transition from constraints-inequalities (if they are present at the problem specification) to constraints-equalities is required. For example, the constraint $p_i \geq a_i$ becomes equality $p_i - a_i - z_i^2 = 0$ when $p_i = a_i + z_i^2$ (here z_i is a new variable). Constraint $p_i \geq p_j$ becomes an equality when $p_j = z_j$, $p_i = a_i + z_i^2$. The double inequality $a_i \leq p_i \leq b_i$ becomes an equality when $p_i \leq b_i + (a_i - b_i)\sin^2 z_i$. Similarly, by use of auxiliary variables, other transformations from constraints-inequalities to constraints-equalities are possible.

Let's consider the method of Lagrange multipliers used for search of the functional $I(\mathbf{p})$ extremum at constraints-equalities $h_i(\mathbf{p})=0$, $i=\overline{1,N_h}$, where $\mathbf{p}=\left(p_1, \ p_2, \ \cdots \ ,p_{N_p}\right)^T$, $N_h < N_p$. As variables $p_1, \ p_2, \ \cdots, \ p_{N_p}$ are connected by N_h relationships, then the number of independent variables appears to be $N_p - N_h$. Therefore, the problem can be reduced to a problem of extremum search for a functional $I(\mathbf{p})$, not of N_p, but of $N_p - N_h$ variables. For this purpose, it is possible in the beginning to express N_h variables from the relations $h_i(\mathbf{p})=0$; for example - the first N_h parameters $p_1, \ p_2, \cdots, p_{N_h}$, by remaining variables $p_{N_h+1}, \ p_{N_h+2}, \ \cdots, \ p_{N_p}$. Further, it will be necessary to substitute the obtained expressions for $p_1, \ p_2, \ \cdots, \ p_{N_h}$ in the functional $I(p_1, \ p_2, \ \cdots, p_{N_p})$ to derive the functional $I_1(p_{N_h+1}, \ p_{N_h+2}, \ \cdots, \ p_{N_p})$, which will be dependent only from $N_p - N_h$ variables. Then, variables at which functional I_1, and, hence, the functional I under condition of constraints reach their extreme value, can be found from the equation

$$\frac{\partial I_1}{\partial p_k}=0,\ k=\overline{N_h+1,N_p}\,.$$

However, this is a rather complicated way to account for constraints on the vector **p** components. To find explicit dependences $p_1,\ p_2,\ \cdots,\ p_{N_h}$ from $p_{N_h+1}, p_{N_h+2},\ \cdots,\ p_{N_p}$, it is necessary to find an analytical solution of the system of nonlinear equations $h_i(\mathbf{p})=0$, $i=\overline{1,N_h}$. This problem is solvable only in rare instances.

The Lagrange method allows us to overcome these difficulties. It involves the introduction of parameters λ_i ($i=\overline{1,N_h}$) - the so-called Lagrange multipliers, and transition from minimization of the functional $I(\mathbf{p})$ to minimization of the expanded functional

$$L(\mathbf{p},\lambda)=I(\mathbf{p})+\sum_{k=1}^{k=N_h}\lambda_k h_k(\mathbf{p})\,.$$

Multipliers $\lambda_1,\lambda_2,\ \cdots,\lambda_{N_h}$ as well as variables $p_1,\ p_2,\ \cdots,\ p_{N_p}$ are arguments of the expanded functional. Thus, the number of unknowns becomes N_h+N_p. To find them, the necessary condition of the expanded functional $L(\mathbf{p},\lambda)$ minimum is written down on variables included in the vector **p**, appended with equations $h_i(\mathbf{p})=0$, ($i=\overline{1,N_h}$). For calculating the derivatives included in the system of N_h+N_p equations

$$\frac{\partial L}{\partial p_k}=0,\ k=\overline{1,N_p},\ h_i(\mathbf{p})=0,\ i=\overline{1,N_h}\,,$$

variables λ are considered as independent values. The vector of variables **p** found from these equations delivers an extremum to the functional $I(\mathbf{p})$ for constraints $h_i(\mathbf{p})=0$, ($i=\overline{1,N_h}$).

Thus, the method of Lagrange multipliers reduces the problem of finding the conditional extremum of the functional $I(\mathbf{p})$ to a problem of finding the unconditional extremum of the expanded functional $L(\mathbf{p},\lambda)$.

Let's apply the Lagrange multipliers method for the search of extremum for the functional $I(p_1,p_2)$, dependent on two variables given the constraint $h(p_1,p_2)=0$. The equation $h(p_1,p_2)=0$ determines a curve on the plane of parameters p_1, p_2, shown by the dashed line (Fig. 2.12). Here, the continuous lines show projections of the equal level lines (constant values) of the functional $I(p_1, p_2)$ on the same plane p_1, p_2.

In the absence of constraints, the functional $I(p_1, p_2)$ extremum is reached at the point *A*, however coordinates of this point do not satisfy the condition $h(p_1, p_2) = 0$ as they do not belong to the curve ℓ. Therefore, the point *A* is not a point of conditional extremum and, accordingly, not a solution of this problem. The solution is the point *B* with coordinates p_{10}, p_{20}. At this point, the curve ℓ is tangent to the line of least values of the functional $I(p_1, p_2)$.

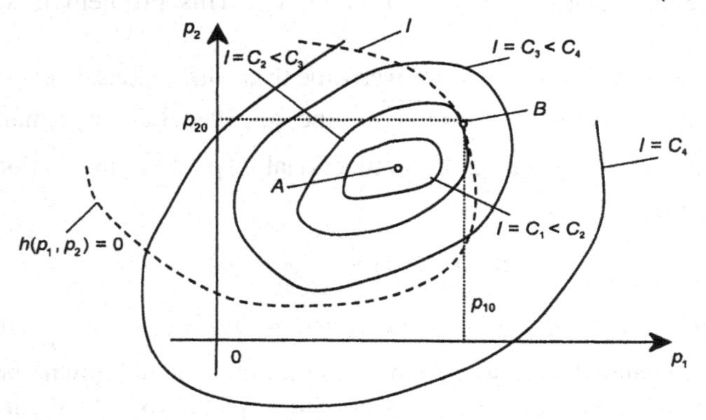

Fig. 2.12. Search of functional extremum under the constraint $h(p_1,p_2)=0$

According to the Lagrange multipliers method an expanded functional $L(p_1, p_2, \lambda) = I(p_1, p_2) + \lambda h(p_1, p_2)$ should be introduced, and coordinates p_{10}, p_{20}, at which the functional $L(p_1, p_2, \lambda)$ reaches its extremum, can be found from the equations $\partial L / \partial p_1 = 0, \partial L / \partial p_2 = 0, h(p_1, p_2) = 0$.

Solutions for some conditional extremum problems, constraints can also be given as inequalities $g_{ym}(\mathbf{p}) \geq 0$, $i = \overline{1, N_g}$ defining a domain in the *N*-dimensional space. For such a definition of constraints, the extremum can be reached both inside this domain and on its boundary. If inside the domain an extremum can be searched as an unconditional one, then to find the function $I(\mathbf{p})$ extreme value on the boundary, the above considered method of Lagrange multipliers can be applied, since conditions $g_i(\mathbf{p}) = 0$, $i = \overline{1, N_g}$ are valid on the domain boundary.

This reasoning may be used for search of the extremum for the functional $I(p_1, p_2)$ of two variables at the constraint $g(p_1, p_2) \geq 0$. If the expression $I(p_1,p_2)$ determines the domain limited by a closed curve ℓ, then the extreme value of $I(p_1,p_2)$ inside this domain can be found by solving the equations $\partial I/\partial p_1 = 0$, $\partial I/\partial p_2 = 0$. On the line ℓ we have equations

$$\partial L/\partial p_1 = 0,\ \partial L/\partial p_2 = 0,\ g(p_1,p_2) = 0,$$

where $L(p_1,p_2,\lambda) = I(p_1,p_2) + \lambda g(p_1,p_2)$.

Problems of extremum search at presence of constraints are solved as well when components of the parameters' vector **p** are represented by unknown functions and their derivatives. In the simplest cases the vector **p** contains a single component which is a function of multiple variables, for example, a function of coordinates $x_1, x_2, \ldots, x_N$. Problems of finding of function $p(x_1, x_2, \ldots, x_N)$, delivering a minimum for the functional $I(p)$, are referred to as variational problems. Let's consider features of application of the Lagrange multipliers method at the solution for such problems. Assume, for simplification, that the desired function $p(x)$ depends on a single coordinate x.

Let the minimized functional contain not only the function $p(x)$, but also its derivative. For such problems, called isoperimetric, the search for an extremum of the functional

$$I(p) = \int_a^b f(p(x), p'_x)dx$$

at a constraint on the function $p(x)$, given as

$$\int_a^b h(p(x), p'_x)dx = d\,,$$

where $p'_x = dp(x)/dx$ and d is a given number, is required. The expanded functional $L(p,\lambda)$ can be written down as

$$L(p,\lambda) = \int_a^b f(p(x), p'_x)dx + \lambda\left(\int_a^b h(p(x), p'_x)dx - d\right).$$

The desired function $p(x)$ is found from the necessary condition of its minimum. That is the condition of equality to zero of the expanded functional variation, i.e. $\delta L = 0$, on the function $p(x)$. Variation of a functional (see Appendix B) is defined as the main (linear) part of its increment at the given variation $\delta p(x)$ of the desired function $p(x)$.

Let's consider an example introducing an expanded functional and finding of the Lagrange multipliers for search of density $\sigma(s)$ of electric charge q, distributed on the surface S of a charged conductor. It is well-known that because of charge distribution on a conductor surface, the electric field energy $W_e = \frac{1}{2}\oint_s \varphi_e \sigma(s)ds$ created by it becomes its minimal value (here φ_e is the electric potential). The objective functional can be written down as $I(\sigma) = W_e = \frac{1}{2}\oint_s \varphi_e \sigma(s)ds$. The electric charge of a conductor is connected to its surface density as $q = \oint_s \sigma(s)ds$. Considering this relation as a constraint, the expanded functional can be given as:

$$L(\sigma) = \frac{1}{2}\oint_s \varphi_e \sigma(s)ds + \lambda \oint_s \sigma(s)ds .$$

From the necessary condition of its extremum (the stationary condition)

$$\delta L(\sigma) = \oint_s (\varphi_e(s) + \lambda)(\delta\sigma)ds = 0 ,$$

by virtue of arbitrariness of the variation $\delta\sigma$, we find $\varphi_e(s) = -\lambda = \text{const}$.

Thus, we arrive at the conclusion that distribution of a charge q on a conductor surface results in constant potential on the conductor surface. And the Lagrange multiplier appears to be proportional to the conductor potential.

In this problem the Lagrange multiplier becomes a constant. Generally, the multipliers $\lambda = \lambda(x_1, x_2, ..., x_N)$ are functions of coordinates. Isoperimetric variational problems are the simplest ones. In other problems the constraints on desired function $p(x)$ are taken into account similarly by the introduction of expanded functionals which are included in the given constraints.

Thus, for search of the objective functional minimum in the presence of constraints on optimization parameters, both in the method of penalty functions and in the Lagrange method, an expanded functional with included constraints is introduced and minimized by methods of unconditional minimization considered in the previous paragraph. In practice for optimization problems, expanded functionals frequently include, simultaneously, penalty functions as well as Lagrange multipliers, taking into account some constraints by means of penalty functions, and others – by means of Lagrange multipliers.

Let's consider an example of the solution of an inverse problem with constraints by the Lagrange multipliers method.

Let's find the currents i_k, $k = \overline{1, N_i}$ of inductively connected coils under the condition of $\sum_{k=1}^{k=N_i} i_k = i_0$, at which the coils' magnetic field energy

$$W_m = \frac{1}{2}\sum_{k=1}^{k=N_i} L_k i_k^2 + \sum_{k=1}^{k=N_i}\sum_{p=1,p\neq k}^{k=N_i} M_{k,p} i_k i_p$$

is minimal. Inductances L_k and mutual inductances $M_{k,p}$, $(k,p=\overline{1,N_i})$ of coils are given parameters. Assuming $I = W_m$ be the objective functional and taking into account $\sum_{k=1}^{k=N} i_k = i_0$ as the constraint, we can introduce an expanded functional:

$$L = I + \lambda\left(\sum_{k=1}^{k=N_i} i_k - i_0\right).$$

From the necessary condition of the minimum $\frac{dL}{di_k} = 0,\ k = \overline{1,N_i}$ and $\sum_{k=1}^{k=N_i} i_k - i_0 = 0$, we derive $N+1$ equations for currents i_k, $k = \overline{1,N_i}$ and for the Lagrange multiplier λ:

$$\begin{cases} L_k i_k + \sum\limits_{\substack{p=1\\ p\neq k}}^{p=N_i} M_{k,p} i_p + \lambda = 0, \quad k = \overline{1,N_i}, \\ \sum\limits_{k=1}^{k=N_i} i_k = i_0. \end{cases}$$

Here, $\Phi_k = L_k i_k + \sum\limits_{\substack{p=1\\ p\neq k}}^{p=N_i} M_{k,p} i_p$ is the magnetic flux linked with the k-th coil. Magnetic field energy of the coils is minimal, if magnetic fluxes of all coils are identical and equal, that is $\Phi_k = -\lambda$, $k = \overline{1,N_i}$. In particular, in the case of two coils $M_{1,2} = M_{2,1} = M$, required currents are found from the equations

$$\begin{cases} L_1 i_1 + M i_2 = -\lambda, \\ L_2 i_2 + M i_1 = -\lambda, \\ i_1 + i_2 = i_0, \end{cases} \quad \text{from which} \quad \begin{cases} i_1 = i_0 \dfrac{L_2 - M}{L_1 + L_2 - 2M}, \\ i_2 = i_0 \dfrac{L_1 - M}{L_1 + L_2 - 2M}. \end{cases}$$

Lagrange multipliers introduced for the account of constraints allow effective calculation of the gradient of the expanded functional on the parameters' vector. It is well-known that calculations of gradients of functionals are the most frequently carried out procedures when solving inverse problems. As it has been shown in Section 2.2, when using gradient methods, finding the functional's gradient demands the highest calculation expense. Therefore, the technique of its calculation determines the efficiency of the entire minimization process as well.

Let's compare various approaches to calculation of the expanded functional gradient at the solution of inverse problems using electric circuits theory.

The circuit equations can be written down as $f_k(\mathbf{p},\mathbf{i})=0, \quad k=\overline{1,N_i}$, or

$$\mathbf{f}(\mathbf{p},\mathbf{i})=0, \tag{2.9}$$

where $\mathbf{p}=\left(p_1, p_2, \cdots, p_{N_p}\right)^T$ and $\mathbf{i}=\left(i_1, i_2, \cdots, i_{N_i}\right)^T$ are vectors of circuit parameters and currents, and $\mathbf{f}(\mathbf{p},\mathbf{i})=\left(f_1(\mathbf{p},\mathbf{i}), \ f_2(\mathbf{p},\mathbf{i}), \ \cdots \ , f_{N_i}(\mathbf{p},\mathbf{i})\right)^T$.

The problem involves finding a parameters' vector **p**, on which the given functional $I(\mathbf{p},\mathbf{i})$ reaches its minimal value. According to the Lagrange multipliers method, an expanded functional should be introduced considering the circuit equations (2.9) as constraints imposed on the desired parameters **p**:

$$L(\mathbf{p},\mathbf{i},\boldsymbol{\lambda})=I(\mathbf{p},\mathbf{i})+\boldsymbol{\lambda}^T\mathbf{f}(\mathbf{p},\mathbf{i}),$$

where $\boldsymbol{\lambda}=\left(\lambda_1, \ \lambda_2, \ \cdots \ , \lambda_{N_i}\right)^T$ are the Lagrange multipliers.

One of the variants of calculating the functional $L(\mathbf{p},\mathbf{i},\boldsymbol{\lambda})$ gradient involves the analytical solving of a system of equations $\mathbf{f}(\mathbf{p},\mathbf{i})=0$ with regard to the vector **i**, with substitution of the obtained solution $\mathbf{i}=\mathbf{i}(\mathbf{p})$ into the expanded functional and its subsequent differentiation. Gradient calculation in this way is unacceptable for the most nonlinear inverse problems, as the analytical solution of Eq. (2.9) is impossible or rather tedious.

Another, most frequently used method of gradient calculation, is its numerical calculation from the following relations

$$\frac{dL(\mathbf{p},\mathbf{i}(\mathbf{p}))}{d\mathbf{p}}=\begin{pmatrix} \dfrac{L(\mathbf{p},\mathbf{i}(\mathbf{p}),\boldsymbol{\lambda}) - L(p_1+\Delta p_1, p_2, \cdots, p_{N_p}, \mathbf{i}(p_1+\Delta p_1, p_2, \cdots, p_{N_p}), \boldsymbol{\lambda})}{\Delta p_1} \\ \dfrac{L(\mathbf{p},\mathbf{i}(\mathbf{p}),\boldsymbol{\lambda}) - L(p_1, p_2+\Delta p_2, \cdots, p_{N_p}, \mathbf{i}(p_1, p_2+\Delta p_2, \cdots, p_{N_p}), \boldsymbol{\lambda})}{\Delta p_2} \\ \vdots \\ \dfrac{L(\mathbf{p},\mathbf{i}(\mathbf{p}),\boldsymbol{\lambda}) - L(p_1, p_2, \cdots, p_{N_p}+\Delta p_{N_p}, \mathbf{i}(p_1, p_2, \cdots, p_{N_p}+\Delta p_{N_p}), \boldsymbol{\lambda})}{\Delta p_{N_p}} \end{pmatrix}.$$

When calculating the gradient by this method it is necessary to solve $N_p + 1$ times the system of Eq. (2.9) for calculation of the vector **i** at various **p**. This demands heavy computing expenses.

Let's consider a way to use Lagrange multipliers for functional L gradient calculation. We shall calculate the derivative of the expanded functional on the vector **p**

$$\frac{dL(\mathbf{p},\mathbf{i},\boldsymbol{\lambda})}{d\mathbf{p}} = \frac{dI}{d\mathbf{p}} + \underbrace{\frac{d\boldsymbol{\lambda}}{d\mathbf{p}}\mathbf{f}}_{\text{is 0 as f=0}} + \left(\frac{d\mathbf{f}}{d\mathbf{p}}\right)^T \boldsymbol{\lambda} = \frac{\partial I}{\partial \mathbf{p}} + \left(\frac{\partial \mathbf{i}}{\partial \mathbf{p}}\right)^T \frac{\partial I}{\partial \mathbf{i}} + \left(\frac{\partial \mathbf{f}}{\partial \mathbf{p}} + \frac{\partial \mathbf{f}}{\partial \mathbf{i}} \cdot \frac{\partial \mathbf{i}}{\partial \mathbf{p}}\right)^T \boldsymbol{\lambda} =$$
$$= \left(\frac{\partial I}{\partial \mathbf{p}} + \left(\frac{\partial \mathbf{f}}{\partial \mathbf{p}}\right)^T \boldsymbol{\lambda}\right) + \left(\frac{\partial \mathbf{i}}{\partial \mathbf{p}}\right)^T \left(\frac{\partial I}{\partial \mathbf{i}} + \left(\frac{\partial \mathbf{f}}{\partial \mathbf{i}}\right)^T \boldsymbol{\lambda}\right). \tag{2.10}$$

Let's find the Lagrange multipliers $\boldsymbol{\lambda}$ from the condition of equality to zero of expression in the brackets in the second summand of Eq. (2.10), that is to determine them from the solution of the system of linear equations

$$\left(\frac{\partial \mathbf{f}}{\partial \mathbf{i}}\right)^T \boldsymbol{\lambda} = -\frac{\partial I}{\partial \mathbf{i}}. \tag{2.11}$$

Then the gradient of the expanded functional can be calculated by means of the following relation:

$$\frac{dL}{d\mathbf{p}} = \frac{\partial I}{\partial \mathbf{p}} + \left(\frac{\partial \mathbf{f}}{\partial \mathbf{p}}\right)^T \boldsymbol{\lambda}. \tag{2.12}$$

This method of calculation of the gradient demands a single solution of the system of linear equations (2.11) and (2.12) that is a serious advantage in comparison with the methods discussed above.

Let's consider an example. In the electric circuit shown in Fig. 2.13 let it be necessary to find the resistance r, at which the functional $I(r,\mathbf{i}) = \left(i_1^2 + i_2^2 - 0.1\right)^2$ reaches its minimal value.

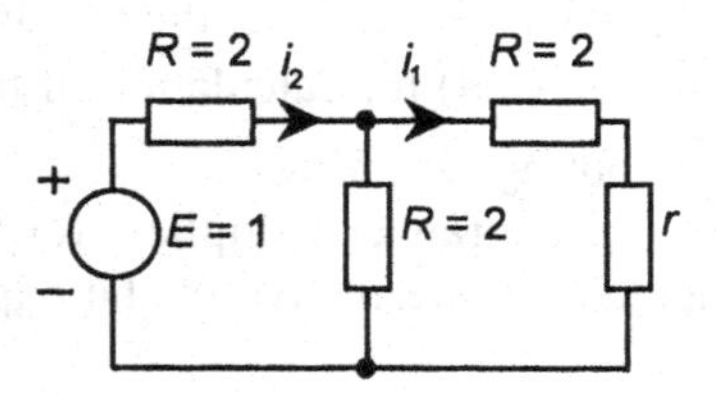

Fig. 2.13. Electric circuit with the resistor r of optimized resistance

To write down Eq. (2.9), we apply the mesh current method:

$$\begin{cases} (4+r)i_1 - 2i_2 = 0, \\ -2i_1 + 4i_2 = 1 \end{cases} \quad \text{or} \quad \begin{cases} f_1(r,\mathbf{i}) = (4+r)i_1 - 2i_2, \\ f_2(r,\mathbf{i}) = -2i_1 + 4i_2 - 1. \end{cases} \tag{2.13}$$

Then the expanded functional is

$$L(r,\mathbf{i},\boldsymbol{\lambda}) = \left(i_1^2 + i_2^2 - 0.1\right)^2 + \lambda_1\left((4+r)i_1 - 2i_2\right) + \lambda_2\left(-2i_1 + 4i_2 - 1\right).$$

According to Eq. (2.11), multipliers λ_1 and λ_2 should satisfy the following equations:

$$\begin{cases} (4+r)\lambda_1 - 2\lambda_2 = -4i_1\left(i_1^2 + i_2^2 - 0.1\right), \\ -2\lambda_1 + 4\lambda_2 = -4i_2\left(i_1^2 + i_2^2 - 0.1\right). \end{cases} \tag{2.14}$$

It should be noted that the left parts of the systems of equations (2.13) and (2.14) are similar. Therefore scopes of work for solution of Eq. (2.13) and (2.14) are practically identical. We shall determine λ_1 and λ_2 from Eq. (2.14), using expression (2.4) for calculation of the desired gradient:

$$\frac{dL(r,\mathbf{i}(r))}{dr} = \frac{\partial I(r,\mathbf{i})}{\partial r} + \sum_{k=1}^{k=2} \lambda_k \frac{\partial f_k(r,\mathbf{i})}{\partial r} = 0 + \lambda_1 i_1 + \lambda_2 0 =$$

$$= -\frac{20i_1^3 + 20i_1 i_2^2 + 10i_2 i_1^2 + 10i_2^3 - 2i_1 - i_2}{5(3+r)} i_1.$$

Eq. (2.13) has a solution $i_1 = \dfrac{1}{6+2r}$, $i_2 = \dfrac{4+r}{2(6+2r)}$ and, hence, the functional gradient is given by

$$\frac{dL(r,\mathbf{i}(r))}{dr} = \frac{-224 + 36r + 32r^2 + 3r^3}{320(3+r)}.$$

From the necessary condition of minimum for the functional $dL(r,\mathbf{i}(r))/dr = 0$, and taking into account that $r>0$, we find $r_{\min} = 2$.

Let's solve a problem similar to the one considered above, but in nonlinear statement. In the electric circuit shown in Fig. 2.13 let it be required to find the value of the parameter a of the resistor's nonlinear characteristic $r = ai_1^2 - 3$, at which the functional $I(r,\mathbf{i}) = \left(i_1^2 + i_2^2 - 0.1\right)^2$ becomes its minimum.

The circuit equation (2.1) and the expanded functional are given by:

$$\begin{cases} f_1(r,\mathbf{i}) = (4 + ai_1^2 - 3)i_1 - 2i_2, \\ f_2(r,\mathbf{i}) = -2i_1 + 4i_2 - 1, \end{cases}$$

$$L(r,\mathbf{i},\boldsymbol{\lambda}) = \left(i_1^2 + i_2^2 - 0.1\right)^2 + \lambda_1\left((1 + ai_1^2)i_1 - 2i_2\right) + \lambda_2\left(-2i_1 + 4i_2 - 1\right).$$

In this case of nonlinear electric circuit, the equation (2.11) for multipliers λ_1 and λ_2 is linear, as before:

$$\begin{cases} (1 + ai_1^2)\lambda_1 - 2\lambda_2 = -4i_1\left(i_1^2 + i_2^2 - 0.1\right), \\ -2\lambda_1 + 4\lambda_2 = -4i_2\left(i_1^2 + i_2^2 - 0.1\right). \end{cases} \tag{2.15}$$

Solving them with respect to λ_1 and λ_2, and substituting the solution into Eq. (2.12), an expression for the expanded functional gradient can be derived:

$$\frac{dL(a,\mathbf{i}(a))}{da} = \frac{\partial I(a,\mathbf{i})}{\partial a} + \sum_{k=1}^{k=2} \lambda_k \frac{\partial f_k(a,\mathbf{i})}{\partial a} = 0 + \lambda_1 i_1 + \lambda_2 \cdot 0 = -\frac{(2i_1 + i_2)\left(0.1 - i_1^2 - i_2^2\right)}{3ai_1}.$$

Then, the problem solution can be obtained by any of the gradient methods: $a^{(n+1)} = a^{(n)} + \alpha \left.\frac{dL}{da}\right|_{a=a^{(n)}}$, where the factor α is defined by choice of the method used. After necessary calculations, we have $a_{opt} = 500$.

Calculation of the gradient of the expanded functional, carried out on each step of descent, requires solution (numerical or analytical) of the following system of nonlinear equations:

$$\begin{cases} f_1(r,\mathbf{i}) = 0, \\ f_2(r,\mathbf{i}) = 0. \end{cases}$$

It is important to note that even in the case of several desired parameters (as in the example considered above, there was a single parameter a), calculation of a gradient will demand only a single solution of a nonlinear problem on each step of descent. Therefore, the considered method of calculation of the gradient, based on the use of Lagrange multipliers, is considerably more efficient than its direct numerical calculation. Its efficiency increases for large numbers of inverse problem desired parameters.

The method of a functional's gradient calculation described above can be applied to any electric circuit. Let, for definiteness, the electric circuit be described by a system of equations using the mesh-current method

$$\mathbf{CRC}^T\mathbf{i} = \mathbf{CE}. \tag{2.16}$$

Here, $\mathbf{C}$ is a matrix of the major loops, $\mathbf{i} = \left(i_1,\ i_2,\ \cdots, i_{N_i}\right)^T$ is the vector of mesh-currents, $\mathbf{E}$ is the vector of sources assumed to be constant, and $\mathbf{R} = \mathrm{diag}\left(r_1,\ r_2,\ \cdots\ , r_{N_p}\right)$, $\mathbf{p} = \left(r_1,\ r_2,\ \cdots\ , r_{N_p}\right)^T$ is the vector of inverse problem parameters. Let the minimized functional be given by $I = I(\mathbf{p},\mathbf{i})$.

The expanded functional can be given as

$$L(\mathbf{p},\mathbf{i},\boldsymbol{\lambda}) = I(\mathbf{p},\mathbf{i}) + \boldsymbol{\lambda}\left(\mathbf{CRC}^T\mathbf{i} - \mathbf{CE}\right). \tag{2.17}$$

Let's denote $\mathbf{f} = \left(\mathbf{CRC}^T\mathbf{i} - \mathbf{CE}\right)$ and Eq. (2.11) for determination of the Lagrange multipliers be

$\left(\dfrac{\partial \mathbf{f}}{\partial \mathbf{i}}\right)^T \boldsymbol{\lambda} = -\dfrac{\partial I}{\partial \mathbf{i}}$, from which $\left(\mathbf{CRC}^T\right)^T \boldsymbol{\lambda} = -\dfrac{\partial I}{\partial \mathbf{i}}$, or $\mathbf{CRC}^T\boldsymbol{\lambda} = -\dfrac{\partial I}{\partial \mathbf{i}}$.

Eq. (2.17) is linear with respect to $\boldsymbol{\lambda}$ and its solution can be written down as

$$\boldsymbol{\lambda} = -\left(\mathbf{CRC}^T\right)^{-1}\frac{\partial I}{\partial \mathbf{i}}.$$

The type of minimized functional defines $\dfrac{\partial I}{\partial \mathbf{i}}$, and its calculation does not involve any difficulties. Components of the gradient of expanded functional are given by

$$\frac{dL}{dr_k} = \frac{\partial I}{\partial r_k} + \boldsymbol{\lambda}^T\frac{\partial \mathbf{f}}{\partial r_k} = \frac{\partial I}{\partial r_k} - \left[\left(\mathbf{CRC}^T\right)^{-1}\frac{\partial I}{\partial \mathbf{i}}\right]^T \mathbf{CR}'_{r_k}\mathbf{C}^{\mathrm{T}}\left(\mathbf{CRC}^{\mathrm{T}}\right)^{-1}\mathbf{CE},$$
$$k = \overline{1, N_p}, \tag{2.18}$$

$$\mathbf{R}'_{r_k} = \frac{\partial \mathbf{R}}{\partial r_k} = \mathrm{diag}(0,\ 0,\ \cdots,\ 0,\ 1,\ 0,\ \cdots\ ,0)$$

$\uparrow$ unit in position k

After transforming Eq. (2.18) in view of symmetry of the matrix $\mathbf{CRC}^T$, necessary conditions of minimum of the expanded functional can be written down as

$$\frac{dL}{dr_k} = \frac{\partial I}{\partial r_k} - \left(\frac{\partial I}{\partial \mathbf{i}}\right)^T \mathbf{b}_k = 0,\ k = \overline{1, N_p}, \tag{2.19}$$

where $\mathbf{b}_k = \left(\mathbf{CRC}^{\mathrm{T}}\right)^{-1} \mathbf{CR}'_{r_k} \mathbf{C}^{\mathrm{T}} \left(\mathbf{CRC}^{\mathrm{T}}\right)^{-1} \mathbf{CE}$.

Here the vectors $\mathbf{b}_k$, $k = \overline{1, N_p}$ can be calculated at the initial stage of the problem solution. In respect to complexity, the calculation of vectors $\mathbf{b}_k$, $k = \overline{1, N_p}$ is the same as of the solution of Eq. (2.16). Modern codes of analytical calculations allow obtaining analytical expressions for $\mathbf{b}_k$, $k = \overline{1, N_p}$, even for problems with several tens of optimized parameters. Use of analytical methods, together with the Lagrange multipliers method, allows simplifying and considerably speeding up calculations of the gradient of the expanded functional and, thus, simplifying and speeding up the inverse problem solution.

Let's consider some examples of search by the Lagrange method of optimum solution for non-stationary problems in electric circuits theory.

We shall consider the problem of optimal charging of a condenser. The condenser in a circuit with series connected sections r and C, when subjected to input voltage $u(t)$, is charged up to voltage u_0 during a period of T. It is required to find such a voltage $u(t)$ at which the energy losses in the resistor are minimal. The circuit input voltage obeys the condition $u(0) = 0$.

As the circuit current is related to the voltage u_c on the condenser as $i = C\frac{du_c}{dt}$, then the objective functional is given by

$$I = \frac{1}{2}\int_0^T \left(C\frac{du_c}{dt}\right)^2 r dt\,.$$

Assuming the circuit equation $rC\frac{du_c}{dt} + u_c = u(t)$ as a constraint, the expanded functional can be written down as

$$L = I + \int_0^T \lambda(t)\left[rC\frac{du_c}{dt} + u_c - u(t) \right] dt = I + \int_0^T \lambda(t)\, rC\frac{du_c}{dt} dt + \int_0^T \lambda(t)[u_c - u(t)]\, dt,$$

where the Lagrange multiplier $\lambda(t)$ is a function of time.

As the desired functions are $u_c(t)$ and $u(t)$, i.e. functions of time, then the necessary conditions of minimum for the functional L are expressed through its variations in the form of $\delta_{u_c} L = 0$ and $\delta_u L = 0$.

For calculation of the variation $\delta_{u_c} L$ we shall transform the expanded functional, performing integration by parts:

$$L = I + \lambda(t) rCu_c\Big|_{t=0}^{t=T} + \int_0^T \left(-rCu_c\frac{d\lambda(t)}{dt} + \lambda(t)u_c - \lambda(t)u(t) \right) dt .$$

Then, the equation $\delta_{u_c} L = 0$ becomes:

$$\delta_{u_c} I + \lambda(T) rC\delta u_c(T) - \lambda(0) rC\delta u_c(0) + \int_0^T \left(-rC\frac{d\lambda(t)}{dt} + \lambda(t) \right) \delta u_c dt = 0 .$$

The variation of objective functional $\delta_{u_c} I = \int_0^T rC^2 \frac{du_c}{dt} \cdot \frac{d}{dt}(\delta u_c) dt$ is also transformed using integration by parts:

$$\delta_{u_c} I = rC^2 \frac{du_c}{dt} \delta u_c \Big|_{t=0}^{t=T} - \int_0^T rC^2 \frac{d^2 u_c}{dt^2} \delta u_c dt .$$

By virtue of arbitrariness of the variation δu_c in the expression $\delta_{u_c} L = 0$, factors at δu_c under integrals should become zero. Hence, we obtain the following equation for the function $\lambda(t)$:

$$rC\frac{d\lambda}{dt} - \lambda = -rC^2 \frac{d^2 u_c}{dt^2} . \qquad (2.20)$$

As $\delta u_c(0)$, $\delta u_c(T)$ are equal to zero, then $\lambda(0)$ and $\lambda(T)$ are arbitrary values, and, in particular, $\lambda(T)=0$ can be assumed. Introducing a variable $t_1=T-t$ in Eq. (2.20), it is possible to proceed to the equation $rC\frac{d\lambda}{dt_1} + \lambda = rC^2 \frac{d^2 u_c}{dt_1^2}$, which should be solved for the initial condition $\lambda(0)=0$.

The necessary condition $\delta_u L = 0$ of minimum for the functional L results in the relation

$$\delta_u L = -\int_0^T \lambda(t)\delta u(t) dt = 0 ,$$

from which follows that condition $\lambda(t)=0$ must hold for the point of minimum of the functional.

For the considered problem the solution of Eq. (2.20) can be easily found: $\lambda(t)=0$ at $\frac{d^2u_c}{dt^2}=0$, i.e. at $\frac{du_c}{dt}=\text{const}$. Thus, the charging current of the condenser should be constant and the circuit input voltage should change under the linear law.

There are several other problems similar to the problems considered above; for example, there is the problem of search of the circuit's input voltage *u(t)*, at which the current in some circuit branch reaches a preset value during a given period *T*. The objective function in this case can be written down as

$$I = 0.5\left(i(t)-i(T)\right)^2.$$

Its variation to current is $\delta I = \left(i(t)-i(T)\right)\delta i$.

If at a transient in the circuit, the power output in resistor *r* over a period of time *T* should be *P(T)*, then the objective function can be written down as

$$I = 0.25\int_0^T \left[i^2(t)r - P(T)\right]^2 dt.$$

The variation of this function to current is $\delta I = r\int_0^T \left[i^2(t)r - P(T)\right]\, i(t)\delta i dt$.

The subsequent course of solution is similar to the one considered above.

This approach can be applied to the general case when the electric circuit is described by a system of state equations:

$$\dot{\mathbf{X}} = \mathbf{AX} + \mathbf{U}(t),\ \mathbf{X}(0) = \mathbf{X}_0,\ 0 \le t \le T,$$

where $\mathbf{X}(t) = \left(x_1(t),\ x_2(t),\ \cdots,\ x_{N_x}(t)\right)^T$ is the vector of condition variables, $\mathbf{U}(t)$ is the vector of sources, and $\mathbf{A}$ is the matrix of the circuit state equations. It is required to find the voltages included in the vector of circuit sources $\mathbf{U}(t)$, at which the functional $I\left(\mathbf{X},\mathbf{U}(t)\right) = \int_0^T \varphi\left(\mathbf{X},\mathbf{U}(t)\right)dt$ reaches its minimal value.

The expanded functional can be given by

$$L = I + \int_0^T \boldsymbol{\lambda}^T\left[\dot{\mathbf{X}} - \mathbf{AX} - \mathbf{U}(t)\right]\, dt = I + \int_0^T \boldsymbol{\lambda}^T\dot{\mathbf{X}}\ dt - \int_0^T \boldsymbol{\lambda}^T\left[\mathbf{AX}+\mathbf{U}(t)\right]\, dt,$$

where $\boldsymbol{\lambda} = \left(\lambda_1(t),\lambda_2(t),\ \cdots,\ \lambda_{N_\lambda}(t)\right)^T$ are the Lagrange multipliers. Integrating by parts we get:

$$L = I + \boldsymbol{\lambda}^T(t)\mathbf{X}(t)\Big|_0^T - \int_0^T \left(\frac{d\boldsymbol{\lambda}^T}{dt}\mathbf{X} + \boldsymbol{\lambda}^T(\mathbf{AX} + \mathbf{U}(t)) \right) dt .$$

The necessary condition of minimum for the expanded functional $\delta_{\mathbf{X}} L = 0$ results in the equation

$$\delta_{\mathbf{X}} I = -\boldsymbol{\lambda}^T(t)\delta\mathbf{X}\Big|_0^T + \int_0^T \left(\frac{d\boldsymbol{\lambda}^T}{dt}\delta\mathbf{X} + \boldsymbol{\lambda}^T \mathbf{A}(\delta\mathbf{X}) \right) dt . \qquad (2.21)$$

Then, $\delta_{\mathbf{X}} I$ can be written down as

$$\delta_{\mathbf{X}} I = \int_0^T \delta_{\mathbf{X}} \varphi(\mathbf{X}, \mathbf{U}) dt = \int_0^T \Phi(\mathbf{X}, \mathbf{U})^T (\delta\mathbf{X}) dt,$$

$$\Phi(\mathbf{X}, \mathbf{U}) = (\varphi_1(\mathbf{X}, \mathbf{U}),\ \varphi_2(\mathbf{X}, \mathbf{U}),\ \cdots\ \varphi_N(\mathbf{X}, \mathbf{U}))^T \ .$$

Equating factors at variations $\delta\mathbf{X}$ in the left and right parts of Eq. (2.21), we get $\frac{d\boldsymbol{\lambda}^T}{dt} + \boldsymbol{\lambda}^T \mathbf{A} = \Phi^T(\mathbf{X}, \mathbf{U}(t))$, or:

$$\frac{d\boldsymbol{\lambda}}{dt} + \mathbf{A}^T \boldsymbol{\lambda} = \Phi(\mathbf{X}, \mathbf{U}(t)) . \qquad (2.22)$$

The initial conditions $\boldsymbol{\lambda}(0)$, $\boldsymbol{\lambda}(T)$ can be obtained equating factors at variations $\delta\mathbf{X}(0)$, $\delta\mathbf{X}(T)$ in the left and right parts of Eq. (2.21). Their form is defined by the form of functional I and by initial conditions $\mathbf{X}_0$.

The necessary condition of minimum for the functional $\delta_{\mathbf{U}} L = 0$ results in the following:

$$\int_0^T \boldsymbol{\lambda}^T(t)\delta_{\mathbf{U}}(\mathbf{AX} + \mathbf{U}(t)) dt = \int_0^T \delta_{\mathbf{U}} \varphi(\mathbf{X}, \mathbf{U}) dt = \int_0^T \phi^T(\mathbf{X}, \mathbf{U}) \delta\mathbf{U} dt . \qquad (2.23)$$

Equating factors at variation $\delta\mathbf{U}$ in both parts of equation (2.23), we get

$$\boldsymbol{\lambda}(t) = \phi(\mathbf{X}, \mathbf{U}(t)) . \qquad (2.24)$$

Equations (2.22) and (2.24) allow calculating functions $\boldsymbol{\lambda}(t)$ and the desired voltage $\mathbf{U}(t)$ at the circuit input.

The sequence of solution of the problem is as follows. Being given initial function $\mathbf{U}^0(t)$, the system of equations (2.22) should be solved to find functions $\boldsymbol{\lambda}(t)$. If condition (2.24) holds at initial voltage $\mathbf{U}^0(t)=\mathbf{U}(t)$, and at the found functions $\boldsymbol{\lambda}(t)$, then the calculation comes to an end. In this case function $\mathbf{U}^0(t)$ determines the desired voltage. If the condition (2.24) does not hold, then the input voltage should be improved according to the applied method, and Eq. (2.22) should be resolved.

As stated above, it has been shown that the Lagrange method allows effectively solving an extensive range of inverse problems in electric circuits theory. Parameters of circuit elements, as well as currents and voltages, being functions of time, can be the desired values. Application of the Lagrange method for solution of inverse problems in the electromagnetic field theory, when processes are described by differential equations in partial derivatives, shall be discussed in more detail in Chapter 4.

2.4 Application of neural networks

One effective method for finding the solution of inverse problems in electrical engineering is based on the use of artificial neural networks. It can be considered as a method of processes mathematical modeling, alternative to classical mathematical analysis methods. Application of artificial neural networks assumes the use of computers for the object's mathematical modeling [7,8].

At solution of problems by the neural networks method, we find the vector $\mathbf{w}$ of output variables by the vector $\mathbf{v}=(v_1,v_2,\cdots,v_{N_v})^T \in S_v$ of input variables (actions) and the vector $\mathbf{p}=(p_1,p_2,\cdots,p_{N_p})^T \in \Pi$ of optimized parameters, without setting up and solving neither the differential nor integral equations connecting them. Here, S_v and Π are the sets of allowable input actions and parameters, respectively. When considering neural networks, we shall assume that the vector p of the optimized parameters is the input vector.

As it was noted in Section 1.1, the solution of inverse problems becomes considerably simpler if an analytical relationship connecting the device's input and output variables is known. Frequently in view of the absence of such relationships, connections between vectors $\mathbf{p}$ and $\mathbf{w}$ are found during the problem solution by repeatedly solving the analysis problem numerically. This procedure, however effectively realized, occupies a major part of an inverse problem's solution time. Furthermore, by the vector **w** of the output variables, a vector $\mathbf{y}=(y_1,y_2,\cdots,y_{N_y})^T$ of criteria can be determined which allows writing down inverse problem's objective functions. In this section, when describing mathematical tools of neural networks it is convenient to assume that the neural network outputs are elements of the vector **y**.

The approach discussed in this section involves prior finding of numerical connections between input variables **p** and criteria **y**. These connections can be written down as

$$\mathbf{y}(\mathbf{p})=\left(y_1(\mathbf{p}),\ y_2(\mathbf{p}),\ \cdots,y_{N_y}(\mathbf{p})\right)^T,\ \mathbf{p}=\left(p_1,\ p_2,\ \cdots,p_{N_p}\right)^T,$$

where **y** is a generally nonlinear vector function determined by the inverse problem conditions, N_p is the number of optimized parameters, and N_y is the number of criteria. As noted above, finding an analytical expression for the function **y**, as a rule, is impossible. However, repeatedly solving analysis problems for a set of various vectors $\mathbf{p}_j$ ($j=\overline{1,K}$), it is possible to find a set of corresponding vectors $\mathbf{y}_j$ ($j=\overline{1,K}$). Resulting sets of vectors p_j и y_j can be considered as a numerical description of the function $\mathbf{y}(\mathbf{p})$.

It is assumed that the set of vectors $\mathbf{p}_j$ ($j=\overline{1,K}$) is a representative one, so that

$$\forall \mathbf{p}\in S_v,\ \ \exists\ \mathbf{p}_j:\ \left\|\mathbf{p}_j-\mathbf{p}\right\|<\varepsilon\left\|\mathbf{p}\right\|, \tag{2.25}$$

where ε is a small value. It is intuitively obvious that such a description of $\mathbf{y}(\mathbf{p})$ will be more accurate, the less ε is. It is also obvious that when increasing N_p and reducing ε, the number K of vectors $\mathbf{p}_j$ necessary to satisfy Eq. (2.25) will increase. Required time for solution of the inverse problem will increase accordingly. Neural networks allow describing the properties of function $\mathbf{y}(\mathbf{p})$ with accuracy, sufficient for practical applications at comprehensible scopes of calculations that is at not too large a K.

Finding of multivariable function $y_k(\mathbf{p})=f(\mathbf{p})=f(p_1,p_2,\cdots,p_{N_p})$ becomes simpler if we can express it as a set of nonlinear one-variable functions. The question of feasibility of such representation for an arbitrary nonlinear function is of fundamental importance for neural networks. It has been shown [9] that any continuous multivariable function f can be exactly represented by superposition of continuous one-variable functions:

$$f(p_1,p_2,\ldots,p_{N_p}) = \sum_{i=1}^{2N_p+1} \alpha_i \left(\sum_{j=1}^{N_p} \beta_{ij}(p_j) \right), \tag{2.26}$$

where α and β are arbitrary one-variable nonlinear functions. Functions α and β effecting exact representation of the function f are rather complex. In practice, approximate representations of the function f are allowable. Therefore bounded nonlinear functions must be used, such as, for example, the sigmoid unipolar function $\varphi(x) = \dfrac{1}{1+e^{-bx}}$, where b is a parameter, the bipolar function $\varphi(x) = \tanh(bx)$, or the unit step function $1(x)$. At their application Eq. (2.26) can be written down as $f = \varphi\left(\sum_{i=1}^{N_v} W_i p_i\right)$, where W_i are constant factors. Thus, y_k, the k-th component of vector $\mathbf{y}$, can be presented as

$$y_k(p_1,p_2,\ldots,p_{N_p}) = \varphi\left(\sum_{i=1}^{N_v} W_{k,i} p_i\right). \tag{2.27}$$

Other nonlinear functions find application in practice; specific features of the inverse problem determine their form. For example, at solution of inverse problems in electric circuits theory can be polynomials or series of special kinds discussed in Section 1.2.

It is well-known, that the expression

$$y_k(\mathbf{p}) = \varphi_k\left(\sum_i W_{k,i} p_i + b_k\right) = \varphi_k(\mathbf{Wp}+\mathbf{b}),$$

where $\mathbf{W} = \begin{pmatrix} W_{1,1} & W_{1,2} & \cdots & W_{1,N} \\ W_{2,1} & W_{2,2} & \cdots & W_{2,N} \\ \vdots & \vdots & \ddots & \vdots \\ W_{M,1} & W_{M,2} & \cdots & W_{M,N} \end{pmatrix}$, $\mathbf{b} = \begin{pmatrix} b_1 \\ b_2 \\ \vdots \\ b_M \end{pmatrix}$,

(similar to Eq. (2.27)), describes the functioning of a living organism nerve cell, called a neuron. The neuron circuit model shown in Fig. 2.14 also reflects properties of summation, transformation and transfer of a nervous pulse (signal). The weight factor $W_{k,i}$ is referred to as the Neural Network Coefficient, the signal b_k as the offset, and the nonlinear function φ_k as the function of neuron activation.

In the human nervous system, the neurons (which number is on the order of 10^{11}, and the number of connections between them is 10^{15}) are connected with each other and participate in reception, processing and transfer of electrochemical signals through the brain's nerve paths. There are several signals at a neuron input whereas at its output there is only one signal, which depends on input signals, properties of communication lines, and on the neuron itself.

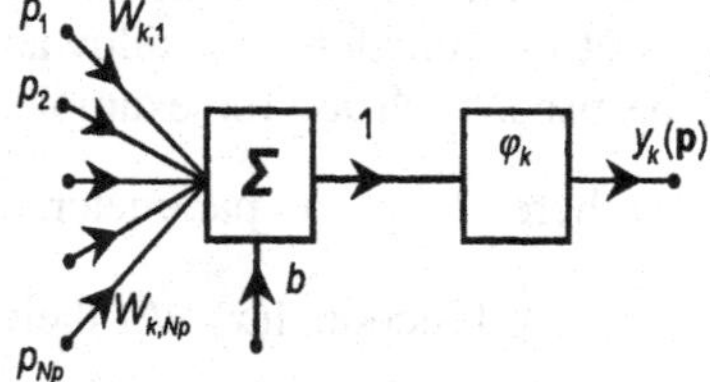

Fig. 2.14. The neuron model

Signal intensity at the neuron input depends on properties of channels (synapses) incoming to the neuron, and also on the quality of the outgoing channel. If the signal at the input of a cell exceeds some threshold level, it results in an avalanche-like signal (pulse) at its output. Though the signal transfer time by a cell is rather long (about 4 milliseconds), the nervous system shows rather high speed of work (for example, at visual recognition) due to parallel transfer of pulses by a large number of interconnected neurons.

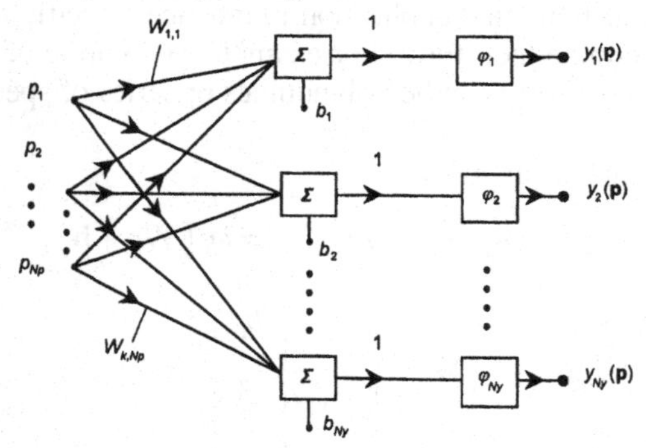

Fig. 2.15. Single layer neural network

It was presumed that the application of artificial neurons would allow constructing computers that operate like a human brain. And though these hopes were not justified (apparently, because of the greater complexity of processes in the human brain than it was presumed), neural networks have allowed solving of a number of problems more effectively than universal computers.

For calculation of all components of the vector **y** on the basis of neuron circuit shown in Fig. 2.14, it is necessary to connect several neurons to form a so-called artificial neural network. The elementary network is formed by the connection of neurons, having at their inputs components p_i of the input vector **p** multiplied by factors $W_{k,i}$ (k is the number of a neuron) (Fig. 2.15).

The neural network shown in Fig. 2.15 is called a single-layer network. However, the majority of practical problems require application of networks containing large numbers of layers. Outputs y_1, y_2, ... y_k of the network shown in Fig. 2.15 can be considered as components of an intermediate vector $\mathbf{y}^1$. Furthermore, it can be considered as the input for the second layer of neurons. In turn, the output $\mathbf{y}^2$ of the second layer can be considered as the input for the third layer, etc. In that case we have a multilayer network. The first layer of the network with input signals v_i is called the input layer. The last layer, forming the network output signals, is the output, and all other layers are called hidden layers. If each of the input vector components of a layer in a network is fed through communication lines to the inputs of all the summing units of that layer, such networks are called fully connected networks. Matrices **W**, as well as offsets **b** for various layers, can differ. Besides, layers can have different activation functions φ.

Generally the number of layers and the number of neurons in each layer can be assumed to be arbitrary. Numbers of input and output signals can differ from the number of neurons in intermediate layers.

Three-layer neural networks with an identical function of activation for every neuron have found practical application. In Fig. 2.16, a three-layer neural network is shown. The hidden layer contains identical nonlinear activation functions φ, and input and output layers do not contain nonlinear functions.

The neural network inputs and outputs shown in Fig. 2.16 are connected by the relationship $\mathbf{y} = \mathbf{V}\varphi(\mathbf{W}\mathbf{p}+\mathbf{b})$, where vector **p** is the input and vector **y** is the output.

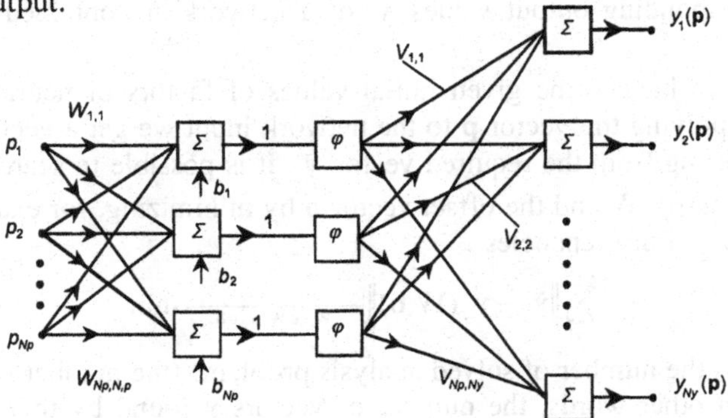

Fig. 2.16. Three-layer neural network

Let's consider the following example. In this problem we wish to find the shape of a polar tip providing a specified distribution of magnetic induction along a given line. At numerical solution the required pole shape is determined by the set of coordinates $p_1, p_2, \cdots, p_{N_p}$ of its surface points, and the distribution of magnetic induction is determined by its values in k points on the given line. At change of coordinate p, any of the points on the pole surface, i.e. at change of the pole shape in vicinity of a single point, the magnetic induction changes at all points of the line. The magnetic induction in k points on the line and coordinates p_n of the pole surface points are connected nonlinearly. This connection is generally not expressed analytically and is not known *a priori*. However, any specified shape of pole surface determines a certain distribution of magnetic induction along the given line. Therefore it is possible to set a relationship between any given set of coordinates of points on the pole surface and a set of values of a magnetic induction in aggregate points on the line. Thus, the magnetic induction **B** column-matrix is connected to the column-matrix **p** of points coordinates on the pole surface by a nonlinear equation $\mathbf{B}=\psi(\mathbf{p})$, similar to the equation $\mathbf{y} = \mathbf{V}\varphi(\mathbf{Wp}+\mathbf{b})$, connecting the inputs and outputs of a neural network.

To find the neural network output vector **y** for any input vector $\mathbf{p} \in \Pi$, it is necessary to calculate the factors of matrices $\mathbf{V},\mathbf{W}$ and $\mathbf{b}$. Calculation of matrices $\mathbf{V},\mathbf{W},\mathbf{b}$, or in other words, the network's adjustment, is referred to as its learning. Matrices $\mathbf{V},\mathbf{W},\mathbf{b}$ are found on the basis of solution of several analysis problems, which allow finding characteristics' vectors **y** by a specified set of parameters' vectors **p** at the device input. Thus, found vectors **y** are considered to be required vectors $\tilde{\mathbf{y}}$ on the neural network output for the same set of parameters **p** on the network input as on the device input. It should be noted that for a number of problems sets, having input parameters **p** and corresponding output values $\tilde{\mathbf{y}}$ of a network are obtained from experiments.

Assume we have some given initial values of factors of neural network matrices. Applying the vector **p** to the network input we get a vector **y** at its output, differing from the required vector $\tilde{\mathbf{y}}$. It is possible to search for factors of the matrix **W** and the offset vector **b** by minimizing, for example, the sum of norms of discrepancies

$$\sum_{j=1}^{j=N} \left\| \tilde{\mathbf{y}}_j - \mathbf{y}_j(\mathbf{W},\mathbf{b}) \right\| \xrightarrow[W_{k,n} \in \mathbf{W},\ b_k \in \mathbf{b}]{} \min .$$

Here, N is the number of solved analysis problems (the number of learning steps) or, in other words, the number of vectors **y** found by their solution. Factors of matrix **V** can be found similarly. Thus, finding of factors of matrices $\mathbf{V},\mathbf{W},\mathbf{b}$, i.e. network learning, is reduced to the search for a minimum of a nonlinear functional.

The specific form of neural network equations connecting vectors **p** and **y** allows applying an effective method of finding the factors of matrices $\mathbf{V},\mathbf{W},\mathbf{b}$. Let's consider this method in more detail. For simplification we assume that only factors of matrix **W** are determined.

Let the weight factors be sought from the condition of minimum for the functional $I(\mathbf{W}) = \|\tilde{\mathbf{y}} - \mathbf{y}(\mathbf{W})\|$, which can be written down as

$$I(\mathbf{W}) = \sum_{h=1}^{N} I^h(\mathbf{W}) = \frac{1}{2}\sum_{h=1}^{N}\sum_{i=1}^{N_y}(\tilde{y}_i^h - y_i^h(\mathbf{W}))^2,$$

where N is the number of learning steps, i.e. the number of pairs of input and output vectors.

Let's apply the gradient method for iterative calculation of the matrix **W**: $\mathbf{W}^{h+1} = \mathbf{W}^h + c\mathbf{e}(\mathbf{W})$, where the vector **e(W)** defines direction in the space of weights **W**, and h is the number of iterations (learning steps). The c is called the learning factor. To find **e(W)** derivatives of objective function on required weights, both the output and hidden layers of the network should be calculated.

Values of weight factors on the learning step h are calculated by means of the following relation:

$$W_{n,k}^{h+1} = W_{n,k}^{h} - \eta\frac{\partial I^h}{\partial W_{n,k}^h} + \alpha(W_{n,k}^h - W_{n,k}^{h-1}),$$

where parameters η, α, determine the rate of iterative process convergence, and are found from empirical relations. The derivative $\partial I^h / \partial W_{n,k}^h$ on the learning step h in the above expression, is given by

$$\frac{\partial I^h}{\partial W_{n,k}} = \frac{\partial}{\partial W_{nk}}\left[\frac{1}{2}\sum_{i=1}^{N_y}(\tilde{y}_i - y_i)^2\right] = \sum_{i=1}^{N_y}(y_i - \tilde{y}_i)\frac{\partial \varphi_n(p_k W_{n,k})}{\partial W_{n,k}}V_{in}.$$

Apparently, the derivative $\partial I^h / \partial W_{n,k}^h$ is proportional to the error $\tilde{\mathbf{y}} - \mathbf{y}$ on the network output, and in this connection this algorithm of weight factors calculation is called the method of error inverse transmission.

Matrix **V** and vector **b** can be calculated similarly as the matrix **W**.

This process of neural network learning is called learning with a teacher since it is assumed at its execution that for each input vector **p** the corresponding vector $\tilde{\mathbf{y}}$ at the network output is known.

The objective function $I(\mathbf{V},\mathbf{W},\mathbf{b})$ can have several minima that considerably complicate the finding of factors of matrices $\mathbf{V},\mathbf{W},\mathbf{b}$, as it requires application of methods of search of the global minimum. Some possible methods of its search are described in this chapter.

The necessary number of input and output sets for calculation of matrixes $\mathbf{V},\mathbf{W},\mathbf{b}$ depends, in particular, on dimensions of vectors **p** and **y** which can reach several hundreds or more. If in a three-layer network the number of inputs is N_p, outputs is N_y, and the number of hidden layer neurons is m, then the total number of unknowns, which should be calculated for network learning, is $(N_p+N_y+1)m$. Factors of matrices $\mathbf{V},\mathbf{W},\mathbf{b}$ can be calculated at the required accuracy for conditions that a sufficient number of corresponding sets of vectors **p** and **y** is given.

After network adjustment, i.e. after determination of $\mathbf{V},\mathbf{W},\mathbf{b}$ matrices, the network should be tested on some set of corresponding input p_i and output y_i $\left(i=\overline{1,N_t}\right)$ data, which had not been used in the weight factors calculations. At testing, matrices $\mathbf{V},\mathbf{W},\mathbf{b}$ should not be corrected. If the determined network outputs differ from the required values no more than a given value, then the network can be assumed as a learned one. If the error

$$\varepsilon\|\tilde{y}\| = \max_{i=\overline{1,N_t}} \|\tilde{\mathbf{y}}_i - \mathbf{y}(\mathbf{p}_i)\|$$

exceeds the given value, it is necessary to return to the learning level to correct the matrix factors on new pairs of vectors $\mathbf{p}_i$ and $\mathbf{y}_i$ (here N_t is the number of test vectors). On reaching the specified error level, the network can be used for solution of inverse problems, i.e. for calculation of criteria vector **y** on a specified vector **p** of input parameters.

Let's consider as an example for the solution of an inverse problem by the method of neural networks, finding of resistances of resistors $R1$-$R4$ in the electric circuit of a *TTL*-logic element (Fig. 2.17). For the circuit shown in the figure below, realizing logic operations NAND, 5 criteria are introduced, among which, in particular, are the allowable level of disturbing voltage in zero condition $\Delta U \geq 0.8\text{V}$, average power consumption $P \leq 20\text{mW}$, and the allowable signal delay $t_d \leq 15\text{ns}$.

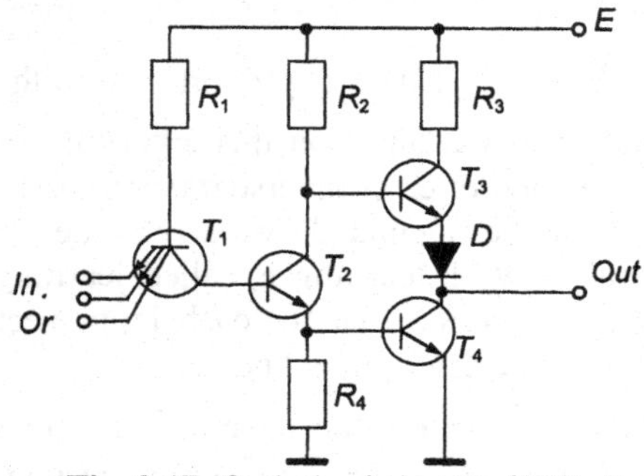

Fig. 2.17. Optimized circuit of TTL-logic

Resistances of circuit resistors are included into the parameters' vector $\mathbf{p} = \left(R_1, R_2, R_3, R_4\right)^T$ as its components. Imposed constraints are $0.5 \le R_1 \le 10$, $0.5 \le R_2 \le 10$, $0.05 \le R_3 \le 1$, and $0.1 \le R_4 \le 2$ (resistances are measured in kOhm). The network learning dependences of criteria from controlled parameters have been defined by analytical expressions. Criteria calculated by use of these expressions were considered as accurate values.

A three-layer neural network was used. The number of network inputs is 4 and the number of outputs is 5. The number of the hidden layer neurons was assumed to be equal to 30. At learning on 200 teaching examples, the root-mean-square error was monotonously decreased and reached the value 0.00132. After learning, the neural network was tested on 300 examples which were not used at learning. At that, the root-mean-square error was 0.035.

Neural networks find application not only for solutions of inverse problems of the type considered above, but also for solutions of so-called classification problems, i.e. problems requiring one to relate an object described by a set of attributes to one of several groups that is closest to it. Such problems arise in various areas of engineering, including electrical engineering.

If in an electric circuit the parameters of one or several elements have exceeded the allowable limits, then it will appear in deviations of the measured voltages and currents from their allowable values. If a fault occurs, it is necessary to compare its attributes to characteristic attributes of groups found at classification. Thus, there is a problem of classification of the fault character, i.e. finding of a type or a group of faults to which it can be related.

One may judge about serviceability of an electric circuit or about the character of a fault on the basis of analysis of measured values, for example, voltages between nodes, branch currents, input resistances or transfer functions. For this purpose the vector of measured values should be related to a certain group describing faults of one type, because various vectors of measured values can correspond to the same fault. Thus, it is necessary to put a dependence between the vector of measured values and one of the groups (classes) of input signals. Further, based upon belonging of a signal to either group, it is possible to draw a conclusion concerning the character of the circuit fault.

This problem can be solved by means of a neural network with an input vector $\mathbf{p} = \left(p_1,\ p_2,\ \cdots, p_{N_p}\right)^T \in \Pi$ of measured values, and an output signal corresponding to the serviceability or to a certain fault of the circuit. To solve this problem by means of a neural network, it is necessary to learn or to train the network at the first stage. At the second stage, by means of the already learned network, it would be possible to place dependency between the input vector of measured values and a certain fault.

Let's consider the process of network learning. This is carried out on the set of input vectors $\mathbf{p} \in \Pi$ with a number that considerably exceeds the number of possible faults. The learning problem of neural network involves allocation of groups of vectors close to each other by some criterion, which define one or several faults of the same kind. It allows a reduction of the information content concerning the properties of input vectors of the network.

Let each of the faults be characterized by an input vector $\mathbf{p} = \left(p_1,\ p_2,\ \cdots, p_{N_p}\right)^T \in \Pi$ of the neural network, with elements that are the electric circuit measured characteristics, for example, node voltages. Assume that vectors $\mathbf{p} \in \Pi$ are normalized so, for example, the norm of each vector is equal to unity. The set Π is partitioned into groups by use of an accepted criterion of closeness of vectors. One of distance measures $d(\mathbf{p}^{(1)}, \mathbf{p}^{(2)})$ between the vectors $\mathbf{p}^{(1)}$ and $\mathbf{p}^{(2)}$ can be used as criteria of closeness: for example, $d_1 = \sum_{j=1}^{N_p} \left(p_j^{(1)} - p_j^{(2)}\right)^2$, $d_2 = \sum_{j=1}^{N_p} \left|p_j^{(1)} - p_j^{(2)}\right|$ or $d_3 = \max_{j=\overline{1,N_p}} \left|p_j^{(1)} - p_j^{(2)}\right|$. The degree of closeness for vectors $\mathbf{p}^{(1)}$ and $\mathbf{p}^{(2)}$ with elements of binary attributes (0 or 1) is defined as the Hamming interval

$$d_H = \sum_{j=1}^{N_p} [p_j^{(2)}(1 - p_j^{(1)}) + p_j^{(1)}(1 - p_j^{(2)})],$$

which is equal to zero only at concordance of vectors.

Another method of distance introduction is the calculation of correlation between vectors. The calculation of distance is carried out from the following relationship:

$$d_C(\mathbf{p}^{(1)}, \mathbf{p}^{(2)}) = \sum_{j=1}^{N_p} \frac{\left(p_j^{(1)} - M_1\right)\left(p_j^{(2)} - M_2\right)}{\sigma_1 \sigma_2},$$

$$M_k = \frac{1}{N_p} \sum_{j=1}^{N_p} p_j^{(k)},\quad \sigma_k = \sqrt{\frac{1}{N_p} \sum_{j=1}^{N_j} \left(p_j^{(k)} - M_k\right)^2},\quad k = 1,2.$$

The preliminary normalizing and centering of each of the vectors is assumed:

$$p_j^{new} = \frac{p_j^{old} - M_{p,old}}{\sigma_{p,old}},\quad j = \overline{1,N_p}.$$

Here, M and σ are the same as above. The index "*old*" specifies that the vector and function components have been taken before and the index "*new*" – after normalization and centering.

Let the set of data $\mathbf{p}_j \in \Pi,\ j=\overline{1,N_j}$ be arbitrarily divided into sets of disjointed groups (classes) $\pi_k : \Pi = \bigcup_{k=1}^{k=N_\pi} \pi_k$. Also let there be in every k-th class a chosen vector $\tilde{\mathbf{p}}_k \in \pi_k$, determining the kernel of the class, that is a vector, having the minimal sum of distances to other vectors of this class.

The best way of partitioning into classes (in terms of the accepted measure of distance) and choice of kernel of each of the classes corresponds a minimum of the following function

$$D(\tilde{\mathbf{p}}_1,\ \tilde{\mathbf{p}}_2,\ \cdots,\tilde{\mathbf{p}}_{N_\pi},\ \pi_1,\ \pi_2,\ \cdots,\pi_{N_\pi}) = \frac{1}{2}\sum_{k=1}^{N_\pi}\sum_{\mathbf{p}\in\pi_k} d(\mathbf{p},\tilde{\mathbf{p}}_k). \tag{2.28}$$

This function depends on the method of partitioning of the set of vectors Π on classes and of the choice of a kernel in each class. Methods realized by neural networks carrying out minimization of the function D are called methods of classification without a teacher. Accordingly, such networks are called self-organizing.

Minimization involves cyclic execution of the following steps.

On the first step, sets of vectors Π are partitioned into classes on the specified kernels $\tilde{\mathbf{p}}_1,\ \tilde{\mathbf{p}}_2,\ \cdots,\tilde{\mathbf{p}}_{N_\pi}$ by the following rule: a vector $\mathbf{p}_j$ is related to a class π_k if the distance $d(\mathbf{p}_j,\tilde{\mathbf{p}}_k)$ between it and the kernel $\tilde{\mathbf{p}}_k$ of this class is less than the distance between $\mathbf{p}_j$ and the kernels of other classes. On the second step, the best kernel $\tilde{\mathbf{p}}_k$ in each class is determined from the condition

$$\sum_{\mathbf{p}\in\pi_k} d(\mathbf{p},\tilde{\mathbf{p}}_k) \xrightarrow[\mathbf{p}\in\pi_k]{} \min. \tag{2.29}$$

The problem solution on the second step depends on the accepted definition of the distance. If, in particular, the distance is $d_1(\mathbf{p}^{(1)},\mathbf{p}^{(2)}) = \sum_{j=1}^{N_p}\left(p_j^{(1)} - p_j^{(2)}\right)^2$, that is the square of the Euclidean distance is assumed, then the kernel delivering a minimum in Eq. (2.29) for the class π_k will be the "center of gravity" vector of this class

$$\tilde{\mathbf{p}}_k = \frac{1}{N_{\pi_k}}\sum_{j=1}^{N_{\pi k}}\mathbf{p}_j,\ \mathbf{p}_j \in \pi_k.$$

On each step of the considered process the function D decreases, which provides its convergence. As a result of its minimization in Eq. (2.16) we have a set of kernels $\tilde{\mathbf{p}}_k$ being "centers" of classes, formed by groups of vectors $\mathbf{p}$.

Let's consider how a neural network realizes this process. Figure 2.18 shows a single-layered neural network containing N neurons. Each of the input vector $\mathbf{p}$ components is applied to all neurons of the network. The output signal of each of the neurons simultaneously is the network output signal. Weights of connections (elements of matrix $\mathbf{W}$) in the beginning of the learning are arbitrary and should be determined. Let's consider the process of self-learning of this network.

In the k-th step let the network be activated by the vector $\mathbf{p}_k$. The following should be calculated for every neuron in the network:

$$g_i = d(\mathbf{p}_k, \mathbf{W}_i^{(k)}),\ \ \mathbf{W}_i^{(k)} = \left(W_{i,1}^{(k)},\ W_{i,2}^{(k)},\ \cdots, W_{i,N_p}^{(k)}\right)^T .$$

Here, $\mathbf{W}_i^{(k)}$ is the weights vector of the neuron with number i. The neuron, possessing the least g_i, is declared the winner-neuron at the k-th step. This technique of choice is called a strategy, at which "the winner gets all". The neighborhood S includes neurons having values g, close to the value g_i of the winner-neuron (see Fig. 2.18). Then the weights of the winner-neuron and neurons within its neighborhood are recalculated (changing direction to the vector $\mathbf{p}_k$) by the following rule (Kohonen's rule [10]):

$$\mathbf{W}_i^{(k+1)} = \mathbf{W}_i^{(k)} + \eta_i^k \left(\mathbf{p}_k - \mathbf{W}_i^{(k)}\right),\ \ i \in S .$$

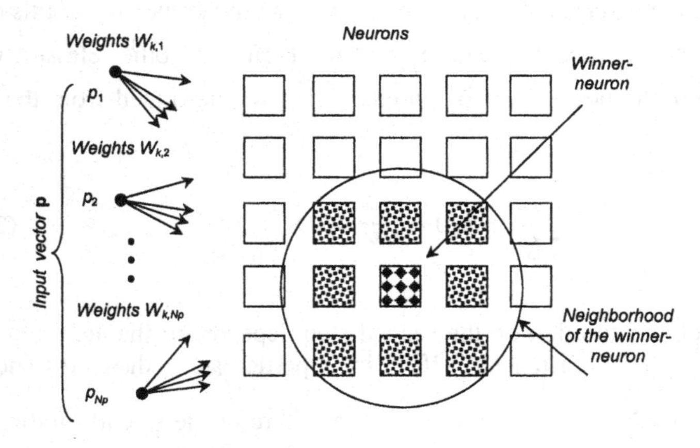

Fig. 2.18. Self-organizing neural network

Here, η_i^k is the learning factor of the i-th neuron at the k-th step. The value of η_i^k decreases as the distance between the i-th neuron and the winner increases. It shall be noted that the "closeness" of neurons "n" and "m" here is understood as closeness of values $g_n = d(\mathbf{p}_k, \mathbf{W}_n^{(k)})$ and $g_m = d(\mathbf{p}_k, \mathbf{W}_m^{(k)})$. It has been shown [11] that the described process of learning is equivalent to minimization of Eq. (2.28). At large numbers of steps, vectors of neuron weights becoming winners on a step more frequently than others define kernels of classes, and their neighborhoods form so-called clusters – groups of neurons, characterizing a certain fault (or some group of faults). Thus, after the learning the neural network will be partitioned into clusters. We shall designate a serial number to each of the clusters, and the fault corresponding to a cluster will get the same serial number.

The neural network learned in such a manner possesses a property to classify vectors acting at its input. To create a new, simpler neural network, we shall replace each of the clusters in the learned network by a single neuron – the kernel of this cluster, that is by the neuron most frequently becoming the winner-neuron in this cluster during learning. For any vector $\mathbf{p}_i$ coming to the input of the new neural network, we shall calculate the value

$$g_k = d(\mathbf{p}_i, \mathbf{W}_k), \quad k = \overline{1, N_k},$$

where $\mathbf{W}_k$, $k = \overline{1, N_k}$ are vectors of weights of neuron connections. As the neural circuit has been learned preliminary, only one of the values g_k will be close to zero. The number of the corresponding kernel (it coincides with the number of the cluster to which the kernel belonged) will determine the number of the class to which the given vector belongs.

It should be noted that vectors $\mathbf{p} \in \Pi$ are normalized, so that, for example, the norm of each of the vectors is equal to unity.

Self-organizing networks are applied for the diagnostics of faults of such electric circuits as active *RC*-filters and for diagnostics of the short-circuit location on a transmission line by measurement of voltages in various points of the line. Such networks give significant effect at forecasting of power systems' performance when the network is preliminary trained on extensive experimental data obtained by analysis of the system performance for many years.

Practical construction and use of neural networks is based substantially on experience and empirical relations. On the first step the network initial configuration is selected, in particular, the number of the hidden layers and the number of neurons. Then the learning of the network is executed, i.e. the weights of neuron connections are found. One of heuristic rules defines the necessary amount of observations, i.e. the number of pairs of input and corresponding output vectors, as a number of 10 times exceeding the number of connections in the network. It is well-known that the number of observations required for learning increases under a nonlinear law with increasing of input vector components number. Experience shows that for the majority of problems, the number of observations lays in limits from several hundreds up to several thousands.

For big errors of a network on the test set of data it is necessary to increase the number of neurons. If the error does not decrease with increasing number of tests, it means that the number of neurons in the network can be reduced.

Special attention should be given on selection of a learning set that should reflect essential properties of the object modeled by means of the neural networks method.

2.5 Application of Volterra polynomials for macromodeling

Neural circuits can be applied in obtaining solutions of devices for macromodeling problems. As noted in Section 1.2, these problems may also be solved by means of a Volterra series or polynomials of split signals. Here, we shall consider the process of macromodeling using a Volterra series

$$w(t)=\sum_{k=1}^{\infty}\int_{-\infty}^{\infty}\cdots\int_{-\infty}^{\infty}h_k(\tau_1,\tau_2,\cdots,\tau_k)\prod_{r=1}^{r=k}v(t-\tau_r)d\tau_1 d\tau_2\cdots d\tau_k,$$

connecting the input $v(t)\in S_v$ and output $w(t)\in S_w$ signals of a nonlinear electric circuit. In this approach, the electric circuit for which the macromodel is created is considered as an unknown operator carrying out the transformation of the set of input signals S_v onto the set of output signals S_w.

A segment of a Volterra series, containing q members, is called Volterra functional polynomial of q degree. For macromodeling it is necessary to define kernels $h_k(\tau_1,\tau_2,...,\tau_k)$, $k=\overline{1,q}$ in frequency or in the time domain. Sets of input and output signals obtained as a result of experiments or calculations are used as initial data for defining the kernels.

Use of Volterra polynomials for operators' approximation is supported by the Frechet theorem [12,13], which deals with the approximation of continuous functionals, and may not be strictly applied for approximation of operators. The possibility of approximation of nonlinear operators by means of functional polynomials is proven more rigidly in [14]. When using the Volterra series, research of their convergence is the most important and intricate part. At that, the question of series convergence is not so specific. The method of Volterra polynomials, the application for macromodeling for the case of divergence of corresponding Volterra series, will be discussed below in an example.

Let's describe the procedure of a nonlinear device macromodeling (Fig. 2.19) by use of experimental data. Assume for simplicity that $q=2$. Then,

$$w[v(\tau),t]=V_1[v(\tau),t]+V_2[v(\tau),t],$$

where

$$V_1[v(\tau),t]=\int_{-\infty}^{\infty}h_1(\tau_1)v(t-\tau_1)d\tau_1, V_2[v(\tau),t]=\int_{-\infty}^{\infty}\int_{-\infty}^{\infty}h_2(\tau_1,\tau_2)v(t-\tau_1)v(t-\tau_2)d\tau_1 d\tau_2.$$

v(t) Nonlinear device w(t)

Fig. 2.19. Arbitrary nonlinear device, for which a macromodel is constructed by the method of a Volterra series

Let's search for the kernels $h_1(\tau_1)$, $h_2(\tau_1,\tau_2)$ of the Volterra polynomials in two stages. For the first stage we shall express V_1, and V_2 through the output signal $w(t)$. For this purpose, actions $\alpha_1 v(t)$, $\alpha_2 v(t)$, where α_1 and α_2 are arbitrary numbers not equal to zero, should be applied by taking turns applying them to the circuit input and measuring the reactions $w_1(t), w_2(t)$. Their linear combination will be given as

$$\beta_1 w_1(t)+\beta_2 w_2(t)=(\beta_1\alpha_1+\beta_2\alpha_2)V_1[v(\tau),t]+(\beta_1\alpha_1^2+\beta_2\alpha_2^2)V_2[v(\tau),t].$$

We may choose numbers β_1, β_2, so that V_1, and V_2 can be determined. To find V_1, we assume

$$\begin{cases}\beta_1^{(1)}\alpha_1+\beta_2^{(1)}\alpha_2=1,\\ \beta_1^{(1)}\alpha_1^2+\beta_2^{(1)}\alpha_2^2=0.\end{cases} \tag{2.30}$$

Then, we have

$$V_1=\int_{-\infty}^{\infty}h_1(\tau_1)v(t-\tau_1)d\tau_1=\beta_1^{(1)}w[\alpha_1 v(t)]+\beta_2^{(1)}w[\alpha_2 v(t)].$$

Similarly, to find V_2, we assume

$$\begin{cases}\beta_1^{(2)}\alpha_1+\beta_2^{(2)}\alpha_2=0,\\ \beta_1^{(2)}\alpha_1^2+\beta_2^{(2)}\alpha_2^2=1.\end{cases} \tag{2.31}$$

Then, we have

$$V_2 = \int_{-\infty}^{\infty}\int_{-\infty}^{\infty} h_2(\tau_1,\tau_2)v(t-\tau_1)v(t-\tau_2)d\tau_1 d\tau_2 =$$

$$= \beta_1^{(2)} w\left[\alpha_1 v(t)\right] + \beta_2^{(2)} w\left[\alpha_2 v(t)\right].$$

Systems of equations (2.30) and (2.31) for any real $\alpha_1 \neq \alpha_2$ and $\alpha_1, \alpha_2 \neq 0$ have unique solutions $\left(\beta_1^{(1)}, \beta_2^{(1)}\right)$ and $\left(\beta_1^{(2)}, \beta_2^{(2)}\right)$, respectively. Therefore, by applying signals $\alpha_1 v(t)$ and $\alpha_2 v(t)$ to the circuit input, then measuring $w\left[\alpha_1 v(t)\right]$ and $w\left[\alpha_2 v(t)\right]$ and setting their linear combinations $\beta_1^{(1)} w\left[\alpha_1 v(t)\right] + \beta_2^{(1)} w\left[\alpha_2 v(t)\right]$, $\beta_1^{(2)} w\left[\alpha_1 v(t)\right] + \beta_2^{(2)} w\left[\alpha_2 v(t)\right]$, we can determine first V_1 and second V_2 members of the Volterra polynomial. This approach can be similarly applied for definition of members V_k, $k = \overline{1,q}$ of Volterra polynomials of any order q.

Let's proceed to the second stage of experimental finding of the Volterra polynomials' kernels. The first stage results allow us to determine experimentally any of the Volterra series members. Let's consider the procedure of finding kernels by the example of determination of the kernel $h_2(\tau_1, \tau_2)$. At first we shall find the Fourier-image of this kernel

$$H_2(j\omega_1, j\omega_2) = \int_{-\infty}^{\infty}\int_{-\infty}^{\infty} h_2(\tau_1, \tau_2)\, e^{-j(\omega_1\tau_1 + \omega_2\tau_2)} d\tau_1 d\tau_2 \ .$$

Then, we apply the following signal at the circuit input:

$$v(t) = U_0\left(\cos\omega_1 t + \cos\omega_2 t\right). \tag{2.32}$$

Identical amplitudes and initial phases for both harmonic components are assumed for simplicity of further calculations. Then, on the basis of definition of the Fourier -image of the kernel of k-th order, we have

$$\int_{-\infty}^{\infty}\int_{-\infty}^{\infty} h_2(\tau_1,\tau_2)v(t-\tau_1)v(t-\tau_2)\, d\tau_1 d\tau_2 =$$

$$= \frac{U_0^2}{4}\int_{-\infty}^{\infty}\int_{-\infty}^{\infty} h_2(\tau_1,\tau_2)\left[e^{j\omega_1(t-\tau_1)} + e^{-j\omega_1(t-\tau_1)} + e^{j\omega_2(t-\tau_1)} + e^{-j\omega_2(t-\tau_1)}\right] \times$$

$$\times\left[e^{j\omega_1(t-\tau_2)} + e^{-j\omega_1(t-\tau_2)} + e^{j\omega_2(t-\tau_2)} + e^{-j\omega_2(t-\tau_2)}\right] d\tau_1 d\tau_2 =$$

$$= U_0^2\left\{ H_2^{1,2}\cos\left[(\omega_1 + \omega_2)t + \angle H_2^{1,2}\right] + H_2^{2,-1}\cos\left[(\omega_2 - \omega_1)t + \angle H_2^{2,-1}\right] + H_2^{1,-1} + H_2^{2,-2} + \right.$$

$$\left. + H_2^{1,1}\cos\left[2\omega_1 t + \angle H_2^{1,1}\right] + H_2^{2,2}\cos\left[2\omega_2 t + \angle H_2^{2,2}\right]\right\},$$

where $H_2^{\pm k,\pm \ell} = |H_2(\pm j\omega_k, \pm j\omega_\ell)|$ and $\angle H_2(\pm j\omega_k, \pm j\omega_\ell)$, $k,\ell = 1,2$ are the amplitude and phase of the Volterra polynomial kernel of the 2nd order in the frequency domain. For the integral calculation, the property of kernel symmetry has been used: $H_2(j\omega_1 j\omega_2) = H_2(j\omega_2 j\omega_1)$.

The first component in this result is of special interest:

$$U_0^2 H_1^{1,2} \cos\left[(\omega_1 + \omega_2)t + \angle H_1^{1,2}\right].$$

Presence of this component shows that if a harmonic signal (see Eq. (2.32)) is applied at the input of the circuit and the harmonic component with frequency $\omega_1 + \omega_2$ is extracted from the output signal (for example, by means of a filter), then it is possible to determine the amplitude and the phase of the Fourier-image of the 2nd order kernel at this frequency. By means of changing frequencies ω_1 and ω_2 of the input signals, it is possible to determine $H_2(j\omega_1 j\omega_2)$ for all frequencies. Similarly, extraction of the Fourier-image of the 3rd order kernel is performed, except a signal containing *three* harmonic components should be applied at the input, etc.

This approach can be used for macromodeling in the frequency domain. Determination of $h_2(\tau_1, \tau_2)$ in the time domain requires the use of Fourier inversion.

The method considered above of the experimental definition of the Volterra polynomials' members imposes rather weak constraints on α_1 and α_2. They should satisfy only the conditions $\alpha_1 \neq \alpha_2$ and $\alpha_1, \alpha_2 \neq 0$. However, it can be shown that the accuracy of macromodeling depends on the choice of α_1 and α_2 values. The macromodel accuracy can be characterized by the root-mean-square error

$$\varepsilon = \frac{1}{N_i} \sum_{i=1}^{N_i} \left(\frac{1}{\max\limits_{t\in[0,T]}^2 \left| w_{circ}\left[v_i(t)\right] \right|} \int_0^T \left(w_{mod}\left[v_i(t)\right] - w_{circ}\left[v_i(t)\right] \right)^2 dt \right)^{1/2}, \qquad (2.33)$$

where $w_{mod}\left[v_i(t)\right] - w_{circ}\left[v_i(t)\right]$ are the output signals of the macromodel and the modeled object.

In the case of $q = 2$, the values $\alpha_1 = 1$, $\alpha_2 = 0.414$ are optimal. At any q, one of these numbers (for example, α_1) can always be assumed equal to 1. Other $q-1$ values can be found from the condition of minimum of approximation error defined by the members of the polynomial of degree $q+1$. At minimization of the root-mean-square error of approximation for $q=3$ and $q=4$, the following sets of numbers are optimal:

$$q=3:\quad \alpha_1=1,\quad \alpha_2=0.5,\quad \alpha_3=0.596,$$

$$q=4:\quad \alpha_1=1,\quad \alpha_2=0.56,\quad \alpha_3=0.644,\quad \alpha_4=0.693.$$

To estimate the quantity of Volterra polynomial members necessary for creating a macromodel displaying the object properties with specified accuracy, in general, is not trivial. If a macromodel created by use of q members does not result satisfactorily then the number of series members should be increased. Moreover, calculations for definition of the first q kernels should be repeated, as the systems of equations similar to Eqs. (2.30) and (2.31) differ for various q.

Let's create the macromodel of the nonlinear electric circuit shown in Fig. 2.20 by means of the Volterra polynomial [15].

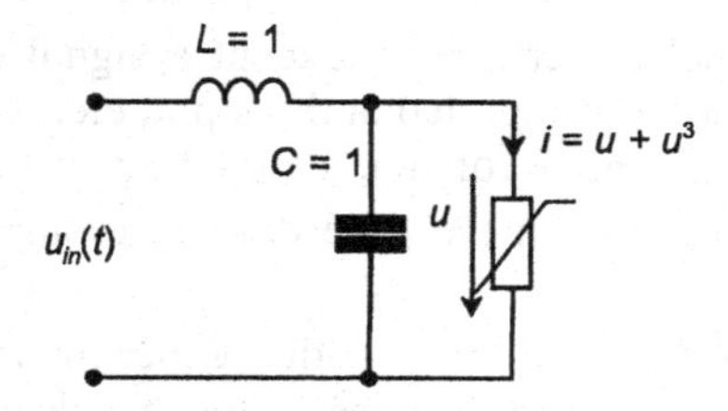

Fig. 2.20. Nonlinear electric circuit for macromodeling

We shall find analytical dependence between the set of input signals $u_{in}(t)=U_0\cos\omega t$ and the corresponding set of circuit responses $y\left[u_{in}(t)\right]=u(t)$ in the range of frequencies $\omega\in(0,1]$ and amplitudes $U_0=(0,5]$. At these amplitudes of external actions and at the given nonlinearity, the Volterra functional series diverge.

We shall use a Volterra polynomial of the 5th degree for macromodeling. As the nonlinear element ampere-volt characteristic is an odd function, then the spectrum of the circuit's response $u(t)$ contains only odd harmonics. Therefore only odd members of series are included into the macromodel:

$$y[\hat{u}_{in}(t)]=\sum_{k=1}^{3}V_{2k-1}\left[u_{in}(\tau),t\right].$$

Actions of Eq. (2.32) type with frequencies in the range $\omega\in(0,1]$ and amplitudes $U_0=(0,5]$ were applied at the circuit input. Corresponding responses $u(t)$ were calculated by numerically solving the system of circuit state equations for each of the input actions. Kernels were calculated for 5 frequencies. Making linear combinations of circuit responses, Volterra polynomial members V_{2k-1}, $k=\overline{1,3}$ were calculated. After calculation of the spectra of the obtained signals, harmonic components were extracted to determine Fourier-images of Volterra kernels.

The root-mean-square error of thus obtained macromodel calculated from Eq. (2.33) is 1.6 % for the specified ranges of amplitude and frequency change at the maximal error of 13 %. To numerically calculate the integral in Eq. (2.33), interpolation of Volterra kernels in the frequency area was used, necessitating Fourier inversion. For macromodeling with specified accuracy, calculations at 5 frequencies prove to be adequate. Already by this simple example, advantages of the Volterra polynomials method in comparison with neural networks, which demand numerous experiments for learning, are appreciable.

A more complex example concerns macromodeling of the operational amplifier circuit (μA741HC) containing 22 transistors. The amplifier connection circuit, used for the calculation of the output signal, is shown in Fig. 2.21.

For macromodeling the following Volterra polynomial was used:

$$\hat{u}_{circ}(t)=\sum_{k=1}^{8}V_{2k-1}\left[u(\tau),t\right]+\sum_{r=1}^{4}V_{2r}\left[u(\tau),t\right].$$

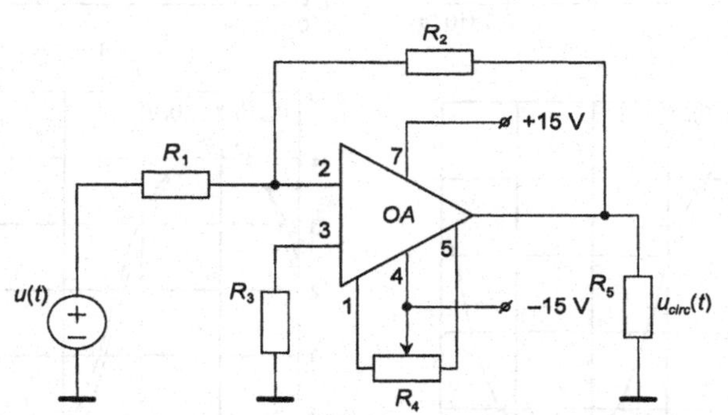

Fig. 2.21. Operational amplifier connection circuit

Modeling was carried out on the set of input signals $u(t)=U\sin\omega t$ $U\in\left(0,0.4\right]$ V, $f\in\left[1,\ 10^{6}\right]$ Hz in the mode of essential nonlinearity (the harmonic factor was up to 30%).

The root-mean-square error of thus obtained macromodel was 0.73% in the specified ranges of amplitude and frequency change for the maximal error of 9.5%.

Visually, the results of modeling for input actions $u_{in,1}(t)=0.4\sin(2\pi\cdot 500t)$, $u_{in,2}(t)=0.32\sin(2\pi\cdot 9\cdot 10^3 t)$, $u_{in,3}(t)=0.3\sin(2\pi\cdot 36\cdot 10^3 t)$ and $u_{in,4}(t)=0.35\sin(2\pi\cdot 600\cdot 10^3 t)$ are shown in Fig. 2.22 a, b, c and d, respectively. Curve 1 characterizes the device response $u_{circ}(t)$ and curve 2 – the model response $\hat{u}_{mod}(t)$.

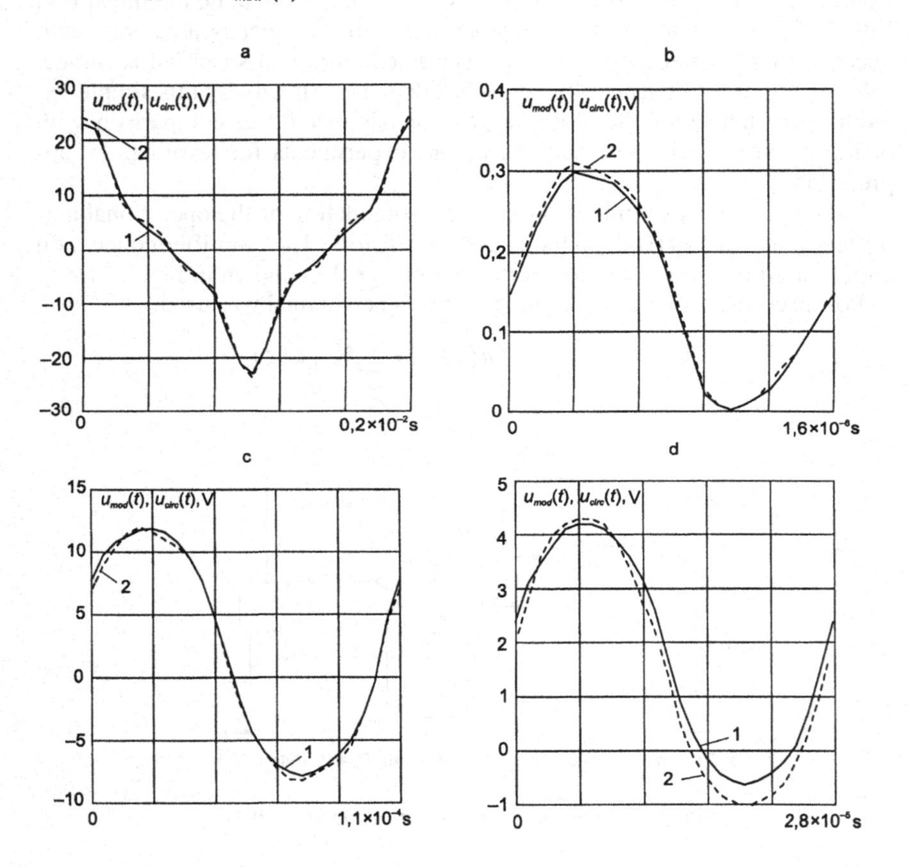

Fig. 2.22. Time responses of the real amplifier (1) and the model (2) for input voltages of different magnitudes and frequencies

2.6 The Search for global minima

Experience in solving inverse problems shows that their objective functionals can have non-unique minimum within the allowable range of change of desired parameters, i.e. they can be non-unimodal. When solving these problems we may wish to find a global minimum. Generally speaking, estimating the possible existence of several extrema for the objective functional is not trivial. This means that in a number of problems, after finding the minimum of the objective functional, it is necessary to check, whether it is global.

When using gradient methods, objective functional's unimodality should be checked during the solution, preceding the search for a global minimum in case several local minima are found. Uses of "soft" methods described below do not give such opportunity, therefore one needs good understanding of inverse problem physical properties to determine the question of its unimodality.

Let's consider some examples.

Assume that it is necessary to find the position coordinate x of a body $(\mu \neq \mu_0)$ having teeth a and b, at which inductance L of the coil C has maximal value (Fig. 2.23).

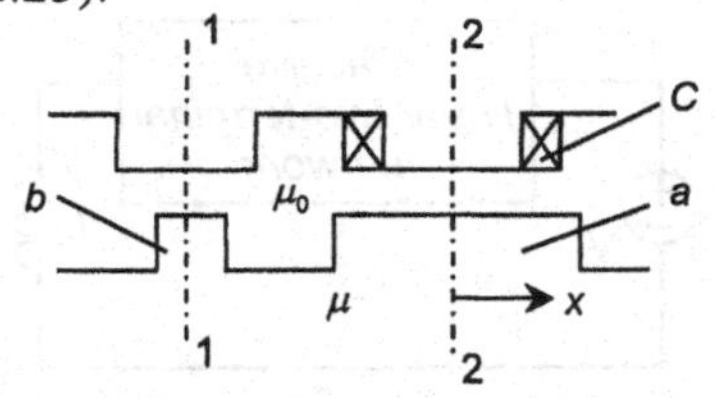

Fig. 2.23. The problem of coil's inductance with non-unique extremum

Evidently, for the case of the coil's axis 2-2 running along the axis 1-1 of the small tooth b, we have a local extremum $L = L_{\max 1}$, as at a small deviation of axis 1-1 from the coil's axis 2-2 the inductance decreases. At the same time, for the case of the coil's axis 2-2 running along the axis of the large tooth (as in the arrangement shown above in Fig. 2.14) we have another extremum of inductance $L = L_{\max 2}$. This simple reasoning allows the assertation that this problem is not unimodal. In practice, teeth a and b may have complex shapes, and determination of the global extremum will demand comparison of inductance extreme values. Therefore, solving of this problem requires finding and subsequent comparison of two local minima.

In this example the objective functional $I(x) = -L$ has no sharp troughs as $|dI/dx|$ (Fig. 2.24) is small. Here the point A corresponds to the local, the point C to the global minimum, and B is the saddle point. Note that points of objective functional minimum in sections with steep slopes, i.e. when $|dI/dx|$ is large in the vicinity of points of minima, and small changes of the parameter x (the coordinate, in this case) near an extremum correspond to large changes of the objective functional, usually do not represent any practical interest. Such solutions of inverse problems are rather sensitive to small changes in parameters and an optimized device with such parameters cannot operate reliably.

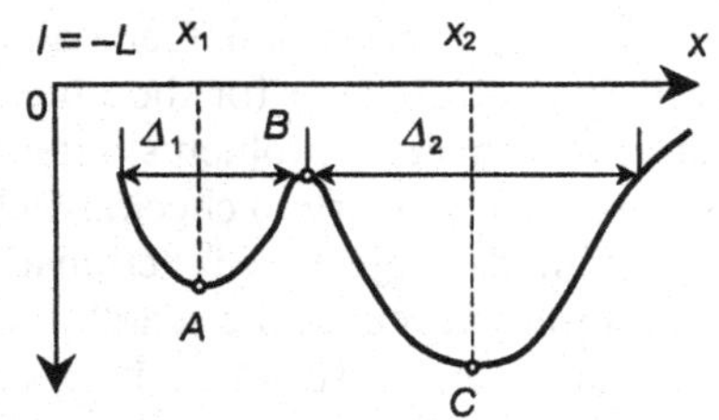

Fig. 2.24. The saddle-point B, points of mimima A,C and their domains of attraction Δx

Let's consider another example. Let, in the circuit shown in Fig. 2.25, it be necessary to find inductance $L \in [L_{\min}, L_{\max}]$, such that the relationship $I(L) = V_L/V$ is maximal for some input signal frequency $\omega = \omega_0$.

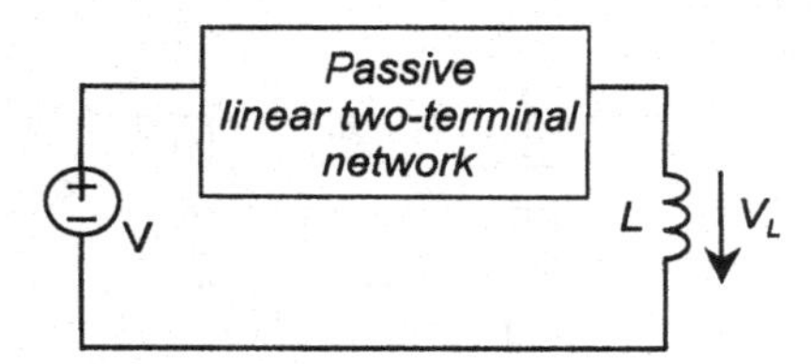

Fig. 2.25. Example of single-extremum problem of inductance L search

It can be shown that this problem is unimodal for any linear two-terminal network. Indeed, let equivalent parameters of the two-terminal network be R_0 and X_0 at the frequency $\omega = \omega_0$. When $X_0 \geq 0$, then $I(L) = V_L/V$ reaches its maximum at $L = L_{\max}$. If $X_0 < 0$ and $|X_0| > R_0$, there is a unique maximum at some $L_{opt} \in [L_{\min}, L_{\max}]$.

Problems of search of global extremum are of great importance for practice that stimulated development of numerous methods of their solution. Even a brief review of these methods lies outside the scope of this book. It shall be noted that the search for a global minimum is much more time-consuming in comparison with the search for a local minimum, and demands significant computing resources.

Let's now consider the most frequently used methods for finding the global minimum, which are most productive for a solution of inverse problems in electrical engineering.

2.6.1. The multistart method and cluster algorithm

Application of gradient methods discussed in Sections 2.2 and 2.3 results in a minimum determined by initial values of desired parameters. Indeed, choosing initial values of the desired parameter from the vicinity Δ_1 of the point x_1 (Fig. 2.24) will lead to the local minimum A. If the initial value x belongs to the segment Δ_2, then a global minimum in point C will be reached. To answer the question whether the found minimum is global, it is necessary to compare it to other minima of the objective functional. Thus, the problem of a global minimum search demands solution of several problems of search of local minima [16,17,18].

One of the most frequently used methods of search of a global minimum is the method of repeated local descents (the method of multistarts). For each descent a new vector of initial parameters, distinct from the previous one, is set. Since the number of objective functional minima at $\mathbf{p}=(p_1,p_2,\cdots,p_n)^T \in \Pi$ is not known in advance, a tentative estimation of the number of initial points (start points) is necessary.

Let the range of definition of the k-th parameter be set by a bilateral inequality $p_{l,k} < p_k < p_{u,k}$. We shall choose N_k equidistant start points on the interval $[p_{l,k}, p_{u,k}]$. Then, the number of start points N necessary for covering of the entire domain Π can be calculated as follows: $N = \prod_{k=1}^{k=n} N_k$.

One can easily see that even for a number of parameters n=10 and an average value $N_k = 10$, the number of start points will be 10^{10}, that is impermissibly large. A valid choice of start points' set is the basic problem for the application of the multistart method. Solution of the problem for a domain's optimal covering may give special grids that allow filling the domain Π in regular intervals.

When filling the domain Π by start points, *a priori* information about the minimized functional can be taken into account. So, for example, it can be known that the objective functional average values are smaller in the subdomain $\pi \in \Pi$ than in subdomain $\eta \in \Pi$. Then, more start points should be placed in the perspective subdomain π than in the subdomain η. It may well be that preferability of a subdomain will be revealed only during the solution. If the objective functional average value in one of the domains appears to be smaller than in others, then the number of points remaining in unpromising domains can be reduced.

A priori information on an optimized device's allowable sensitivity to the change of parameters can also be used in the estimation of the start points' arrangement density.

Indeed, the distance between the adjacent points of the parameters' vector and the change of objective functional are connected through the Lipschitz constant Λ:

$$|I(\mathbf{p}_1) - I(\mathbf{p}_2)| \le \Lambda \|\mathbf{p}_1 - \mathbf{p}_2\|,$$

$$\mathbf{p}_1 = [p_1, \; \cdots, \; p_k, \; \cdots, \; p_n]^T, \quad \mathbf{p}_2 = [p_1, \; \cdots, \; p_k + \Delta, \; \cdots, \; p_n]^T.$$

The Lipschitz constant in this case defines the sensitivity of the objective functional to changes of the k-th parameter that can be estimated on the basis of *a priori* information. Let's consider the procedure of such estimation in more detail.

Let $\Delta p_{\min} = \| \mathbf{p}_1 - \mathbf{p}_2 \|$ be the accuracy at which technical, technological or operational conditions allow maintaining the device parameters. In other words, vectors $\mathbf{p}_1^*$ and $\mathbf{p}_2^*$ are indistinguishable from the point of view of the device performance, if the following inequality holds:

$$\|\mathbf{p}_1^* - \mathbf{p}_2^*\| < \Delta p_{\min}.$$

Let $\Delta I_{\max}$ be the maximum deviation of the functional from its optimal value. For changes of the functional more than $\Delta I_{\max}$, the device ceases to satisfy its functionality. Then, the relation $\hat{\Lambda} = \Delta I_{\max} / \Delta p_{\min}$ can be considered as an upper estimate for the Lipschitz constant Λ. Here, $\Delta I_{\max}$ and $\Delta p_{\min}$ reflect *a priori* concepts about the device quality, its manufacturing technological possibilities or its service conditions.

It was noted, when considering the examples shown in Figs. 2.23-2.24, that inverse problems' solutions corresponding to sharp troughs of the objective functional are quite sensitive to small changes of parameters, and devices created on the basis of such solutions do not represent practical interest. Estimation of the Lipschitz constant allows estimating the minimal distance between initial points of descent:

$$\|\mathbf{p}_1 - \mathbf{p}_2\|_{\min} \approx \frac{|I(\mathbf{p}_1) - I(\mathbf{p}_2)|}{\hat{\Lambda}}.$$

Indeed, let initial points of descent be chosen from this condition. Inserting one more point $\mathbf{p}^*$ between them we shall assume that this point belongs to the domain of attraction of the minimum $\mathbf{p}_{opt}$, to which points $\mathbf{p}_1$ and $\mathbf{p}_2$ do not belong. That is, choosing of $\mathbf{p}_1$ and $\mathbf{p}_2$ does not allow finding this minimum. However, by virtue of the above estimation, the minimum $\mathbf{p}_{opt}$ is in a sharp trough of the objective functional and its search does not represent any interest.

Development of the idea of multistart results in the so-called cluster algorithm, classed among random search algorithms. A random-number generator assigns start points in the domain Π. The quantity of points is approximately equal for each of the components of the parameters' vector **p**. In the process of movement from start points to local minima it appears that some points are concentrated, i.e. they converge towards the same minimum. At detection of such concentrations (clusters) only a single point of a cluster is kept and all other points are discarded.

Note that the multistart method demands significant computing expenses in case of a large dimensionality for the vector **p**. Search algorithms for clusters in many-dimensional space are also rather complex and ineffective in the sense that they start to "reveal" clusters only at significant number of steps of descent with various initial parameters' vectors **p**. Therefore it is expedient to carry out multistart search algorithms for a global minimum on parallel working computers, as processes of search of various minima are independent from each other.

2.6.2 "Soft" methods

Besides gradient methods, such as the method of the quickest descent, the method of adjoint gradients, the method of Newton, etc. - so-called "soft" methods of optimization, are also applicable. For search of a global extremum these methods use models of natural processes that explain the terminology accepted at their description. Here we shall consider the "method of simulated annealing" [19] and one of evolutionary methods, the so-called genetic algorithm.

The gradient methods of optimization demand calculation of functional derivatives by the vector of unknown parameters. To move to an extremum, these methods use the information on the functional's behavior in the vicinity of the current value of the parameters' vector **p**. As has been noted above, one of the important properties of inverse problems is the non-differentiability of the functional in view of the fact that optimization parameters can only have discrete values. The "soft" methods of a minimum search do not require calculation of functional gradients, hence their essential advantage.

The genetic algorithm operates simultaneously with an aggregate group of parameters' vectors $\mathbf{p}_i, i=\overline{1,N}$ covering, in the beginning of the process, all the space of allowable values of optimized variables. The algorithm of this method results in step-by-step concentration of the set of operating points near the best extremum point among those revealed during calculations. At that, there is no need to calculate gradients of variables. The method of simulated annealing assumes calculations with one operating point, but also does not require calculation of gradients when searching for an extremum.

The simulated annealing method

The annealing method [20] is classed among random search methods. If the inverse problem current solution is in the vicinity Δ_1 (Fig. 2.24) of a local minimum x_A of the objective functional, then during the solution there is a possibility to pass over the saddlepoint B to find the global minimum. Hence, this method allows increasing of the functional during minimization.

The author of this method (Metropolis, 1953) has suggested using an algorithm for search of the functional global minimum simulating the process of controlled annealing. In this method, the minimized functional is interpreted as the energy of a cooling down body. At slow controlled annealing, thermal balance is established at each temperature T and the body energy corresponding to this temperature will be minimal. Thus, one can reach a global minimum of energy by lowering the body temperature smoothly. In the case of quick cooling, irregularities of structural and internal pressures are formed. Therefore, the body energy will be higher than at slow cooling, which corresponds to convergence to a local minimum.

Let the solution of an inverse problem be reduced to minimization of a functional:

$$I(\mathbf{p}) \xrightarrow[\mathbf{p}\in\Pi]{} \min ,$$

where Π is the set of allowable values of the parameters' vector. Then we set the following constants: $T = T_{\max} > 0$, $0 < \alpha < 1$, and choose an arbitrary vector $\mathbf{p} \in \Pi$. We shall further consider this vector $\mathbf{p}$ as the problem is current solution. The annealing method algorithm involves iteration of the following actions until the constant T does not become less than a given small number $T_{\min}$:

1. Choice of a new arbitrary vector $\mathbf{p}_1 \in \Pi$;

2. Calculation of $\Delta = I(\mathbf{p}_1) - I(\mathbf{p})$, $\xi = e^{-\Delta/T}$ and generation of a random number r from the interval [0,1];

3. Transition to the new current solution, according the following rule:

- if $\Delta \leq 0$ or ($\Delta > 0$ and $\xi > r$) - accept $\mathbf{p}_1$ as the new current solution ($\mathbf{p} = \mathbf{p}_1$);
- if $\Delta > 0$ and $\xi \leq r$ - ignore the vector $\mathbf{p}_1$ and keep $\mathbf{p}$ as the current solution.

4. Reduction of the constant T, ($T_1 = \alpha T$).

In the beginning of the search when the vector $\mathbf{p}$ is far from optimum, the temperature $T = T_{\max}$ has the greatest value and ξ is close to unity. Therefore, the probability of choosing a current solution that increases the functional is high. Acceptance of such solutions corresponds to movement towards the saddlepoint B, instead of towards the minimum A (see Fig. 2.24). As we approach a global minimum, the temperature decreases and the probability of functional increase falls. For a reliable search of a global minimum by the annealing method, a rather slow decrease in temperature ($\alpha \approx 1$) and large numbers of iterations and calculations of objective functional are required.

Key parameters of the annealing method are the initial T_{max} and the final T_{min} values of temperature, as well as the factor α of its reduction. They are selected based upon reasons of reliability of finding the global minimum. In most cases the choice is made empirically or by results of preliminary numerical experiments. Correct choice of parameters T_{max}, T_{min} and α can considerably speed up the solution search process.

Obviously, at high initial temperatures the speed of change of the objective functional will be insignificant. The high temperature part of the process does not provide essential advancement towards a minimum. On the other hand, choice of small initial T_{max} reduces the probability of finding the global minimum.

Small values of the final temperature T_{min} considerably delay the final low temperature stage of the minimum search process. To prevent excessive computing expenses, the solution process can be finished if at K consecutive temperature reductions there is no reduction of the objective functional. Typical values of K are 5÷50.

In the case of minimization of functionals with unknown quantity and distribution of local minima, application of the annealing method is preferable in comparison with the multistart method. Reliable localization of the global minimum in the multistart method requires dense covering of the set Π for allowable values of the vector $\mathbf{p}$ by start points, which is inconvenient in the case of, for example, large dimensionality of $\mathbf{p}$.

In the annealing method selection of only three parameters (T_{max}, T_{min}, α) is required. It shall also be noted that the annealing method does not demand calculation of gradients of the objective functional, which makes its application for the solution of inverse problems very effective.

However, at solution of inverse problems with unimodal objective functionals the annealing method essentially lacks efficiency (concerning the number of the functional calculations necessary for search of a minimum) in comparison with gradient methods.

Genetic algorithm

When using gradient methods and the annealing method at each moment during calculations there is only one current parameters' vector $\mathbf{p}$, corresponding to the best value of the functional at that moment. Evolutionary methods operate simultaneously with a group of vectors **p**, called a population [1,21,22]. The extremum search process models the process of evolution of living organisms based on Darwin's evolutionary theory concerning heredity, variability and natural selection. We shall consider the genetic algorithm in more detail, as it will be used for solution of inverse problems discussed in this book.

Further, we shall use commonly accepted terminology for description of a genetic algorithm and consider it in more detail.

Each point in the space of optimized variables is defined by a vector $\mathbf{p}=(p_1,p_2, \cdots,p_n)^T \in \Pi$ of these variables, referred to within the framework of accepted terminology as a chromosome. Each variable p_i in a chromosome acts as a separate gene. The minimized functional $I(\mathbf{p})$ is referred to as fitness-function and serves in the genetic algorithm as the parameter of vitality (fitness) of a given individual. An individual is virtually an association of "chromosome + fitness-function" (Fig. 2.26a) and can be presented in the form of $(\mathbf{p},I(\mathbf{p}))$. The set of all individuals $(\mathbf{p}_i,I(\mathbf{p}_i))$, $i=\overline{1,N}$ forms a population.

Chromosomes (vectors **p**) of each individual in an initial population are selected randomly from the set Π. Further, the genetic algorithm involves iteration of the following actions:

- selection of all individuals in a population according to the values of their fitness-function;
- individuals with best values of fitness-function form the parents' generation, and those with the worse values do not participate further in the process;
- on the basis of the parents' generation, descendants are produced, forming a new population jointly with parents.

A graphic representation of the algorithm is shown in Fig. 2.26b. It simulates natural selection of individuals by certain attributes (survival of the most adapted). Such organization of the process of global extremum search is common for all evolutionary methods. The individual possessing the best fitness-function at a given step is considered as the current solution of the problem. The process of extremum search comes to the end when the population has ceased to change. It occurs when values of genes in individual chromosomes and values of fitness-functions of parents and descendants remain constant within the specified accuracy. In that case, we may say that the genetic algorithm has led to an extremum. The best individual, that is an individual with the minimal value of fitness- function, corresponds to the problem's solution.

This way of producing a new generation makes up the basis of the genetic algorithm and is further referred to as style of evolution. Various styles of evolution, as well as selection of concrete style parameters, define the genetic algorithm modification. When realizing a style of evolution, significant freedom of actions is allowable, however three main principles of evolutionary development must be observed:

- the principle of natural selection realized by preservation of the most adapted;
- individuals, owing to which the new generation on its total vitality (the total value of fitness-functions of individuals) should surpass the previous one (or not be worse);
- the principle of inheritance of best properties, owing to which useful attributes transferred from parents to descendants, are fixed in generation as a whole, the principle of variability, which provides genes changing in new generations.

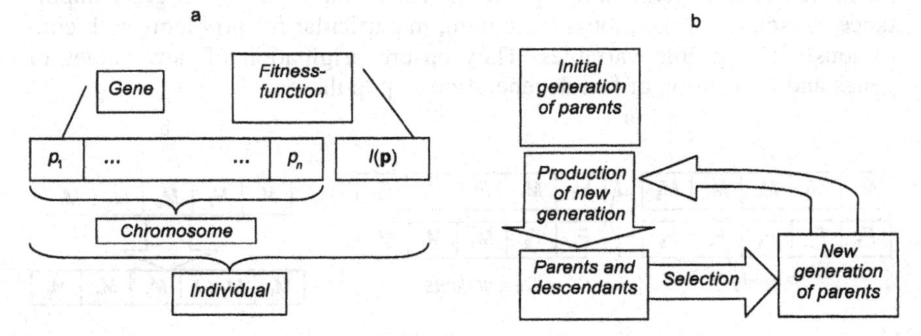

Fig. 2.26. Representation of properties of an individual and its fitness-function in the chromosome **a**) and the block diagram of genetic algorithm **b**)

In classic genetic algorithm, production of descendants is carried out by means of the so-called crossover (crossbreeding) of a pair of parental individuals, when chromosomes of parents are severed in casual points and thus obtained pieces are sewn crosswise in chromosomes of two descendants (Fig. 2.27a). Choice of parents (from all parents-individuals in the current generation) is carried out randomly or by the so-called "roulette rule" when the probability P_i of choice of i-th individuals as a parent depends on the value of its fitness-function and is determined by the following formula:

$$P_i = \frac{I_{\max} - I_i}{\sum_{k=1}^{N} I_{\max} - I_k},$$

where N is the number of individuals in the population and $I_{\max}$ is the worst value of fitness-function among all individuals of the population. Descendants originating as result of a crossover, bear chromosomal sets of both the first and the second parents. In this way, the mechanism of inheritance and securing useful attributes in descendants is realized.

New genes in chromosomes of descendants (i.e. changes of values of parameters' vector **p** elements) occur when mutations are used. Descendants are exposed to mutations after a crossover. At mutation some genes in an individual's chromosome are changed by a random variable δ (Fig. 2.27b). After mutation a check of belonging of the new individual to the population should be carried out (as previously noted, the condition of **p** belonging to Π is checked). If the individual does not belong to the population then the mutation procedure should be redone with another δ. The probability μ of mutations of each gene is a parameter of the evolution style.

To increase the overall performance of genetic algorithm some of the descendants are produced without crossover, i.e. by direct mutations of a single parent (so-called "gemmating" process). These mutations are of great importance for search of the global extremum, in particular for problems with continuously changeable variables. They ensure origination of new values of genes and prevention of fast degeneration of population.

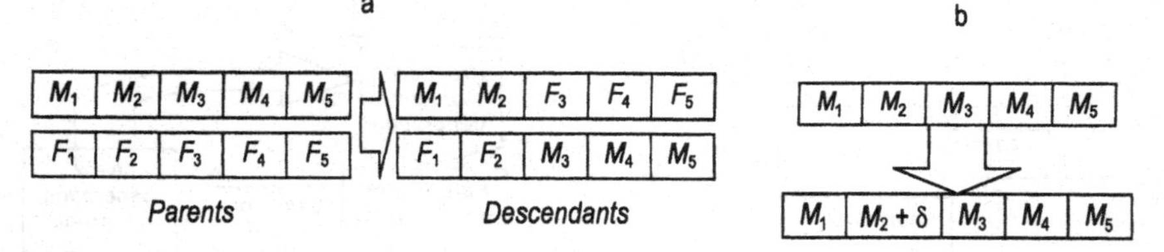

Fig. 2.27. Block diagrams of crossover **a**) and gene mutation procedures **b**)

Let's consider the performance of the genetic algorithm by an example to search for the global minimum for the following functional:

$$I(p_1, p_2) = p_1^2 + p_2^2 + 5p_1p_2 - 10^2\left[\cos\left(p_1 - \frac{p_2}{6}\right) + \cos\left(\frac{p_2}{3} - 1\right)\right], \quad p_1, p_2 \in [-10, 10],$$

which has several local minima in the parameters' vector definitional domain.

Assume the size of population is 10 individuals. Fig. 2.28a shows equal level lines of the functional (the fitness-function) $I(p_1, p_2)$ and the arrangement of randomly chosen individuals of the first generation. One can easily see that individuals are located in domains of attraction of several local minima. Then, we execute an iteration of the genetic algorithm. The arrangement of individuals of the new (second) generation is shown in Fig. 2.28b. All individuals of the second generation have concentrated in domains of attraction of only two minima, one of which is the global minimum. After execution of the following, third step (Fig. 2.28c), all individuals are located only in the domain of attraction of the global minimum. However, it is apparent that chromosomes (parameters' vectors) of individuals are not identical; therefore it is necessary to continue iterating. After execution of the fourth iteration (Fig. 2.28d), chromosomes of all individuals in the population have practically coincided, and the global minimum is found.

This example shows that the genetic algorithm does not require *a priori* data on minimized functional properties, such as availability of local extrema or estimations of dimensions of domains of attraction of extrema. Calculations of minimized functional derivatives are also not required. These properties of the genetic algorithm make it rather convenient for finding the functionals' global extrema.

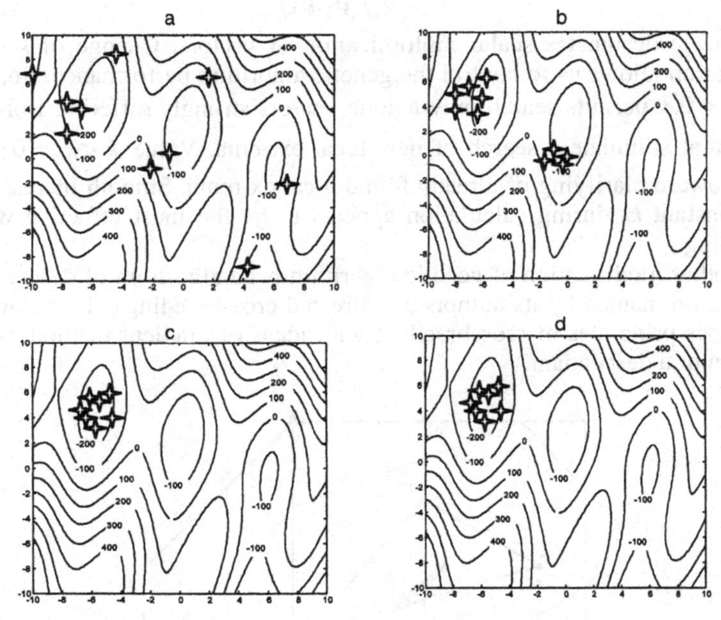

Fig. 2.28. The fitness-function relief and arrangement of best individuals of the first **a**), second **b**), third **c**) and fourth **d**) generations

Some modifications of the genetic algorithm which increases its efficiency are considered below.

Let's allocate a group among the parents (usually 10÷15% of their number) showing best fitness-functions – the so-called elite. Furthermore, the individuals included in the elite group will be chosen as parents at least once. Such style of evolution supplements the above considered "roulette rule" and provides faster distribution of the current population's best individuals throughout the population, speeding up convergence of the algorithm. On the other hand, it increases the risk of population degeneration before finding the global extremum. At degeneration an individual and its descendants drive out less viable at present, but potentially perspective individuals during a small number of generations. Therefore, use of this modification of genetic algorithm is expedient at the last stages of its execution.

To prevent degeneration the following modification of the genetic algorithm can be used. On each step individuals with close values of chromosomes are excluded from the population. As a measure of closeness of individuals, acts parameter k determined by the following expression:

$$k = \frac{\langle \mathbf{p}_1, \mathbf{p}_2 \rangle}{\sqrt{\langle \mathbf{p}_1, \mathbf{p}_1 \rangle \langle \mathbf{p}_2, \mathbf{p}_2 \rangle}} \in [0,1].$$

Here $\langle , \rangle$ designates scalar multiplication of vectors. Change of k during calculation allows us to control the genetic algorithm performance. So, value $k < k_0 = 0.8$ permits search only among vectors strongly differing from each other that stimulate a search of new local extrema. Value $k < k_0 = 0.99$ directs toward clarifying of already found local extrema. Smooth increasing of the constant k_0 during calculation appears to be the most effective way of searching.

Another modification of genetic algorithm is an algorithm of descendants' production, named by its authors as "directed crossbreeding". This algorithm combines principles of crossbreeding with ideas of gradient methods at production of a descendant.

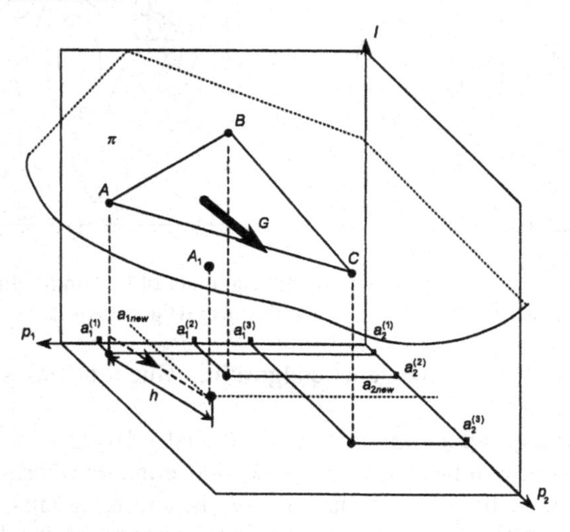

Fig. 2.29. Graphical interpretation of combined genetic + gradient method of optimum search

Let there be some population of individuals $(\mathbf{p}_i, I(\mathbf{p}_i)), i=\overline{1,N}, N>>2$. We shall describe the algorithm of directed mutation for three individuals at a number of genes in a chromosome of each individual equal to two ($\mathbf{p}_i = \left(p_1^{(i)}, p_2^{(i)}\right)^T$). At that, the algorithm provides visual graphic interpretation. Figure 2.29 shows three randomly chosen individuals $(\mathbf{p}_1, A), (\mathbf{p}_2, B)$, and $(\mathbf{p}_3, C)$, where $A = I(\mathbf{p}_1)$, $B = I(\mathbf{p}_2)$, and $C = I(\mathbf{p}_3)$. It is possible to pass a unique plane π through the points *A*, *B* and *C*. Genes of all three individuals define the direction of antigradient **G** in the plane π, and in this sense it can be considered as the result of their crossbreeding.

Three descendants can be formed in various ways. For example each individual can generate a single descendant as shown in Fig. 2.29 for the individual $(\mathbf{p}_1, A)$, which generates the descendant $((p_{1,new}, p_{2,new}), A_1)$. The step value *h*, i.e. the descendant's shift in the antigradient direction **G** with respect to the parental individual, is chosen from the relationship $h = \Delta \cdot \rho$, where ρ is a random variable distributed in regular intervals over the interval [0,1], and Δ is a parameter of the "directed mutation" method.

Another way of a descendant production by three individuals can be as follows. A search of an individual with the minimal value of fitness-function is carried out from the center of mass of the triangle *ABC* in the antigradient direction. This search can be executed, for example, by the gradient method. The resulting individual is assumed further as the single descendant of three parents. Other algorithms of generation of descendants, using estimations of the minimized functional gradient, are also possible.

The difference of this approach from gradient methods should be emphasized. Individuals from a population for crossbreeding are chosen randomly and, generally, are not close to each other. Therefore, in particular, at early stages of population development, not the local value of the functional gradient in some point is estimated, but some large-scale inclination of the extremum search domain. At final stages when the majority of individuals are located in a domain with a unique extremum, this approach becomes a version of the gradient method. Thus, positive features of the gradient method and the genetic algorithm are combined. This algorithm of crossbreeding is normally generalized for the case of an arbitrary number of genes in chromosomes.

Another direction for improvement of genetic algorithm is its realization on multiple processor computers. The number of parameters on which the extremum is searched is assumed that to be very large, and use of a uniprocessor computer will lead to long and tedious calculations. Let *M* processes on a solution of a single problem work simultaneously. Then it will be possible to make regular exchanges of groups of best individuals generated by parallel processes. It is possible and probable that evolution in various populations will go in different ways. Therefore, thus organized computing process is not equivalent to the process of the problem solution by use of a uniprocessor computer with M-multiplied size population. The total expenses of resources of a computer for executing of M parallel processes are less than that of performance of a single process with M-multiplied size population. At the same time, speeds of convergence of both processes are practically identical.

Efficiency of application of the genetic algorithm in many respects depends on the choice of its parameters. Our experience shows that if the number of optimization problem's parameters is equal to n, it requires setting the size of the population equal to $(50 \div 150)n$, and the number of parents – of the order of $(10 \div 20)n$. At gemmating, it is expedient to set large probabilities of mutation (0.1÷0.5) and small values of mutation (~5%). Thus, the descendant chromosome will have some new genes close to the parent genes. For the descendants produced by crossover, on the contrary, it is expedient to subject a small number of genes (~10%) in a chromosome to mutations, however the mutation value should be sufficiently large (~50%). Thus, there will be individuals in the population having genes considerably differing on value from genes of the whole population. These recommendations are not rigid and concern problems with $n < 100$. Optimal values of parameters also essentially depend on the features of the problem.

Comparing gradient methods and the genetic algorithm, the efficiency of the latter at solution of problems with discrete parameters and with large number of local extrema shall be noted. At solution of the majority of inverse problems in electrical engineering, it is necessary to search for a global extremum in view of discreteness of optimization parameters. Besides, for many practical problems, accuracy in 2÷3 significant digits is sufficient. For solutions of such problems the use of the genetic algorithm is preferable. Due to these features of the genetic algorithm, the circle of optimization problems solved with its help has recently extended appreciably.

A disadvantage of the genetic algorithm is the slow convergence near an extremum. Therefore, in cases when high-accuracy solutions are required, it should be expedient at first to obtain solutions near extrema by means of the genetic algorithm. Then, using them as initial approximation, more accurate solutions can be obtained by the gradient method.

References

1. Collette, Y., and P.Siarry (2004). *Multiobjective Optimization: Principles and Case Studies*. Berlin: Springer.
2. Chiampi, M., C.Ragusa, and M.Repetto (1996). Fussy approach for multiobjective optimization in magnetics. *IEEE Trans on Mag*, vol 32, no3.
3. Ross, T.J. (2004). *Fuzzy Logic with Engineering Applications*. Chichester: John Wiley and sons.
4. Reklaitis, G.V., A.Ravindran, and K.M.Ragsdell (1983). *Engineering Optimization*. New York: John Wiley and sons.
5. Rakitski, Yu.V., S.M.Ustinov, and I.G.Chernorutski (1979). *Numerical methods for the solution of stiff systems*. Moscow: Nauka.
6. Korn, G.A., T.M.Korn (2000). *Mathematical Handbook for Scientists and Engineers: Definitions, Theorems, and Formulas for Reference and Review*. New York: Dover Publication.
7. Gilev, S.E., A.N.Gorban, and E.M.Mirkes (1990). Several methods for acceleration the training process of neural networks in pattern recognition. USSR Academy of Sciences, Siberian Branch, Institute of Biophysics, Preprint N146B, Krasnoyarsk.
8. Mueller, B., and J. Reinhardt (1990). *Neural networks*. Berlin: Springer–Verlag.
9. Kolmogorov, L.N. (1957). Representation of continuous multivariable functions as a superposition of continuous one-variable functions and adding. *DAN, USSR*, vol 114, no5,953-956.
10. Kohonen, T. (1995). Self-organizing maps. Berlin: Springer – Verlag.
11. Ritter, H., and K. Schulten (1986). On the stationary state of the Kohonen Self-organizing sensory mapping. *Biological Cybernetics*, vol. 54, 234-249.
12. Frechet, M. (1910). Sur les fonctionnelles continues. *Ann. De l'Ecole Normale Sup 3-me ser*, vol 27.
13. Rugh, W.J. (1981). *Nonlinear System Theory: The Volterra/Wiener Approach*. Baltimore, MD: Johns Hopkins Univ Press.
14. Baesler, I., and I.K. Daugavet (1993). Approximation of Nonlinear Operators by Volterra Polynomials. *Amer Math Soc Transl (2)*, vol 155, 47-57.
15. Soloviova, E.B.(2001). Methods of macromodeling of nonlinear circuits, synthesis of operators and approximations of signals. Ph.Thesis, St.Petersburg State Electrotechnical University.
16. Bomze, I.M., T.Csendes, R.Horst, and P.M.Pardalos (Eds.) (1997). *Developments in Global Optimization*. Dordrecht, The Netherlands: Kluwer.
17. Horst, R., and P.M.Pardalos (Eds.) (1995). *Handbook of Global Optimization*. Dordrecht, The Netherlands: Kluwer.
18. Torn, A., and Zilinskas (1989). *A Global Optimization*. NY:Springer-Verlag.
19. Van Laarhoven, P.J., and E.H. Aarts (1987). *Simulated Annealing: Theory and Applications*. Dordrecht, The Netherlands: Kluwer.

20. Jonson, D., C.Aragon, and C.Schevon (1998). Optimization by annealing: an experimental evaluation. Part I: graph partitioning, *Operations Research.* vol 37, 865-892.
21. Bach, T. (1996). *Evolutionary Algorithms in Theory and Practice*. Oxford: Oxford University Press.
22. Goldberg, D.E. (1989). *Genetic Algorithms in Search, Optimization, and Machine Learning*. MA: Reading, Addison-Wesley.

Chapter 3. Methods of Solution of Stiff Inverse Problems

In this chapter the so-called stiff inverse problems and methods obtaining their solution will be discussed. We will begin with several examples of stiff problems which will be considered in order to become familiar with their basic properties, and subsequently give definitions of problems of such a type. Further, in Sections 3.2 and 3.3, two basic principles will be introduced which provide a basis for solving stiff problems. Specifically, the principle of quasi-stationarity of derivatives, and the principle of repeated measurements. In Section 3.4, problems of diagnostics of sinusoidal current circuits that are typical inverse problems in circuit theory will be discussed. Conditions, at which these problems should be considered as stiff will be discussed as well. In Section 3.5, a new effective method of diagnostics stiff problems solution will be introduced and illustrated by results of numerical solution of some problems. In Section 3.6, the problem of localization of one or several perturbation sources in an electric circuit by results of measurement of voltages in circuit nodes located remotely with respect to the perturbation sources will be discussed.

3.1 Stiff inverse problems

The purpose of solution of the extensive class of inverse problems is the creation of mathematical models for devices or processes and the determination of parameters of these mathematical models by means of numerical or physical experiments. Creating a mathematical model that reflects the modeled device (or process) in the best way requires simultaneous consideration of all (or most of) known factors that influence its functioning. At the initial stages of an inverse problem solution, it is usually impossible to estimate the importance of either “weak actions”, “small parameters” or “minor alterations” of created mathematical models. Unreasonable neglect of “small quantities” can lead to “throwing out significant details”, and the creation of inadequate models.

The property of stiffness of mathematical models is a consequence of the inclusion of factors in them that differ by their degree of influence. Further research, which may give better understanding of specific properties and processes occurring in modeled devices, allow in many cases reducing the model stiffness by means of reasonable exclusion of factors of little importance. However, for some devices and processes stiffness is an inherent property reflecting the basic principles of their functioning. Being aware of abstractness of this reasoning, we shall illustrate the aforesaid by several examples below.

One may obtain some insight of a stiff mathematical model, and thus of the corresponding stiff inverse problem, when modeling a device that simultaneously includes "rapidly" and "slowly" varying quantities. As an example, let's consider the problem of mathematical modeling of a device by its experimentally determined transient conductivity $y_e(t)$. Assume there is *a priori* information that physical processes in this device correspond to processes in the electric circuit shown in Fig. 3.1a. Then, the purpose of inverse problem solution should be the definition of parameters R, L and C. Such inverse problems are referred to as problems of electric circuits' parametrical synthesis. For simplicity we assume that $y_e(t)$ (in Fig. 3.1b) is an aperiodic function with initial conditions equal to zero. Half-fall time of $y_e(t)$ is designated as $t_{1/2}$.

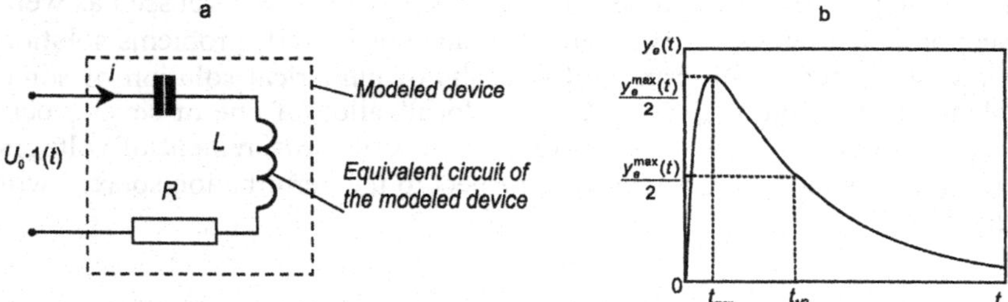

Fig. 3.1. The modeled device, its equivalent circuit **a**) and the transient characteristic of the modeled device **b**)

Transient conductivity $y_m(t)$ of the series circuit R, L and C considered as a mathematical model, is related to the input current $i(t)$ by the following expression: $y_m(t)=i(t)/U_0$. An analytical expression for the current $i(t)$ can easily be derived by the solution of the linear differential equation of the transient process in the RLC-circuit:

$$\frac{d^2i}{dt^2}+\frac{R}{L}\frac{di}{dt}+\frac{1}{LC}i=\frac{U_0}{RLC}. \tag{3.1}$$

The solution of Eq. (3.1) can be written down as:

$$i(t) = A_1 e^{\lambda_1 t} + A_2 e^{\lambda_2 t} = U_0 y_m(t).$$

Constants A_1 and A_2 are defined from the input conditions $i(0)=0$, $u_C(0)=0$:

$$\begin{cases} i(0) = A_1 + A_2 = 0, \\ L\left.\dfrac{di}{dt}\right|_{t=0} = L(A_1\lambda_1 + A_2\lambda_2) = U_0 - u_C(0) = U_0, \end{cases} \Rightarrow \quad \begin{aligned} A_1 = A = \frac{U_0}{L(\lambda_1 - \lambda_2)},\ A_2 = -A, \\ y_m(t) = \frac{A}{U_0}(e^{\lambda_1 t} - e^{\lambda_2 t}). \end{aligned}$$

Here the roots of the characteristic equation λ_1 and λ_2, corresponding to the differential equation (3.1), are given by

$$\lambda_{1,2} = -\frac{R}{2L} \pm \sqrt{\frac{R^2}{4L^2} - \frac{1}{LC}}.$$

The definition of parameters R, L and C for the inverse problem of the devices' mathematical model can be written down as:

$$\int_0^\infty \left(y_e(t) - \frac{A}{U_0}(e^{\lambda_1 t} - e^{\lambda_2 t}) \right)^2 dt \xrightarrow[R,L,C]{} \min. \tag{3.2}$$

The result of solution of Eq. (3.2) will be values of R, L and C, which can be used for the calculation of the roots of the characteristic equation λ_1, λ_2 and the constant A. Figure 3.2 illustrates this solution.

Later on we may need the following properties of the solution: $y_e(t)$ rise time t_{max} up to its maximum value close to the time constant $\tau_1 = |1/\lambda_1|$ of an exponent with maximum modulo index, and $y_e(t)$ half-fall time $t_{1/2}$ close to the time constant $\tau_2 = |1/\lambda_2|$.

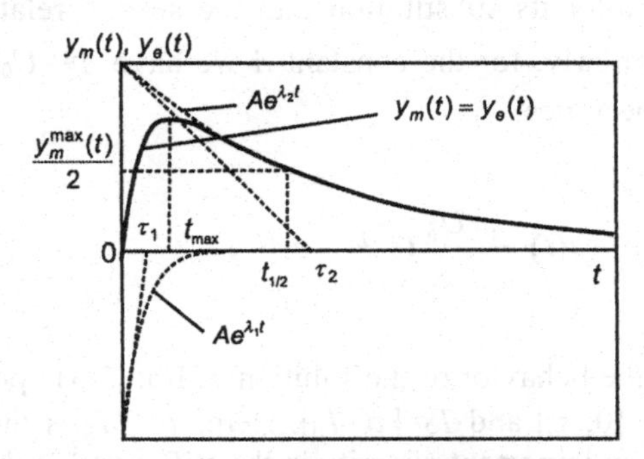

Fig. 3.2. The transient characteristic of the modeled device

Methods of solution of the problems similar to Eq. (3.2), and difficulties arising in that connection, were discussed in Chapter 1. These methods will also be used in subsequent chapters. In this section, the case for which the transient characteristic $y_e(t)$ of the modeled device has "fast"" and "slow" components is of interest, and will be considered below.

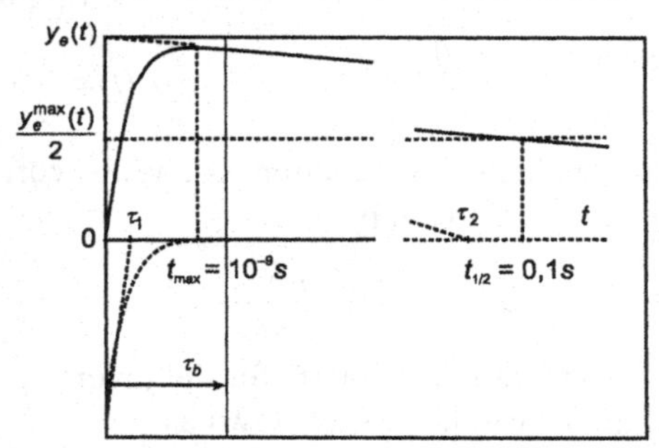

Fig. 3.3. The modeled device's transient characteristic shown on different time intervals

Let the transient characteristic $y_e(t)$ reach its maximum value y_e^{max} at $t_{max}=10^{-9}$ s as shown in Fig. 3.3, with distortion of the time scale for clarity. The half-fall time $t_{1/2}$ of $y_e(t)$ is 0.1 s. In this case $t_{max} << t_{1/2}$, therefore $\tau_1 << \tau_2$; this is the same as $|\lambda_1| >> |\lambda_2|$. Here, the obtained relations between t_{max}, $t_{1/2}$ and time constants τ_1 and τ_2 have been used. Then, according to Vieta's theorem for the roots of the characteristic equation (3.1), we have:

$$\lambda_1 + \lambda_2 = -R/L, \qquad \lambda_1 \lambda_2 = 1/LC.$$

Following [1], from the first relationship when $|\lambda_1| >> |\lambda_2|$, we have $\lambda_1 \cong -R/L$, and after its substitution into the second relationship, we find $\lambda_2 \cong -1/RC$. Similarly, for the constant A we have $A=-U_0/R$, and the solution to Eq. (3.1) becomes:

$$i(t) = -\frac{U_0}{R}(e^{-\frac{R}{L}t} - e^{-\frac{t}{RC}}). \tag{3.3}$$

Let's consider the behavior of the solution of Eq. (3.3) upon two sequential time intervals $T_1=[0, \tau_b]$ and $T_2=[\tau_b, T_0]$. Here, $\tau_b << T_0$ is the duration of the boundary layer – an important quantity in the stiff systems theory. We choose the duration of the boundary layer τ_b so that the Eq. (3.1) solution within the boundary layer, i.e. upon the interval T_1, will be characterized by a fast variation of current. The duration of the interval T_1 for this problem can be chosen equal to $\tau_b=(3\text{-}5)\cdot\tau_1 \approx 10^{-9}$ s (see Fig. 3.3). The interval T_2, lying beyond the boundary layer, is characterized by a slow variation of current and comes to its end along with the completion of the transient process.

For the case under consideration, it is possible to assume with a high degree of accuracy that $e^{\lambda_2 t} = e^{-t/RC} = 1$ on the interval T_1 within the boundary layer, and $e^{\lambda_1 t} = e^{-tR/L} = 0$ on the interval T_2 outside the boundary layer. Then,

$$i(t) = -\frac{U_0}{R}(e^{-\frac{R}{L}t} - e^{-\frac{t}{RC}}) = \begin{cases} t \le \tau_b, & i_1(t) = \frac{U_0}{R}(1 - e^{-\frac{R}{L}t}) \\ t > \tau_b, & i_2(t) = \frac{U_0}{R} e^{-\frac{t}{RC}} \end{cases}. \tag{3.4}$$

Relationships (3.4) show that fast and slow processes can be separated for stiff problems, which considerably simplifies their solution. Indeed, an analytical solution of above stiff inverse problem can be obtained for the condition that characteristic points of $y_e(t)$ dependence are known. Let $t_{\max} = 10^{-9}$s, $t_{1/2} = 0.1$s, $y_e(t_{\max}) = y_e^{\max} = 0.1$Sm, $y_e(t_{1/2}) = y_e^{\max}/2 = 0.05$Sm. Then

$$R = \frac{1}{y_e^{\max}} = 10\text{kOhm}, \quad C = y_e^{\max}\frac{t_{1/2}}{\ln 2} = 1.44\text{mF}, \quad L = \frac{1}{y_e^{\max}}\frac{t_{\max}}{20} = 50\mu\text{H}.$$

It should also be noted that the form of expressions for currents $i_1(t)$ and $i_2(t)$ in (3.4) allows giving the following simple physical interpretation of the stiff inverse problem solution. Properties of the initial circuit, on the interval T_1 within the boundary layer, coincide with properties of a simpler circuit consisting of a series-connected resistor R and inductor L. Outside the boundary layer on the interval T_2, properties of the initial circuit coincide with properties of an RC-circuit, which is simpler than the initial one. Thus, during the inverse problem solution, a correction of the initially accepted mathematical model has been carried out. This correction becomes possible because of the stiffness of the initial problem.

Splitting stiff problems into two or more simpler, and non-stiff problems is one of the effective ways of their solution. Let's consider this method of solution in more detail.

Following [2], we shall give a rigorous definition of stiff systems. Suppose there is a system of nonlinear differential equations of the form

$$\frac{d\mathbf{y}(t)}{dt} = \mathbf{f}(t,\ \mathbf{y})\ ,\ t \in [0,\ T_0]\ , \qquad \mathbf{y}(0) = \mathbf{y}_0;$$
$$\mathbf{y}(t) = \left(y_1(t)\ \ldots\ y_m(t)\right)^T\ ,\ \mathbf{f}(t\ ,\ \mathbf{y}) = \left(f_1(t,\mathbf{y})\ \ldots\ f_m(t,\mathbf{y})\right)^T. \tag{3.5}$$

For a stiff system of differential equations, values of the solution derivatives' norm outside the boundary layer $\tau_b \ll T_0$ is much smaller than inside of it:

$$\left\|\left.\frac{d\mathbf{y}(t)}{dt}\right|_{t\geq\tau_b}\right\| \ll \left\|\left.\frac{d\mathbf{y}(t)}{dt}\right|_{t<\tau_b}\right\| \Leftrightarrow \exists_{N\gg1}: \left\|\left.\frac{d\mathbf{y}(t)}{dt}\right|_{t\geq\tau_b}\right\| \approx \left\|\left.\frac{1}{N}\frac{d\mathbf{y}(t)}{dt}\right|_{t<\tau_b}\right\|,\ N\gg1.$$

After linearizing the right member of Eq. (3.5) in the vicinity of the initial point, we have:

$$\mathbf{f}(t,\mathbf{y}) = \mathbf{f}(t,\mathbf{y}_0) + \frac{\partial \mathbf{f}}{\partial \mathbf{y}}(\mathbf{y} - \mathbf{y}_0) + \ldots.$$

Values of vector $\mathbf{y}(t)$ components' derivatives at $t \in [0,\ \tau_b]$ may reach up to $W \cdot \max|y_k(t)|$, where W is a number satisfying the inequality $0 < W \leq \|\partial\mathbf{f}/\partial\mathbf{y}\|$ and $\partial\mathbf{f}/\partial\mathbf{y}$ is the Jacobi matrix.

The system of differential equations (3.5) is identified as a stiff one if at any vector of input conditions $\mathbf{y}_0$ there will be such numbers $\tau_b \ll T_0$ and $0 < W \leq |\partial\mathbf{f}/\partial\mathbf{y}|$, $N \gg 1$ that assume the following inequalities:

$$\left|\frac{dy_k}{dt}\right|_{t\geq\tau_b} \leq \frac{W}{N} \max_{t\in[0,\ T_0]} |y_k(t)|,\ \ k = \overline{1,m}. \tag{3.6}$$

It is important that the concept of stiffness of a system of differential equations is connected with the interval $t \in [0,\ T_0]$, on which its solution is searched. A system of equations that is stiff on the interval $t \in [0,\ T_0]$ is not stiff on the subinterval $t \in [0,\ \tau_b]$. Applying the definition considered above to the system of linear differential equations

$$\frac{d\mathbf{y}}{dt} = \mathbf{A}\mathbf{y},\ \ \mathbf{y} \in \mathbf{R}^m,\ \ t \in [0, T_0],$$

we obtain the following conditions for the stiff system matrix $\mathbf{A}$ eigenvalues:

$$|\lambda_k(\mathbf{A})|\ e^{\mathrm{Re}\,\lambda_k(\mathbf{A})\tau_b} \leq \frac{L}{N},\quad L = \max_k |\lambda_k(\mathbf{A})|,\quad N \gg 1,\quad \tau_b \ll T_0,\quad k = \overline{1,m}. \tag{3.7}$$

Let's estimate the stiffness of state equations of the circuit shown in Fig. 3.1. The state matrix eigenvalues are $\lambda_1=-10^3\ \mu s^{-1}=-10^9\ s^{-1}$ and $\lambda_2=-1\ \mu s^{-1}=-10^6\ s^{-1}$. The search interval for its solution is $t\in[0,\ 10\ \mu s]$. We assume that the constant W is equal to the module of maximum eigenvalue $W=|\lambda_1|=10^9$. For the boundary layer we have: $\tau_b=5\tau_{\min}=5|1/\lambda_1|=5\cdot 10^{-9}\ s^{-1}$. We can see by choosing $N=100$ that this system is stiff, since the following inequalities are valid:

$$|\lambda_1|e^{\lambda_1\tau_b}=10^9\cdot e^{-5}\cong 6.7\cdot 10^6\le\frac{W}{N}=\frac{10^9}{100}=10^7;$$

$$|\lambda_2|e^{\lambda_2\tau_b}=10^6\cdot e^{-5\cdot 10^{-3}}\cong 10^6\le\frac{W}{N}=10^7.$$

This problem is classed among the so-called singularly perturbed problems. Indeed, from expressions $|\lambda_1|>>|\lambda_2|$, $\lambda_1\cong -R/L$ and $\lambda_2\cong -1/RC$ follows $RC>>LC=\mu$, therefore Eq. (3.1) can be rewritten in the form of:

$$\mu\frac{d^2i}{dt^2}+RC\frac{di}{dt}+i=\frac{U_0}{R},$$

where μ is a small parameter. Presence of a small parameter in the equation or in the system of equations is a characteristic feature of singularly perturbed problems. This singularity allows splitting linear problems into several simpler ones quite easily as it has been done above. In most cases singularly perturbed problems are stiff; however they by no means exhaust the variety of stiff problems. A system of equations can be stiff as well, in cases when it is not possible to select a small parameter. To prove it, we shall consider the following example:

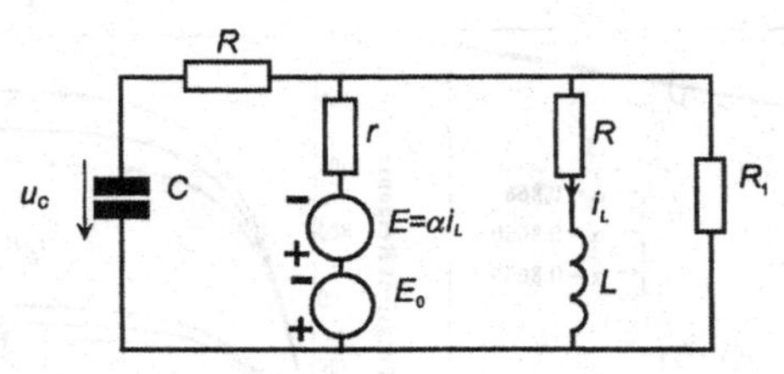

Fig. 3.4. The electric circuit, for which the problem of parametrical synthesis is solved

Let's once again state the parametrical synthesis problem for the electric circuit shown in Fig. 3.4. Assume that R=470 Ohm, r=270 Ohm, C=4 pF, and E_0=1 V are known values. Suppose determination of parameters α, L, R_1 by measured state variables $i_L(t)$ and $u_C(t)$ is required. Assume the exact solution is known: α^*=0.866 kOhm, L^*=1.0 μH, R_1^*=1 kOhm. The system of state equations describing the transient process is given by:

$$\frac{d}{dt}\begin{bmatrix} i_L \\ u_C \end{bmatrix} = \frac{1}{\rho}\begin{bmatrix} \frac{R(\alpha R_1 - 2R_1 r - rR - R_1 R)}{L} & \frac{rR_1}{L} \\ \frac{R_1(\alpha - r)}{C} & -\frac{r + R_1}{C} \end{bmatrix}\begin{bmatrix} i_L \\ u_C \end{bmatrix} + \frac{1}{\rho}\begin{bmatrix} \frac{E_0 R_1 R}{L} \\ \frac{E_0 R_1}{C} \end{bmatrix},$$

$$\rho = rR_1 + RR_1 + rR$$

or

$$\frac{d}{dt}\begin{bmatrix} i_L \\ u_C \end{bmatrix} = \mathbf{A}\begin{bmatrix} i_L \\ u_C \end{bmatrix} + \mathbf{b}, \qquad (3.8)$$

where **A** is the state equations matrix and **b** is the vector of sources.

Dependences $i_L(t)$ and $u_C(t)$ at initial conditions $i_L(0)$=100 mA, $u_C(0)$=100 V are shown in Fig. 3.5a for the time interval of T_1=[0, 50 ns], and in Fig. 3.5b for the time interval T_2=[0, 20 μs]. Analysis of these dependences allows defining the boundary layer duration as approximately $\tau_b \approx 10$ ns, thereby dividing the transient process in two parts:

- boundary layer part $t<\tau_b$ with fast-changing state variables $i_L(t)$, $u_C(t)$ and large values of derivatives $|i'_L(t)|$ and $|u'_C(t)|$;

- outside of the boundary layer part $t > \tau_b$ with slow-changing state variables $i_L(t)$, $u_C(t)$.

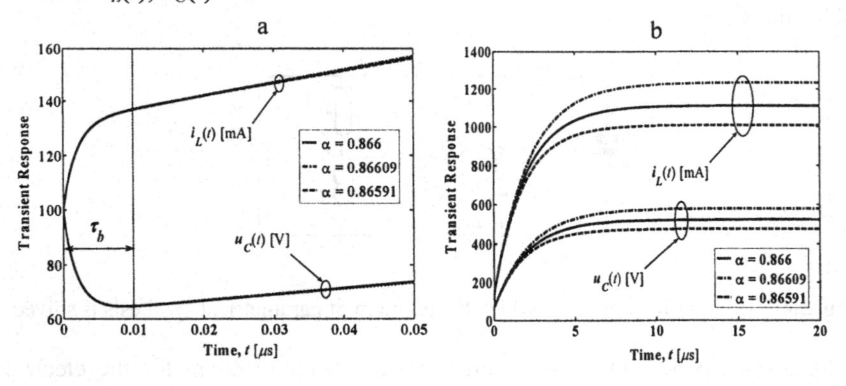

Fig. 3.5. Dependences $i_L(t)$ and $u_C(t)$ near a boundary layer **a**) and outside of the boundary layer **b**) for various values of gain α

This property of the problem ensures that it is a stiff problem. We may assert that the matrix **A** is ill-conditioned, that is $|\lambda_{\max}(\mathbf{A})| >> |\lambda_{\min}(\mathbf{A})|$. Indeed, it is easy to find that **A** is a degenerate matrix (i.e. its eigenvalue is zero), if

$$\alpha_0 = R + r + rR/R_1 . \tag{3.9}$$

The value $\alpha=\alpha^*=0.866$kOhm that was specified in the problem conditions is close to the critical value $\alpha_0|_{r=270,\, R=470,\, R_1=R_1^*1000} = 0.8669\,\text{kOhm}$. Therefore matrix **A** should be close to a degenerate one, and one of its eigenvalues should be close to zero. Indeed, matrix **A** eigenvalues are $\lambda_1\approx-512\ \mu s^{-1}$ and $\lambda_2\approx-0.5\ \mu s^{-1}$.

However, the property of stiffness found for this problem does not allow making simplifying assumptions for the purpose of deriving relations between required parameters or splitting it into simpler and less stiff problems. It is also impossible to select a small parameter. Here, we do not intend to assert that splitting the initial problem is basically impossible. We only draw attention to the situation typical for practice, when simplification of the initial problem requires application of special methods. Such methods will be discussed in the following paragraphs of this chapter. Here, we shall continue the solution of the above problem to demonstrate difficulties arising at the solution of stiff problems.

Let's proceed to solution of the inverse problem. Let $i_L^{\exp}(t)$ and $u_C^{\exp}(t)$ be experimentally found in points $t = t_m,\ m = \overline{1,M}$. We shall search for parameters α, L, R_1 from the condition of minimum of the following functional:

$$F(\alpha, L, R_1) = \max(\delta_u, \delta_i) \xrightarrow[\{\alpha, L, R_1\}]{} \min, \tag{3.10}$$

where $\delta_u = \dfrac{1}{M}\sqrt{\sum\limits_{m=1}^{M}\left(\dfrac{u_C(t_m) - u_C^{\exp}(t_m)}{u_C^{\exp}(t_m)}\right)^2}$, $\delta_i = \dfrac{1}{M}\sqrt{\sum\limits_{m=1}^{M}\left(\dfrac{i_L(t_m) - i_L^{\exp}(t_m)}{i_L^{\exp}(t_m)}\right)^2}$,

$i_L(t_m) = i_L(t, \alpha, L, R_1)\big|_{t=t_m}$ and $u_C(t_m) = u_C(t, \alpha, L, R_1)\big|_{t=t_m}$.

These values are obtained from solution of Eq. (3.8) at some current values of desired parameters α, L, R_1 and at input conditions $i_L^{\exp}(t)\big|_{t=0} = i_L(0)$ $u_C^{\exp}(t)\big|_{t=0} = u_C(0)$. This approach to solve parametrical synthesis problems in its various modifications is one of the basic methods frequently used in practice. Let's estimate possibilities of its application for solution of stiff problems.

In practice, solution of the stiff system of Eq. (3.8) is carried out numerically. We shall use an analytical solution of this system of equations to calculate the functional $F(\alpha,L,R_1)$. This approach will allow eliminating the influence of the error of numerical integration of Eq. (3.8) when solving the parametrical synthesis problem. The solution is given by:

$$\begin{cases} i_L(t) = i_L(\infty) + A_1 e^{\lambda_1 t} + B_1 e^{\lambda_2 t} \\ u_C(t) = u_C(\infty) + A_2 e^{\lambda_1 t} + B_2 e^{\lambda_2 t}, \end{cases} \tag{3.11}$$

where λ_1 and λ_2 are the eigenvalues of matrix **A**, and $(i_L(\infty), u_C(\infty))^T = -\mathbf{A}^{-1}\mathbf{b}$ is the vector of steady-state values of state variables. The constants are equal to:

$$A_1 = \frac{(a_{11} - \lambda_2)\Delta i_L + a_{12}\Delta u_C}{\lambda_1 - \lambda_2}, \quad B_1 = -\frac{(a_{11} - \lambda_1)\Delta i_L + a_{12}\Delta u_C}{\lambda_1 - \lambda_2},$$

$$A_2 = \frac{(a_{22} - \lambda_2)\Delta u_C + a_{21}\Delta i_L}{\lambda_1 - \lambda_2} \quad B_2 = -\frac{(a_{22} - \lambda_1)\Delta u_C + a_{21}\Delta i_L}{\lambda_1 - \lambda_2}, \tag{3.12}$$

where $\Delta i_L = i_L(0) - i_L(\infty)$, $\Delta u_C = u_C(0) - u_C(\infty)$.

Minimization of Eq. (3.10) was carried out by the Nelder-Mead simplex method [3] for 50 various initial approximations $\left(\alpha^{(0)}, L^{(0)}, R_1^{(0)}\right)$. These initial approximations were chosen at random fashion from the given search intervals of parameters: $\alpha^{(0)} \in [\alpha_{\min}, \alpha_{\max}]$, $L^{(0)} \in [L_{\min}, L_{\max}]$ and $R_1^{(0)} \in [R_{\min}, R_{\max}]$. At their choice, the condition $\alpha^{(0)} < \alpha_0\left(R_1^{(0)}\right)$ has been superimposed. The latter allows negative eigenvalues to remain in the domain of matrix **A** and to use formulas (3.11) - (3.12) for the circuit response calculation. In each case the solution was considered to be found, if for two subsequent iterations the increment of parameters in corresponding units did not exceed 10^{-10}.

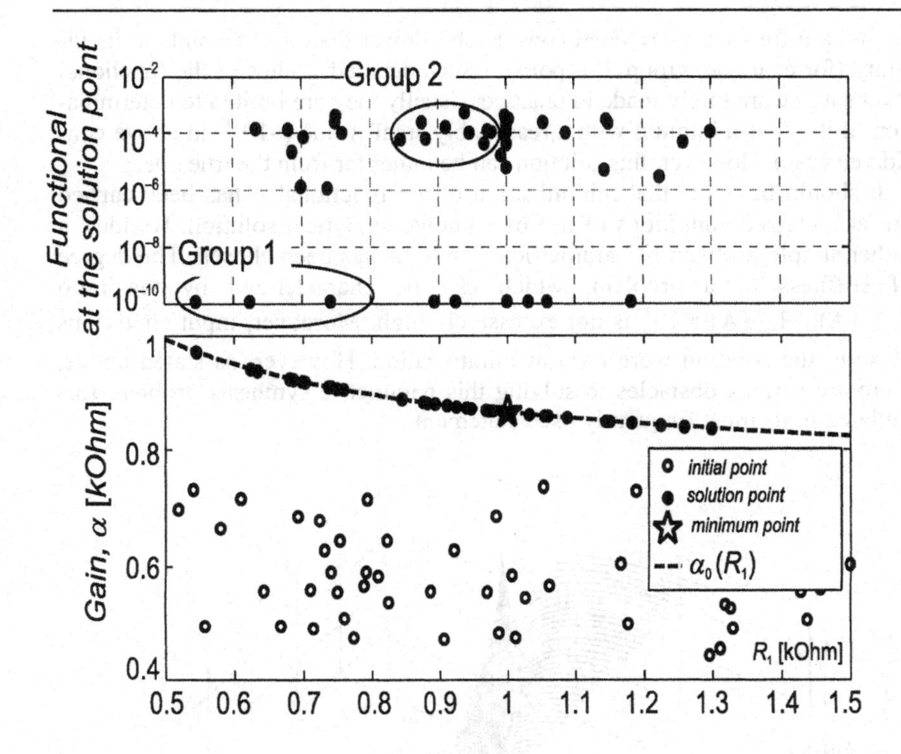

Fig. 3.6. Results of numerical solution of the problem (3.10) from various entry conditions for gain α. Approximate solutions incorporated in Group 1 correspond to large values of minimized functional F, but are arranged near the solution of (3.10). Approximate solutions incorporated in Group 2 are characterized by a small value of F, but they are arranged far from the solution of (3.10)

Figure 3.6 shows results of the numerical experiment. Let's consider them in more detail. Symbol ☆ marks off the exact solution of the problem (3.10), showing the functional F global minimum. Initial approximations (symbol °) at some value $L \in [L_{\min}, L_{\max}]$ and corresponding solutions of the problem (3.10) (symbol •) are shown on the plane α, R_1 in the lower part of the figure. This part of the figure shows that all obtained solutions are located on the curve $\alpha_0(R_1)$ and that the majority of them are far from the correct solution of the problem.

Minimum values of functional F for each of the obtained solutions are shown in the upper part of the figure. It is obvious that values of the functional in points distant from the global minimum (for example, group 1 of points) are frequently arranged considerably lower than in the points in its vicinity (for example, group 2 of points). Such detailed studies of the functional level surface are rarely made in practice. Usually they are limited to determination of the first minimum with a reasonably small value ($\sim 10^{-10}$ in above considered case). However, this solution can be rather far from the true one.

It should be noted that minimization of the functional F has been carried out at idealized conditions of use of a known analytical solution. Besides, a rather simple problem of parametrical synthesis has been chosen. The degree of stiffness of a problem, which can be characterized by the ratio $|\lambda_{max}(\mathbf{A})|/|\lambda_{min}(\mathbf{A})| \approx 10^3$ is not excessively high. Moreover, input conditions close to the solution were used at minimization. However, as stated above, there are serious obstacles to solving this parametric synthesis problem formulated in its most frequently used statement.

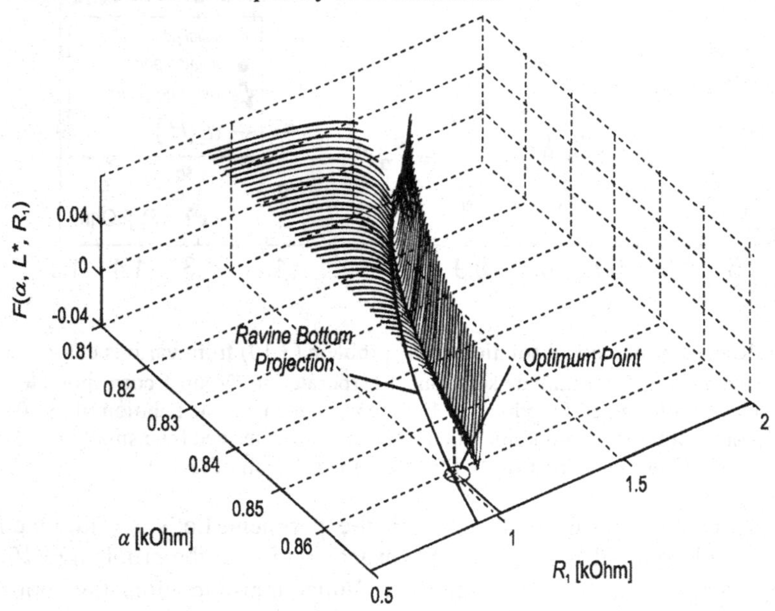

Fig. 3.7. Geometry of the minimized functional F in coordinates α, R_1 at L=L^* and the projection of the ravine bottom on the plane α, R_1

Problems arising at solution are caused by the fact that level surfaces of the functional F are of ravine form in the space of variables α, R_1. To illustrate, the level surface of the functional F is shown in Fig. 3.7 for values α^*=0.86 kOhm and L^*=1 μH. The initial problem at α^*=0.866 kOhm is characterized by a "deeper" ravine that is rather complicated to present graphically. The point of minimum is at the bottom of a ravine with very steep walls. At that, the functional is changing insignificantly along the ravine bottom even at significant changes of parameters. For this reason some of above obtained solutions are close enough to the point of true minimum, however they give large values of the functional. They are located on the steep wall of the ravine. Note that the projection of the ravine bottom is a curve. It is important that the functional F also have ravine structure in space of variables α, L with a ravine directed along the coordinate axis L. Thus, the functional of the above considered inverse problem is ravine type, and the problem is stiff.

This example leads to an important overall conclusion: when the subject (in this case, an electric circuit) of parametrical synthesis, diagnostics or mathematical modeling has the property of stiffness, then the corresponding inverse problem possesses this same property.

Let's introduce, following [2], the definition of a ravine functional. A smooth functional $F(\mathbf{x})$, $\mathbf{x} \in \mathbf{R}^m$ (m-dimensional Euclidean space) is called ravine, if there is an area $G \in \mathbf{R}^m$, where eigenvalues of Hesse matrix $\mathbf{H}''(\mathbf{x})$, ordered in decreasing of absolute values in any point $\mathbf{x} \in G$, satisfy the inequalities:

$$\lambda_1(x) >> \left| \min_i \lambda_i(x) \right|. \tag{3.13}$$

If the Hesse matrix $\mathbf{H}''(\mathbf{x})$ of the functional is positively defined (all its eigenvalues are greater than zero), the inequality (3.13) is equivalent to the condition of ill-conditionality of the Hessian:

$$\Theta = \frac{\lambda_{\max}(\mathbf{H}'')}{\lambda_{\min}(\mathbf{H}'')} >> 1,$$

where Θ is the spectral number of the Hesse matrix. For further use it will be convenient to give one more equivalent definition of the ravine functional using its trajectory of the quickest descent. The trajectory of the functional $F(\mathbf{x})$ is quickest descent is described by the following system of differential equations:

$$\frac{d\mathbf{x}}{d\xi} = -F'(\xi), \quad \mathbf{x}(0) = \mathbf{x}_0. \tag{3.14}$$

A smooth functional $F(\mathbf{x})$ is ravine if the system of differential equations (3.14) is stiff.

Many works are dedicated to methods of minimization of ravine functionals but their discussion lies outside the scope of this book. We shall emphasize only that established methods of minimization are ineffective for ravines of arbitrary form. For example, the coordinate-wise descent method allows solving problems with a linear ravine arranged along one coordinate axis. The modified Hook-Jives method of configurations and Rozenbroke method of coordinate rotation can be used for nonlinear, one-dimensional ravines. The simplex method chosen for solution of the above problem is effective for solution of problems with many dimensional nonlinear ravines (at a small number of variables). It does not use the functional gradient; therefore, its effectiveness does not depend on the relation of derivatives along and across the ravine bottom. However, as apparently, this method also gives bad results in view of weak convergence within the ravine domain. And this in spite of the fact that an analytical solution for the circuit response was used, the search area of parameters was limited by a small area in the vicinity of the problem solution, and initial approximations were close to the solution.

It is necessary to note some other difficulties arising from solution of stiff inverse problems. Usually (for example, for circuits described by differential equation of third order) it is not possible to write down an analytical expression for the functional as has been done above. Then the functional values have to be calculated numerically. So, for example, to calculate the functional (3.10), numerical solution of the system of differential equations (3.8) would be required. Since it is a stiff system for its integration application of special methods, demanding high computing power is necessary. Therefore, this elementary example already indicates the urgency of development of new methods that allow creating effective computing procedures considering characteristic properties of stiff problems. Such methods are discussed below in Section 3.2.

In conclusion, we shall consider one more important property of stiff inverse problems. Further, we shall consider, as is the case in practical problems, that values of variables $i_L^{\exp}(t_m)$ and $u_C^{\exp}(t_m)$ are measured with some relative error $\rho \cdot \Delta_{\max}$, where $\rho \in [-1,1]$ is a random variable and $\Delta_{\max}$ is the relative measurement error describing the accuracy class of measuring instruments.

Assume that after measuring $i_L^{\exp}(t_m)$ and $u_C^{\exp}(t_m)$ by an error $\Delta_{\max}$=0.01%, it was possible to solve the problem (3.10) by means of some method of minimization. Assume that the obtained value of dependent source coefficient was $\alpha_{\exp}$=0.8660000. It is natural to suppose that $\alpha_{\exp}$ was determined with a relative error no larger than $\Delta_{\max}$. Thus, the inverse problem solution will be written down in the form of $\alpha_{\exp}$=0.86600±0.00009, that at first sight can be considered to be a good result. Let's estimate the response changes for the circuit shown in Fig. 3.4 for variation of α within this range. The transient process at found values of parameters is shown in Fig. 3.5 by dashed and dash-dot lines.

This is an unexpected result, which is characteristic for stiff problems. The error of modeled process reproduction by its mathematical model outside the boundary layer is ~10% and exceeds hundreds of times the error Δ_{max}=0.01%, by which experimental data $i_L^{exp}(t)$, $u_C^{exp}(t)$ have been obtained. Therefore, the mathematical model created as a result of solution of the inverse problem is not adequate to the real subject, and the obtained solution is unacceptable.

In this case, the reason of poor adequacy of the model involves strong distortion of the small eigenvalue λ_2 of matrix **A** when setting its elements with a small error. This, in turn, is a corollary that in an ill-conditioned matrix **A,** the information on its small eigenvalue lies in low-order digits of its elements. And it is these orders which are distorted at setting of the matrix elements. So, if the exact value is α=0.86600 kOhm, eigenvalues are λ_1=–512.6128415 μs^{-1}, λ_2=–0.5063187795 μs^{-1}, and at α=0.86609 kOhm they are λ_1=–512.6146800 μs^{-1}, λ_2=–0.4556853442 μs^{-1}. Thus, at an error 0.01% for the parameter α, the error for the large eigenvalue is ~0.00036%, and for the small one it is 10%. For this reason, when reproducing the process by its model, a maximum error of the order of 10% is observed.

The maximum error of a model can be estimated as follows. The conditionality number of the matrix **A** is $\Theta(\mathbf{A}) = |\lambda_1|/|\lambda_2| \approx 1500$. It is always possible to choose such a scale (such system of measurement units) at which the module of the larger eigenvalue λ_1 is unity. Then, $\|\mathbf{A}\|$ will also be of the order of unity, and the module of the small eigenvalue will be $|\lambda_2| = 1/1500 = 6.7 \cdot 10^{-4}$. Let the relative change of matrix elements' values be δ_a=10^{-4}. Then the absolute change of the elements' values will be $\Delta a_{i,j} = \|\mathbf{A}\| \cdot \delta_a \approx \delta_a = 10^{-4}$. Considering that the information on the small eigenvalue λ_2 lies in low-order digits of matrix elements $a_{i,j}$, the relative error for λ_2 will be $\delta_\lambda = \Delta a_{i,j} / |\lambda_2| = \Delta a_{i,j} \cdot \Theta(\mathbf{A}) \approx 15\%$. This is in close fit with the results of a numerical experiment.

3.2 The principle of quasistationarity of derivatives and integrals

In this section we shall consider one of the main principles of effective solution of stiff inverse problems – the principle of quasi-stationarity of derivatives [2]. We shall state it in regard to stiff systems of linear differential equations. At the end of this section, it will be shown how this principle can be used when searching a solution of ill-conditioned systems of algebraic equations to which inverse problems are reduced.

Let's consider an inhomogeneous stiff system of linear differential equations whose Jacobi matrix is negative-definite, that is, one having only real, negative eigenvalues:

$$\frac{d\mathbf{y}(t)}{dt}=\mathbf{Ay}+\mathbf{b},\ t\in[0,\ T],\quad \mathbf{y}(0)=\mathbf{y}_0,\quad \mathbf{y}(t)=\left(y_1(t)\ \ldots\ y_m(t)\right)^T,\quad \mathbf{b}=\text{const.} \tag{3.15}$$

In this most simple case, matrix $\mathbf{A}$ is ill-conditioned as follows from the condition of stiffness of the system of equations (3.16). Let us arrange matrix $\mathbf{A}$ eigenvalues $\lambda(\mathbf{A})$ by decreasing absolute values. We shall assume that they can be divided in two groups, by their values λ_i, $i=\overline{1,k}$ and λ_i $i=\overline{k+1,m}$, according to the following inequalities

$$|\lambda_{\max}|=|\lambda_1|\geq|\lambda_2|\geq\cdots\geq|\lambda_k|>>|\lambda_{k+1}|\geq|\lambda_{k+2}|\geq\cdots\geq|\lambda_m|=|\lambda_{\min}|, \tag{3.16}$$

where $\lambda_{\max}$ and $\lambda_{\min}$ are maximum and minimum eigenvalues by absolute value, respectively.

According to the definition of stiff systems stated in the previous section, there is a boundary layer with duration τ_b on the interval of solution $t\in[0,T]$. It separates intervals of fast and slow changing of variables $\mathbf{y}(t)$. Typical dependence of component $y_p(t)$ of the vector $\mathbf{y}(t)$ is shown in Fig. 3.8: in linear (with distortion of the time scale (Fig. 3.8a)) and logarithmic (Fig. 3.8b) scales.

Obviously, the solution behavior inside and outside of the boundary layer is very distinctive. The derivative $y_1'(t)$ outside the boundary layer has considerably smaller modulo value than inside it. Similarly, for all components of the vector $\mathbf{y}(t)$, the following inequalities are valid:

$$\left\| \left. \frac{dy_i(t)}{dt} \right|_{t \geq \tau_b} \right\| << \left\| \left. \frac{dy_i(t)}{dt} \right|_{t < \tau_b} \right\|, \qquad i = \overline{1,m}.$$

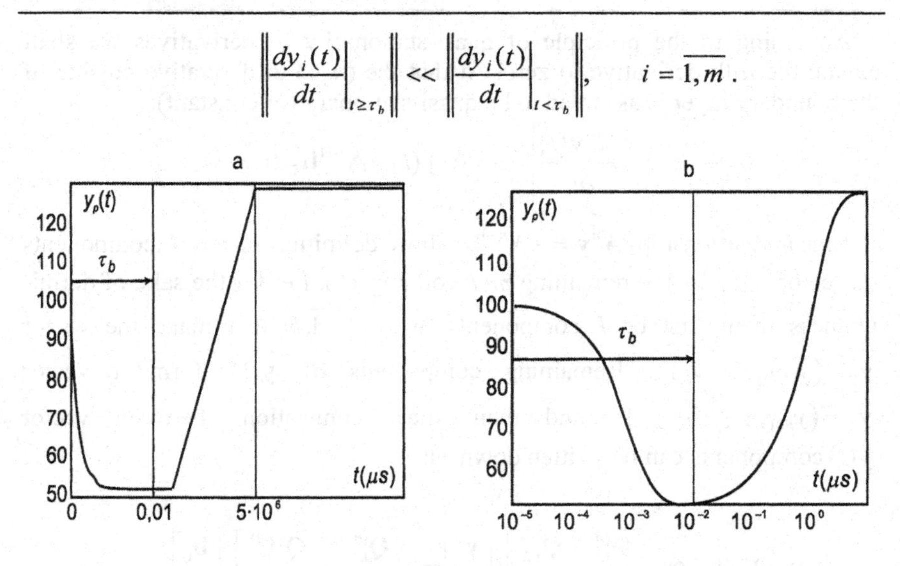

Fig. 3.8. Typical dependence from time for one of the variables of a stiff system of differential equations in linear **a**) and logarithmic **b**) scales

This property of the solution that is characteristic for stiff problems underlies the principle of quasi-stationarity of derivatives. According to this principle the solution of a stiff system of differential equations (3.15) of m-th order outside the boundary layer can be described with high accuracy by solution of a less stiff system of differential equations of $m-k$ order and by a system of linear algebraic connections of k-th order.

Linear connections are defined from the condition of quasi-stationarity of $(n-1)$-th derivative of the solution outside the boundary layer. At that, the quasi-stationarity or constancy of $(n-1)$-th derivative of the solution means that its n-th derivative is zero. Actually, the $(n-1)$-th derivative outside of the boundary layer varies very slowly in comparison with its variation inside the boundary layer. It is "almost" constant or quasi-stationary, therefore the corresponding principle name has been defined by this feature.

The principle of quasi-stationarity of derivatives is constructive in the sense that it not only postulates simplification of describing the stiff system's solution behavior outside the boundary layer, but also shows the way to do it. Let's differentiate by t the system of equations (3.15) n times:

$$\frac{d^2\mathbf{y}}{dt^2} = \mathbf{A}\frac{d\mathbf{y}}{dt} = \mathbf{A}^2\mathbf{y} + \mathbf{A}\mathbf{b}, \quad \cdots \quad \frac{d^n\mathbf{y}}{dt^n} = \mathbf{A}^{n-1}\frac{d\mathbf{y}}{dt} = \mathbf{A}^n\mathbf{y} + \mathbf{A}^{n-1}\mathbf{b}.$$

According to the principle of quasi-stationarity of derivatives we shall equate the n-th derivative to zero. At that the $(n-1)$-th derivative outside of the boundary layer is assumed to be quasi-stationary (or constant):

$$\left.\frac{d^n\mathbf{y}(t)}{dt^n}\right|_{t>\tau_b}=\mathbf{A}^n\mathbf{y}(t)+\mathbf{A}^{n-1}\mathbf{b}=0.$$

Linear relationships $\mathbf{A}^n\mathbf{y}=-\mathbf{A}^{n-1}\mathbf{b}$ allow definition of any k components of vector $\mathbf{y}(t)$ by the remaining $m-k$ components. Let for the sake of definitiveness them first be k components $y_i(t),\ i=\overline{1,k}$ that make the vector $\mathbf{y}_1=(y_1,y_2,\cdots,y_k)^T$. Remaining components of $\mathbf{y}(t)$ form a vector $\mathbf{y}_2=(y_{k+1},y_{k+2},\cdots,y_m)^T$ and then linear connections between vector $\mathbf{y}(t)$ components can be written down as:

$$\mathbf{A}^n\mathbf{y}=-\mathbf{A}^{n-1}\mathbf{b},\ \text{or}\quad \begin{bmatrix}\mathbf{Q}_{1,1}^{(n)} & \mathbf{Q}_{1,2}^{(n)}\\ \mathbf{Q}_{2,1}^{(n)} & \mathbf{Q}_{2,2}^{(n)}\end{bmatrix}\cdot\begin{bmatrix}\mathbf{y}_1\\ \mathbf{y}_2\end{bmatrix}=-\begin{bmatrix}\mathbf{Q}_{1,1}^{(n-1)} & \mathbf{Q}_{1,2}^{(n-1)}\\ \mathbf{Q}_{2,1}^{(n-1)} & \mathbf{Q}_{2,2}^{(n-1)}\end{bmatrix}\cdot\begin{bmatrix}\mathbf{b}_1\\ \mathbf{b}_2\end{bmatrix},\tag{3.17}$$

$$\mathbf{Q}_{1,1}^{(n)}\mathbf{y}_1+\mathbf{Q}_{1,2}^{(n)}\mathbf{y}_2=-\mathbf{Q}_{1,1}^{(n-1)}\mathbf{b}_1-\mathbf{Q}_{1,2}^{(n-1)}\mathbf{b}_2,\ \text{from which}$$

$$\mathbf{y}_1=-\left(\mathbf{Q}_{1,1}^{(n)}\right)^{-1}\left(\mathbf{Q}_{1,2}^{(n)}\mathbf{y}_2+\mathbf{Q}_{1,1}^{(n-1)}\mathbf{b}_1+\mathbf{Q}_{1,2}^{(n-1)}\mathbf{b}_2\right)=\mathbf{L}^{(n)}\mathbf{y}_2+\boldsymbol{l}^{(n)}.$$

Matrix $\mathbf{L}^{(n)}$ and vector $\boldsymbol{l}^{(n)}$ are defined from relations:

$$\mathbf{L}^{(n)}=-\left(\mathbf{Q}_{1,1}^{(n)}\right)^{-1}\mathbf{Q}_{1,2}^{(n)},\qquad \boldsymbol{l}^{(n)}=-\left(\mathbf{Q}_{1,1}^{(n)}\right)^{-1}\left(\mathbf{Q}_{1,1}^{(n-1)}\mathbf{b}_1+\mathbf{Q}_{1,2}^{(n-1)}\mathbf{b}_2\right).\tag{3.18}$$

In Eq. (3.18) $\mathbf{Q}^{(0)}=\mathbf{1}$. As will be shown later the value n=2÷4 is sufficient for qualitative definition of linear connections.

By substitution of the vector $\mathbf{y}_1$ in the system of equations (3.15), the order of the latter is reduced to $m-k$:

$$\frac{d\mathbf{y}(t)}{dt}=\mathbf{A}\mathbf{y}+\mathbf{b},\quad \text{or}\quad \frac{d}{dt}\begin{bmatrix}\mathbf{y}_1\\ \mathbf{y}_2\end{bmatrix}=\begin{bmatrix}a_{1,1} & a_{1,2}\\ a_{2,1} & a_{2,2}\end{bmatrix}\cdot\begin{bmatrix}\mathbf{y}_1\\ \mathbf{y}_2\end{bmatrix}+\begin{bmatrix}\mathbf{b}_1\\ \mathbf{b}_2\end{bmatrix},\tag{3.19}$$

$$\frac{d\mathbf{y}_2(t)}{dt}=a_{2,1}\mathbf{y}_1+a_{2,2}\mathbf{y}_2+\mathbf{b}_2=a_{2,1}\left(\mathbf{L}^{(n)}\mathbf{y}_2+\boldsymbol{l}^{(n)}\right)+a_{2,2}\mathbf{y}_2+\mathbf{b}_2=\mathbf{D}^{(n)}\mathbf{y}_2+\mathbf{d}^{(n)}.$$

Matrix $\mathbf{D}^{(n)}$ and vector $\mathbf{d}^{(n)}$ are defined from relations:

$$\mathbf{D}^{(n)} = a_{2,1}\mathbf{L}^{(n)} + a_{2,2}, \qquad \mathbf{d}^{(n)} = a_{2,1}l^{(n)} + \mathbf{b}_2 .$$

Equations of linear connections (3.17), together with the reduced system of differential equations (3.19), describe the solutions of the stiff system (3.15) outside the boundary layer.

It has been shown in the previous section that the system of differential equations describing a transient process in the electric circuit, shown in Fig. 3.4, is stiff. Let's obtain a reduced system of equations and linear connections outside the boundary layer. Substituting parameters' values R=0.47 kOhm, r=0.27 kOhm, α=0.866 kOhm, C=4 pF, L=1 μH, E_0=1 V in the system of equations (3.8), we have (here the current is measured in milliamperes and the voltage in volts):

$$\frac{d}{dt}\begin{bmatrix} i_L \\ u_C \end{bmatrix} = \underbrace{\begin{bmatrix} -146.8716, & 311.45461 \\ 171.8768, & -366.2475 \end{bmatrix}}_{\mathbf{A}} \underbrace{\begin{bmatrix} i_L \\ u_C \end{bmatrix}}_{\mathbf{y}} + \underbrace{\begin{bmatrix} 542.1617 \\ 288.3839 \end{bmatrix}}_{\mathbf{b}}, \; \mathbf{y}(0) = \begin{bmatrix} i_L \\ u_C \end{bmatrix}_{t=0} = \begin{bmatrix} 10^2 \\ 10^2 \end{bmatrix} = \mathbf{y}_0 \, . \quad (3.20)$$

Matrix **A** eigenvalues are $\lambda_1 \approx -512\ \mu s^{-1}$ и $\lambda_2 \approx -0.5\ \mu s^{-1}$. In this problem the matrices $\mathbf{Q}_{k,p}^{(n)}\ k,p=1,2$, $\mathbf{L}^{(n)}$, $\mathbf{D}^{(n)}$ and vectors $l^{(n)}\ \mathbf{d}^{(n)}$, appearing in Eqs. (3.16)-(3.19), are scalars. Therefore they will be designated further as $Q_{k,p}^{(n)},\ k,p=1,2,\ L^{(n)},\ D^{(n)},\ \ell^{(n)},\ d^{(n)}$.

Let's define linear connections between variables $i_L(t)$ and $u_C(t)$ at various n.

Let n=1, then it corresponds to the assumption of quasi-stationarity of the variables' vector outside the boundary layer. Then the linear connection (3.17)-(3.18) is defined from the first equation of the following system of equations:

$$\frac{d\mathbf{y}(t)}{dt} = 0, \;\Rightarrow\; \mathbf{A}\mathbf{y} = -\mathbf{b} .$$

Substituting numerical values in Eq. (3.20), we have:

$$a_{1,1}i_L + a_{1,2}u_C = -b_1, \quad \text{from which} \quad i_L = \underbrace{\left(-a_{1,2}/a_{1,1}\right)}_{L^{(1)}} u_C + \underbrace{\left(-b_1/a_{1,1}\right)}_{\ell^{(1)}} =$$

$$= \frac{-311.4546}{-146.8768}u_C + \frac{-542.1617}{-146.8768} = 2.1206u_C + 3.6914.$$

Value n=2 corresponds to the assumption of quasi-stationarity of the first derivative of variables' vector outside the boundary layer. The linear connection is defined from the first equation of the following system of equations:

$$\frac{d^2\mathbf{y}(t)}{dt^2}=0, \qquad \text{from which} \qquad \mathbf{A}^2\mathbf{y}=-\mathbf{A}\mathbf{b},$$

then:

$$i_L=\underbrace{\left(-\left(Q_{1,1}^{(2)}\right)^{-1}Q_{1,2}^{(2)}\right)}_{L^{(2)}}u_C+\underbrace{\left(-\left(Q_{1,1}^{(2)}\right)^{-1}\left(a_{1,1}b_1+a_{1,2}b_2\right)\right)}_{\ell^{(2)}}=2.1279u_C-0.1357.$$

Let n=3, then it corresponds to the assumption of quasi-stationarity of the second derivative of variables' vector outside the boundary layer. The linear connection is defined from the first equation of the following system of equations:

$$\frac{d^3\mathbf{y}(t)}{dt^3}=0, \qquad \text{from which} \qquad \mathbf{A}^3\mathbf{y}=-\mathbf{A}^2\mathbf{b},$$

then:

$$i_L=\underbrace{\left(-\left(Q_{1,1}^{(3)}\right)^{-1}Q_{1,2}^{(3)}\right)}_{L^{(3)}}u_C+\underbrace{\left(-\left(Q_{1,1}^{(3)}\right)^{-1}\left(Q_{1,1}^{(2)}b_1+Q_{1,2}^{(2)}b_2\right)\right)}_{\ell^{(3)}}=2.1279u_C-0.1395.$$

Comparing expressions for coefficients of linear connections derived at n=2 and n=3, we can note that they are very close. Obviously, in this particular problem coefficients of linear connection can also be defined from the second equation with the same result.

We shall use the above derived linear connections to reduce the dimensionality of the system of equations (3.20). As a result, we shall obtain systems of algebro-differential equations that give an approximate description of the process beyond the boundary layer:

$$n=1, \quad \frac{du_C}{dt}=-1.7672u_C+922.8497, \quad i_L=2.1206u_C+3.6914,$$

$$n=2, \quad \frac{du_C}{dt}=-0.5076u_C+265.0629, \quad i_L=2.1279u_C-0.1357, \tag{3.21}$$

$$n=3, \quad \frac{du_C}{dt}=-0.5063u_C+264.4116, \quad i_L=2.1279u_C-0.1395.$$

Solution components containing the factor $e^{\lambda_1 t}$ have been damped outside the boundary layer, so the solution contains only components with the factor $e^{\lambda_2 t}$. Therefore, the quality of obtained results is characterized by closeness of the coefficient at u_C to the value $\lambda_2 \approx -0.5063$ in the differential equation.

It is apparent from the expressions (3.21) that this condition sufficiently holds at n=2 and n=3. This indicates that the assumption about quasi-stationarity of the vector of first (n=2) and the second (n=3) derivatives outside the boundary layer was correct.

Exact and approximate solutions of the system of equations (3.20) in logarithmic scale are shown in Fig. 3.9a and b. The constant of integration in the solution of approximated differential equations (3.21) was defined with use of the exact value of the variable $u_C(t)$ at $t=\tau_b$. Dependences $\Delta i_L(t)$ and $\Delta u_C(t)$ of relative errors of approximated solutions of the system of equations (3.20) are shown in Fig. 3.9c and d. Evidently, the coincidence of exact and approximated solutions is already good at n=2.

Let's consider the solution of the system of equations (3.15) inside the boundary layer. We shall use the following expression for any given component $y_p(t)$ of solution $\mathbf{y}(t)$ of the system of equations (3.15):

$$y_p(t) = y_p(\infty) + \sum_{i=1}^{i=k} \alpha_{i,p} e^{\lambda_i t} + \sum_{i=k+1}^{i=m} \alpha_{i,p} e^{\lambda_i t}, \qquad p = \overline{1,m},$$

where $\alpha_{i,p}$ are constants defined by initial conditions. Assume that inequalities (3.16) are valid for matrix $\mathbf{A}$ eigenvalues. The second sum in Eq. (3.21) varies slightly in the interval of the boundary layer. Therefore, it can be approximated by a linear function. At the same time, the first sum decreases exponentially to zero (or to rather small value).

Let's apply the principle of quasi-stationarity of derivatives "in reverse" and calculate not the derivative, but the integral of the variables' vector. For this purpose we shall introduce a new variable $\tau = t/T$ where $T = 5/|\lambda_{k+1}|$, and then integrate both parts of Eq. (3.15) over the bounds 0 to τ:

$$\frac{1}{T}\int_0^{\tau} \frac{d\mathbf{y}}{d\tau} d\tau = \int_0^{\tau} (\mathbf{A}\mathbf{y} + \mathbf{b}) d\tau = \mathbf{A}\int_0^{\tau} \mathbf{y} d\tau + \mathbf{b}\tau,$$

from which $$\int_0^{\tau} \mathbf{y} d\tau = \frac{\mathbf{A}^{-1}}{T}(\mathbf{y} - \mathbf{y}_0) + \mathbf{y}_{\infty}\tau.$$

Here $\mathbf{y}_{\infty} = -\mathbf{A}^{-1}\mathbf{b}$ is the vector of steady-state values of variables.

Let's calculate the second integral of the variables' vector:

$$\int_0^{\tau}\int_0^{\tau} \mathbf{y} d\tau d\tau = \frac{\mathbf{A}^{-1}}{T}\int_0^{\tau} \mathbf{y} d\tau - \frac{\mathbf{A}^{-1}}{T}\mathbf{y}_0\tau + \frac{\mathbf{y}_{\infty}\tau^2}{2} =$$

$$= \mathbf{A}^{-2}T^{-2}(\mathbf{y} - \mathbf{y}_0) - \mathbf{A}^{-1}T^{-1}(\mathbf{y}_0 - \mathbf{y}_{\infty})\tau + \frac{\mathbf{y}_{\infty}\tau^2}{2}.$$

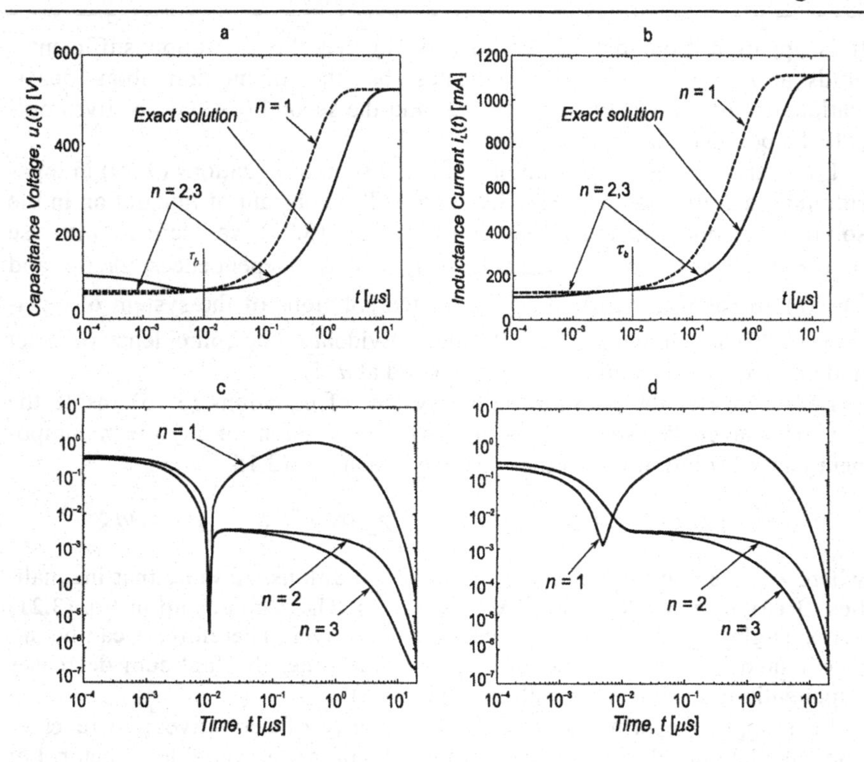

Fig. 3.9. Dependences $u_C(t)$ - **a**) and $i_L(t)$ - **b**) and their approximation expressions obtained by means of linear connections at various n for $t>\tau_b$, and errors of approximation $\Delta u_C(t)$ - **c**) and $\Delta i_L(t)$ - **d**)

Similarly, the n-th integral of the variables' vector is given by:

$$\int_0^\tau\int_0^\tau\cdots\int_0^\tau \mathbf{y}(d\tau)^n = \mathbf{A}^{-n}T^{-n}(\mathbf{y}-\mathbf{y}_0) - \left(\mathbf{A}^{-n+1}T^{-n+1}\tau - \frac{\mathbf{A}^{-n+2}T^{-n+2}\tau^2}{2} - \frac{\mathbf{A}^{-1}T^{-1}\tau^{n-1}}{(n-1)!}\right)(\mathbf{y}_0-\mathbf{y}_\infty) + \frac{\mathbf{y}_\infty\tau^n}{n!}. \quad (3.22)$$

Since $\tau T \le \tau_b \approx 5/|\lambda_k| << T$, we have $\tau << 1$ inside the boundary layer. Therefore, we can neglect all members in expression (3.22) except for the linear ones. Assuming that the (n+1)-th integral of the solution is quasi-stationary, we equate the n-th integral of the solution to zero. Then multiplying both parts of the resulting equality by T^n, we obtain a relationship for linear connections inside the boundary layer:

$$\mathbf{A}^{-n}(\mathbf{y}-\mathbf{y}_0)-\mathbf{A}^{-n+1}(\mathbf{y}_0-\mathbf{y}_\infty)t=\mathbf{0}. \qquad (3.23)$$

The assumption about equality to zero of the integral (3.22) used for deriving linear connection (3.23), is well justified, which can be seen by the example of system (3.15) when m=2. One of the solutions of the system inside the boundary layer under condition of (3.16), mentioned above, can be given by:

$$y_1(\tau)\approx y_1(\infty)+ae^{\lambda_{\max}T\tau}+bT\tau+c,$$

where a, b and c are some numbers, and $T=5/|\lambda_{\min}|$ is the approximate transient time. After integrating the solution n times:

$$\int_0^\tau\int_0^\tau\cdots\int_0^\tau y_1(dt)^n=\frac{a\left(e^{\lambda_{\max}t}-1\right)}{\lambda_{\max}^n T^n}-\sum_{k=1}^{n-1}\frac{a\tau^k}{\lambda_{\max}^{n-k}T^{n-k}k!}+\frac{y_1(\infty)+c}{n!}\cdot\tau^n+\frac{bT\tau^{n+1}}{(n+1)!}$$

it can be seen that at $|\lambda_{\max}|>>|\lambda_{\min}|$ and $\tau\le\tau_b=5/|\lambda_{\max}|$ the first two members are close to zero, as $|\lambda_{\max}^n T^n|=5\left(|\lambda_{\max}|/|\lambda_{\min}|\right)^n>>1$. Remaining members are small, as $\tau<<1$, i.e. the n-fold integral is quasi-stationary.

We shall use Eq. (3.23) to define the linear connection between variables $i_L(t)$ and $u_C(t)$ in equations (3.20) for various n. Desired linear connection between variables inside the boundary layer is given by:

$$i_L(t)=\alpha u_C(t)+\beta+\gamma t, \qquad t\le\tau_b.$$

Coefficients of this dependency derived from relations (3.23) at various n are listed in Table 3.1.

Data in Table 3.1 show that accuracy of definition of linear connections grows rapidly with increasing n, therefore assuming n=2÷4 is sufficient for practical purposes. Given this one can easily see that factor β can be calculated by known input conditions as $\beta=i_L(0)-\alpha u_C(0)$, or in general form

$$\beta=y_1(0)-\sum_{i=2}^{k}\alpha_i y_i(0).$$

Table 3.1. The coefficients of linear connection between variables $i_L(t)$ and $u_C(t)$ inside the boundary layer

n	α		β		γ	
	Value	Relative error (%)	Value	Relative error (%)	Value	Relative error (%)
1	-0.850394	0.1	185.0394	0.1	716.5354	3.2
2	-0.851570	10^{-4}	185.1570	10^{-4}	693.9369	10^{-4}
3	-0.8515709	10^{-8}	185.15709	10^{-8}	693.9926	10^{-5}

Let's construct a simplified model of the system on the interval of boundary layer. For this purpose we shall substitute the linear connection obtained for n=3 into the initial system of equations (3.22). As a result, we have:

$$n=3, \quad \frac{du_C}{dt} = -512.6128u_C + 3.2112\cdot 10^4 + 1.1969\cdot 10^5 t,$$
$$i_L = -0.8515709u_C + 185.15709 + 693.9926t. \qquad (3.24)$$

Solution components with the factor $e^{\lambda_2 t}$ are small on the interval of boundary layer, so the solution contains only components with the factor $e^{\lambda_1 t}$. Therefore, the quality of obtained results is characterized by closeness of the coefficient at u_C to the value $\lambda_1 = -512.6128415\,\mu s^{-1}$ in the differential equation. It is apparent that the approximation (3.24) at n=3 well describes the process in the initial circuit.

It is obvious that $u_C(t)$ can be excluded instead of $i_L(t)$ from the initial system of differential equations in a similar way. For this purpose, the above obtained linear connections should be written down in the form of

$$u_C(t) = \begin{cases} a_1 i_L(t) + b_1 + c_1 t = -1.174300 i_L(t) + 217.4300 + 815.0050t, & t \le \tau_b \\ a_2 i_L(t) + b_2 = 0.469941 i_L(t) + 0.065546 & t > \tau_b. \end{cases} \qquad (3.25)$$

Then, for the current $i_L(t)$ we have the following approximations:

$$\frac{di_L}{dt} = \begin{cases} -512.6128 i_L(t) + 68261.75 + 253837t, & t \le \tau_b \\ -0.506319 i_L(t) + 562.5764, & t > \tau_b. \end{cases}$$

The exact curve of voltage on the capacitance (solid line) and its approximations (dash and dash-dot lines) obtained according to linear connections (3.25) are shown in Fig. 3.10. It can be seen that the linear connections (3.25) describe the exact solution on the whole time interval with high accuracy. The approximation of solution inside the boundary layer, derived when neglecting the member $c_1 t$ in Eq. (3.25), is shown in Fig. 3.10 (dotted line). In this case, the error of approximation in the vicinity of τ_b is significant.

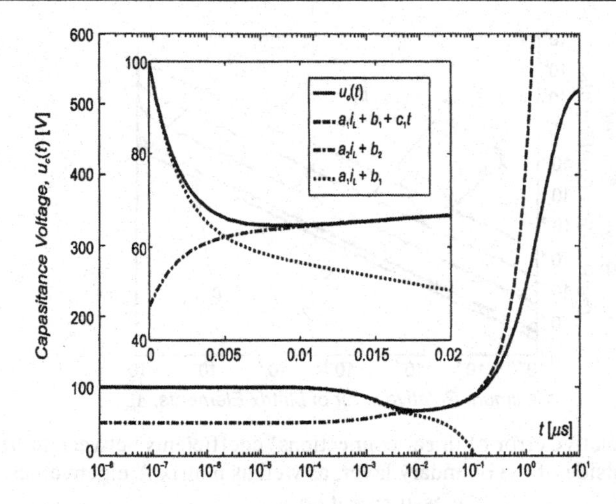

Fig. 3.10. Dependence $u_C(t)$ and its approximation expressions within and outside of the boundary layer

The principle of quasi-stationarity of derivatives has allowed describing the behavior of solution of the stiff problem with high accuracy (3.20), both inside and outside the boundary layer. Simplified models describing the behavior of stiff problem inside the boundary layer (3.24) and outside of it (3.21) have also been created. For definition of linear connections "exact" values of the system of differential equations' matrix coefficients were used. In practice, when solving inverse problems, particularly problems of diagnostics, identification and parametrical synthesis, parameters of diagnosed (identified) devices are found on the basis of experimental data. Therefore, the problem of influence of experimental data errors on accuracy of definition of linear connections considered below is of significant interest.

Linear connections were defined above by means of Eqs.(3.17) and (3.23), in which matrix **A** of the system of equations (3.15) was used. Let this matrix's elements be found experimentally with an error Δ_{max}. Figure 3.11 shows dependences of error for the coefficients of linear connections (3.17) and (3.23) from Δ_{max} for the circuit shown in Fig. 3.4. When deriving these dependences, exactness of initial conditions $\mathbf{y}_0$ and vector of sources **b** have been assumed. Curves on Fig. 3.11 have been obtained as average results of 100 calculations. It is apparent that coefficients a_1, a_2, b_1 and b_2 are defined with an error not larger than Δ_{max}, whereas the coefficient c_1 is defined with an error exceeding Δ_{max} by an order.

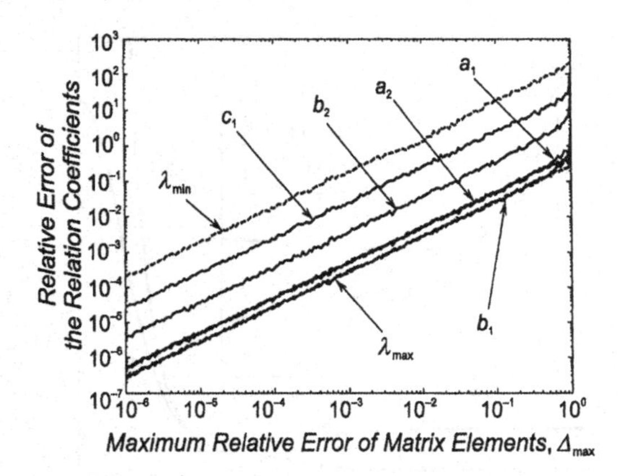

Fig. 3.11. The relative error of linear connections' coefficients between $u_C(t)$ and $i_L(t)$ within and outside of the boundary layer, as well as matrix **A** eigenvalues versus measurement error Δ_{max}

Thus, even this elementary example shows that there are significant distinctions in the stability of linear connections' coefficients to errors of matrix **A** elements. To find properties of stability in general, we shall present matrix **A** of the system (3.15) in the form of $\mathbf{A} = \mathbf{P\Lambda P}^{-1} = \mathbf{P\Lambda Q}^T$, where **P** and **Q** accordingly are matrices of right and left eigenvectors $\mathbf{p}_i$, $\mathbf{q}_i$, $i = \overline{1,m}$, and $\mathbf{\Lambda} = \mathrm{diag}(\lambda_1, \ldots, \lambda_m)$ is the matrix of eigenvalues. In general form, the linear connection outside the boundary layer is given by $\mathbf{A}^n\mathbf{y} = -\mathbf{A}^{n-1}\mathbf{b}$, then:

$$\lambda_1^n\left(\sum_{i=1}^{m}\mathbf{p}_i\left(\frac{\lambda_i}{\lambda_1}\right)^n\mathbf{q}_i^T\right)\mathbf{y} = -\lambda_1^{n-1}\left(\sum_{i=1}^{m}\mathbf{p}_i\left(\frac{\lambda_i}{\lambda_1}\right)^{n-1}\mathbf{q}_i^T\right)\mathbf{b}.$$

According to condition (3.16), $(\lambda_i/\lambda_1)^{n-1} \approx 0, i > k$ can be assumed for the exponent n=3-5. Then, for linear connection we have:

$$\left(\sum_{i=1}^{k}\mathbf{p}_i\left(\frac{\lambda_i}{\lambda_1}\right)^n\mathbf{q}_i^T\right)\mathbf{y} = -\frac{1}{\lambda_1}\left(\sum_{i=1}^{k}\mathbf{p}_i\left(\frac{\lambda_i}{\lambda_1}\right)^{n-1}\mathbf{q}_i^T\right)\mathbf{b}. \tag{3.26}$$

Each of the equations (3.26) represents a linear connection of the form:

$$y_1 = \sum_{i=2}^{m} a_{2,i}y_i + b_2.$$

Therefore, it is necessary to estimate the sensitivity of coefficients $a_{2,i}, i = \overline{2,m}$ and b_2 to the error Δ_{max} of the definition of matrix **A** elements.

Let the perturbed matrix of the system be $\tilde{\mathbf{A}} = \mathbf{A} + \mathbf{F}$, where $\mathbf{F}$ is the matrix of absolute errors, and $\|\mathbf{F}\| \leq \Delta_{\max} \|\mathbf{A}\|$. According to perturbations theory an estimation of distortion of matrix eigenvalues can be performed for the ordered eigenvalues on the basis of the Bauer-Fike theorem [4]:

$$\left|\tilde{\lambda}_i - \lambda_i\right| \leq \theta(\mathbf{P}) \cdot \|\mathbf{F}\|_2 \leq \theta(\mathbf{P}) \cdot \Delta_{\max} \cdot \|\mathbf{A}\|_2, \tag{3.27}$$

where $\tilde{\lambda}_i$ is the eigenvalue of matrix $\tilde{\mathbf{A}}$, $\theta(\mathbf{P}) = \mu_{\max} / \mu_{\min}$ is the spectral number of conditionality of the right eigenvectors' matrix, μ_i is the matrix singular number, and $\|\cdot\|_2 = \mu_{\max}$ is the matrix spectral norm. θ in a bi-logarithmic scale is proportional to the parameter $N_a = \sqrt{\|\mathbf{A}\|_F^2 - \sum_{i=1}^{m} |\lambda_i|^2}$ that describes the matrix asymmetry (here $\|\cdot\|_F$ is the Euclidean (Frobenius) norm of the matrix).

From the estimation (3.27) follows that for matrixes with small parameter N_a, the relative change of dominating eigenvalues $\lambda_1, \lambda_2, .., \lambda_k$ is of the order $\Delta_{\max}$. At the same time small eigenvalues $\lambda_{k+1}, \lambda_{k+2}, .., \lambda_m$ are distorted to a greater extent, since $|\lambda_{k+1}| << |\lambda_{\max}| \leq \mu_{\max}$.

Right and left eigenvectors are defined as solutions for the following system of equations

$$(\mathbf{A} - \lambda_i \mathbf{1})\mathbf{p}_i = \mathbf{0}, \quad (\mathbf{A} - \lambda_i \mathbf{1})^T \mathbf{q}_i = \mathbf{0},$$

which have ill-conditioned matrixes. Therefore, a small difference of matrix $\left(\tilde{\mathbf{A}} - \tilde{\lambda}_i \mathbf{1}\right)$ elements from elements of the correct matrix $(\mathbf{A} - \lambda_i \mathbf{1})$ leads to a significant error in solution of these systems of equations. However, as shown in [5], the error vector is practically parallel to the desired eigenvector of matrix $\mathbf{A}$ and has no influence upon the latter's direction. Therefore, eigenvectors corresponding to dominant eigenvalues are stable against errors of setting of the elements of the system matrix $\mathbf{A}$. From Eq. (3.26), it follows that linear connection coefficients depend only on dominant eigenvectors $\mathbf{p}_i, \mathbf{q}_i, i = \overline{1,k}$. Therefore, *connection coefficients corresponding to $t > \tau_b$, are stable against errors of matrix* $\mathbf{A}$ *elements.*

Similarly, rewriting the expression (3.23) for linear connection inside the boundary layer in the form of:

$$\mathbf{A}^{-n}(\mathbf{y} - \mathbf{y}_0 - \mathbf{b}t) = \mathbf{A}^{1-n}\mathbf{y}_0 t,$$

we have:

$$\left(\sum_{i=k+1}^{m}\mathbf{p}_i\left(\frac{\lambda_m}{\lambda_i}\right)^n\mathbf{q}_i^T\right)(\mathbf{y}-\mathbf{y}_0-\mathbf{b}t)=\lambda_m\left(\sum_{i=k+1}^{m}\mathbf{p}_i\left(\frac{\lambda_m}{\lambda_i}\right)^{n-1}\mathbf{q}_i^T\right)\mathbf{y}_0t. \quad (3.28)$$

Each of equations (3.28) represents a linear connection of the form:

$$y_1=\sum_{i=2}^{m}a_{1,i}y_i+b_1+c_1t. \quad (3.29)$$

As is apparent from expression (3.28), connection coefficients $a_{1i}, i=\overline{2,m}$, b_1 and c_1 in Eq. (3.29) are defined by small modulo eigenvalues of matrix **A** and corresponding eigenvectors. According to Eq. (3.27) and subsequent reasoning concerning eigenvectors, one may conclude that generally all connection coefficients *inside the boundary layer* are unstable to errors of matrix **A** elements.

However, this is not always true. In a number of cases that are important for practical purposes, some coefficients of linear connection (3.29) are stable against errors of matrix elements. These cases correspond to systems with a strongly *separated spectrum of eigenvalues*. For such systems $(\lambda_m/\lambda_i)^n=1$ at i=m and $(\lambda_m/\lambda_i)^n\approx 0, i=\overline{k+1,m-1}$. Therefore, the accuracy of connection (3.29) coefficients $a_{1,i}$, $i=\overline{2,m}$ and b_1 is defined only by the variation of eigenvectors included in expression (3.28). Variation of eigenvectors at distortion of matrix **A** elements can be estimated as follows [5]. Assume that λ_r and $\tilde{\lambda}_r$ are eigenvalues of matrices **A** and $\tilde{\mathbf{A}}$, $\mathbf{p}_r$ and $\tilde{\mathbf{p}}_r$ are corresponding characteristic vectors, and $\gamma=\min_{i\neq r}\left|\lambda_i-\tilde{\lambda}_r\right|$ is the minimal distance between $\tilde{\lambda}_r$ and the remaining part of matrix **A**'s spectrum. Then:

$$\left|\sin\angle(\mathbf{p}_r,\tilde{\mathbf{p}}_r)\right|\leq\|\mathbf{F}\|_2/\gamma.$$

Usually the distortion of matrix **A** maintains the separation of its spectrum, so $\|\mathbf{A}\|_2/\gamma\leq 1$ $\left|\sin\angle(\mathbf{p}_r,\tilde{\mathbf{p}}_r)\right|\leq\Delta_{\max}$. The last condition is valid for any stiff system of the second order, and in particular, for the system considered in this paragraph. In this example, the coefficients a_1 and b_1 of the connection (3.29) are stable against errors of matrix **A** elements as it follows from Fig. 3.11. Generally, even if small eigenvalues do not form a sufficiently compact group, coefficients $a_{1,i}, i=\overline{2,m}$ and b_1 of the connection (3.29) also appear to be stable against errors of matrix **A** elements.

Let's consider the stability of coefficient c_1 to errors of matrix **A** elements. As is apparent from Eq. (3.28), the expression for c_1 involves a summand, which is proportional to λ_m and varies strongly at small variation of elements of **A**. The contribution of this summand into the magnitude of c_1 depends on the relation $\lambda_m \cdot \|\mathbf{y}_0\| / \|\mathbf{b}\|$. For example, the electric circuit shown in Fig. 3.4 is of order 0.1. Therefore, the aggregate error of c_1 is approximately an order less than $\lambda_{\min}$ definition error (Fig. 3.11).

In spite of the fact that the coefficient c_1 of connection (3.23) is defined with a greater error, it affects the solution behavior only in the vicinity of τ_b (see Fig. 3.10). For example, in the considered problem at a 10% error in definition of the coefficient c_1, the greatest error in reproduction of the transient process curve is reached at $t=\tau_b$ and is equal to 1%.

In conclusion, of this section we shall apply the principle of quasi-stationarity of derivatives to the solution of inverse problems that can be reduced to solution of systems of linear algebraic equations. In inverse problems of such type coefficients of a system of equations are determined, for example, by measured currents and voltages of an electric circuit at various conditions of its performance. The vector of unknowns is formed by circuit parameters that are subject to definition. Similar problems will be considered in Sections 3.4-3.5 in more detail.

Let matrix coefficients $g_{k,p} \in \mathbf{G}$ $f_k \in f$, $k,p=\overline{1,m}$ of a system of equations be determined as a result of a series of experiments:

$$\mathbf{Gx} = \mathbf{f}, \tag{3.30}$$

where the solution vector **x** is the solution of some inverse problem. Assume that matrix **G** is ill-conditioned, then the solution of inverse problem by a "direct method", by using the solution of the system of equations (3.30) (for example, by Gauss method or by means of inversion of matrix **G**), will be obtained by a far greater error than the error of input data [6]. The following relation can be used for a crude estimation of the relative error Δ_X when solving the stiff inverse problem (3.30) by means of solution of the system of equations:

$$\Delta_X \le \Delta_{\max} \frac{\|\mathbf{G}\|_F \cdot \|\mathbf{G}^{-1}\|_F}{m}, \quad \|\mathbf{G}\|_F = \sqrt{\mathrm{trace}(\mathbf{G} \cdot \mathbf{G}^T)}.$$

Here $\Delta_{\max}$ is the relative error of experimental definition of matrix **G** coefficients, m is the matrix dimension, and trace is the matrix trace function (the sum of its diagonal elements). Let, for example, Δ_{max}=0.01. Let matrix **G** be congruent with matrix **A** from Eq. (3.20). Then m=2 and:

$$\Delta_X \le \Delta_{max} \frac{\|\mathbf{A}\|_F \cdot \|\mathbf{A}^{-1}\|_F}{m} = 0.01 \cdot 541.9 \approx 6, \quad \text{or} \quad \Delta_X \le 600\%!$$

As this estimation shows, for solution of stiff inverse problems by the "direct method", experimental data with very high accuracy are required. Here some general aspect of the problem should be noted. The solution of the system of equations (3.30) by virtue of the ill-conditionality of matrix **G** is determined by its small eigenvalues. Because of experimental and therefore not so accurate definition of matrix **G** elements, *it does not contain trustworthy information about the small eigenvalues*. This paradoxical statement should be understood as follows. When changing parameters of one or several elements of **G** on a value no more than the measurement error, small eigenvalues of **G**, and therefore the solution of Eq. (3.30) will change considerably.

Let's return to the principle of quasi-stationarity of derivatives and to benefits that it can give for solution of stiff inverse problems. Instead of the problem (3.30) we shall consider the more general problem:

$$\frac{d\mathbf{x}}{dt} = -\mathbf{G}\mathbf{x} + \mathbf{f}, \quad \mathbf{x}(0) = \mathbf{x}_0 \tag{3.31}$$

with solution that coincides with the solution of Eq. (3.30) at $t \to \infty$ (we assume that matrix **G** is positively defined). The system of equations (3.31) satisfies the definition of stiff system of the differential equations, and therefore use of the principle of quasi-stationarity of derivatives is possible. Linear connections between the vector **x** components beyond the boundary layer are given by:

$$\mathbf{G}^n \mathbf{x} = \mathbf{G}^{n-1} \mathbf{f}\,. \tag{3.32}$$

Linear connections can be easily found by experimentally determined matrices **G** and **f** for any n. As was shown above, these *linear connections are defined with the same accuracy at which the experimental data have been obtained*. Therefore relationships (3.32) contain important information on the stiff inverse problem. Using this information, repeated measurements in slightly modified conditions can be performed and the inverse problem can be solved on the basis of thus obtained experimental data. The idea of this approach, formulated by authors of [2] under the name of "principle of repeated measurements", is considered in Section 3.4.

3.3 Using linear relationships for solving stiff inverse problems

As shown in the previous section, data received as a result of experimentation ("primary" data) can be used to determine linear connections between variables in a problem. Irrespective of the degree of stiffness in a problem, the degree of accuracy of connections to be found is the same as the accuracy of measurements that have been carried out.

Here we shall continue the study of the problem of synthesis of an equivalent circuit for the circuit shown in Fig. 3.4 that we started in Section 3.1. The problem consists in definition of parameters α, L and R_1 by relationships $i_L(t)$ and $u_C(t)$ on the interval $[0,T]$ obtained as a result of measurements. It has been shown that these problems cannot be solved using traditional methods, instead the use of linear connections (3.18) and (3.23) gives rise to alternative methods for solutions of inverse problems, discussed below.

Let's proceed according to the following sequence. At first we define coefficients of linear connections and then use them to solve the inverse problem specified. According to Eq. (3.25), we shall write down linear connections for $i_L(t)$ and $u_C(t)$ as follows:

$$u_C(t) = \begin{cases} a_1 i_L(t) + b_1 + c_1 t, & t \le \tau_b \\ a_2 i_L(t) + b_2, & t > \tau_b. \end{cases}$$

The problem of determining the coefficients a_1, b_1, c_1, a_2, b_2 can be formulated as

$$F(\mathbf{x}) = \sum_{m=1}^{M} \left(u_C^{\text{ex}}(t_m) - u_C(t_m) \right)^2 \xrightarrow[\mathbf{x}]{} \min, \tag{3.33}$$

where $\mathbf{x} = (a_1, b_1, c_1, a_2, b_2)^T$ is the vector of unknown coefficients of linear connections and

$$u_C(t_m) = u_C(t, \mathbf{x})\big|_{t=t_m} = \begin{cases} a_1 i_L^{\text{ex}}(t_m) + b_1 + c_1 t_m, & t_m \le \tau_b \\ a_2 i_L^{\text{ex}}(t_m) + b_2, & t_m > \tau_b. \end{cases}$$

The problem (3.33) does not require a solution of a stiff system of differential equations. For its solution, we shall apply the least-squares method. Let's search the solution of (3.33) for time intervals inside and outside the boundary layer. Let $t_m \in [0, \tau_b]$ for $m \le K$ and $t_m \in [\tau_b, T]$ for $m = \overline{K+1, M}$, that is, the first K points of experimental curves lay inside the boundary layer, and the remaining points are outside of it. The following matrixes can be formed:

$$\mathbf{W}_1 = \begin{bmatrix} i_L(t_1) & 1 & t_1 \\ i_L(t_2) & 1 & t_2 \\ \vdots & \vdots & \vdots \\ i_L(t_K) & 1 & t_K \end{bmatrix}, \mathbf{u}_1 = \begin{bmatrix} u_C(t_1) \\ u_C(t_2) \\ \vdots \\ u_C(t_K) \end{bmatrix}, \mathbf{W}_2 = \begin{bmatrix} i_L(t_{K+1}) & 1 \\ i_L(t_{K+2}) & 1 \\ \vdots & \vdots \\ i_L(t_M) & 1 \end{bmatrix}, \mathbf{u}_2 = \begin{bmatrix} u_C(t_{K+1}) \\ u_C(t_{K+2}) \\ \vdots \\ u_C(t_M) \end{bmatrix}.$$

When $K \geq 3$ and $M - K \geq 2$, matrixes $\mathbf{W}_i^T \mathbf{W}_i$, $i = 1,2$ are nonsingular. The problem (3.33) has a unique solution, which according to the least-squares method is given by

$$\mathbf{x}_i = \left(\mathbf{W}_i^T \mathbf{W}_i\right)^{-1} \mathbf{W}_i^T \mathbf{u}_i,\ i = 1,2,\ \mathbf{x}_1 = (a_1, b_1, c_1)^T,\ \mathbf{x}_2 = (a_2, b_2)^T. \quad (3.34)$$

Dependences of coefficient errors for a_1, b_1, c_1, a_2, b_2 versus measurement error calculated from Eq. (3.34) are shown in Fig. 3.12. These dependences are derived by averaging over 100 calculations. Further, all dependences obtained for the random experimental error Δ_{max} will also be given as averaged over 100 calculations.

One can easily see from Fig. 3.12 that error of coefficient b_2 considerably exceeds the experimental error. Loss of accuracy (relative to the experimental error) is of three orders. Reasons of this were discussed in the previous section. They are caused by the smallness of b_2 in comparison with a_2, which in turn is a display of the stiff properties of the problem.

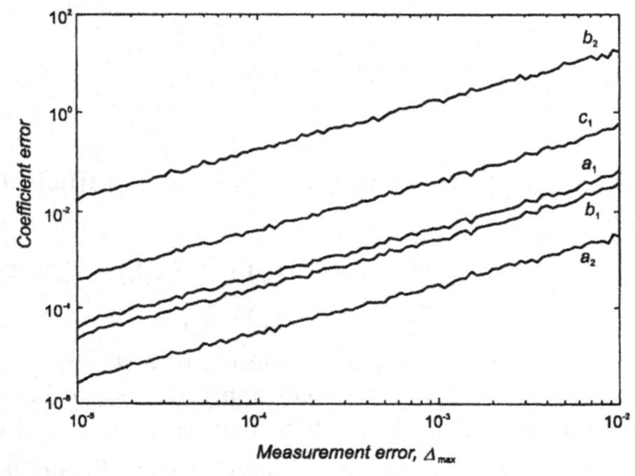

Fig. 3.12. The relative error of linear connections' coefficients between $u_C(t)$ and $i_L(t)$ within and outside of the boundary layer versus measurement error Δ_{max}

To determine parameters α, L and R_1 one needs only three coefficients of linear connection. Let them be a_1, a_2 and c_1. Let's designate coefficients derived from Eq. (3.34) as $a_1^{ex}, a_2^{ex}, c_1^{ex}$, emphasizing the fact that they have been found experimentally. Expressions connecting coefficients $a_1^{ex}, a_2^{ex}, c_1^{ex}$ with desired parameters α, L, R_1, can be found from Eq. (3.18) and (3.23) by use of condition equations (3.8). At that, the value of n in these formulas is assumed to be 2. Large values of n can be required only for the purpose of full elimination of small eigenvalues' influence, that is to provide the condition $(\lambda_i/\lambda_1)^{n-1} \approx 0$, in Eq. (3.26) and (3.28). For the problem under consideration a comprehensible solution can already be obtained at n=2. Analytical expressions for coefficients of connections are given by

$$\begin{aligned}
&a_1(\mathbf{x}) = -D^{-1}L\left(925.93LR_1^2 + 67.50L + 500LR_1 + R_1^2(\alpha - 0.27)\right) \\
&a_2(\mathbf{x}) = 10^{-3}D^{-1}\left[(3.27\alpha^2 - 6.61\alpha + 33.38)R_1^2 + (1-\alpha)R_1 + (\alpha - 270)LR_1^2 + 0.05\right] \\
&c_1(\mathbf{x}) = D^{-1}\left[L\left(925.93\alpha R_1^2 - 685.19R_1^2 - 31.73 - 302.50R_1\right)i_L(0) + \right. \\
&\left. + R_1(\alpha R_1 - 0.13 - 0.74R_1)u_C(0) + (925.93L + 0.47)R_1^2 + 250.00(\alpha+1)LR_1\right], \\
&D = 10^{-3}R_1(1.90R_1 - 1.88\alpha R_1 + 0.24 + 270L + LR_1), \quad \mathbf{x} = (\alpha, L, R_1)^T.
\end{aligned} \tag{3.35}$$

Here, $i_L(0)$ and $u_C(0)$ are also defined by experimental dependences.

Thus, the new problem for definition of the required parameters α, L, R_1 becomes:

$$\min_{\mathbf{x}} F(\mathbf{x}), \quad F(\mathbf{x}) = \left\| \begin{matrix} a_1(\mathbf{x}) - a_1^{exp} \\ a_2(\mathbf{x}) - a_2^{exp} \\ c_1(\mathbf{x}) - c_1^{exp} \end{matrix} \right\|. \tag{3.36}$$

The following important advantages of problem (3.36) should be emphasized:

- the accuracy of coefficients $a_1^{exp}, a_2^{exp}, c_1^{exp}$ found by experimental data weakly depends on stiffness of equations (3.8). It is defined only by the accuracy of the measuring equipment;
- the solution of Eq. (3.36) does not require multiple integration of a stiff system of state equations (3.8) as was necessary for the solution of Eq. (3.10).

Regardless of these essential simplifications, the problem (3.36) nevertheless is complex enough owing to ill-conditionality of the functional **F** Hesse matrix. Let's consider its further simplification by use of the inverse problem property of stiffness.

Stiffness of the initial inverse problem is conditioned on diverse influence of desired parameters on transient in the circuit. Considering two characteristic time zones - inside and outside of the boundary layer, may help to estimate the influence of either parameter on the process in these zones. Estimation of sensitivity of the system of Eq. (3.8) matrix **A** eigenvalues λ_1 and λ_2 to modification of parameters α, L, R_1 will be sufficient for this purpose.

Let's introduce normalized variables $\xi_i \in [0,1]$, $i=1,2,3$:

$\alpha=(\alpha_{max}-\alpha_{min})\xi_1+\alpha_{min}, L=(L_{max}-L_{min})\xi_2+L_{min}, R_1=(R_{max}-R_{min})\xi_3+R_{min}$, where quantities with indexes "min" and "max" limit ranges of definition of α, L, R_1.

To estimate the sensitivity of λ_1 and λ_2 we shall use the average value of their derivative on normalized variables in the whole domain of their definition. Thus, averaged elements of matrix **J** are calculated as follows:

$$J_{i,j}=\frac{\partial}{\partial \xi_i}\int_0^1\int_0^1\int_0^1 \lambda_j d\xi_1 d\xi_2 d\xi_3 \in \mathbf{J},\quad i=\overline{1,3},\quad j=1,2.$$

In the problem under consideration, matrix **J** is given by:

$$\begin{matrix} & \lambda_1 & \lambda_2 & \\ \mathbf{J}= & \left[\begin{matrix} 0.1 & 1 \\ 1 & 0.7 \\ 0.05 & 0.25 \end{matrix}\right. & \left.\right] & \begin{matrix} \xi_1 \\ \xi_2 \\ \xi_3 \end{matrix} \end{matrix}.$$

Here, elements of each column are divided by the maximum element of that column.

Apparently, maximum eigenvalue λ_1 depends weakly on ξ_3. Hence, modification of ξ_3 has only a slight influence on transients inside the boundary layer. Therefore the problem of (3.10) could be divided into two problems: for the interval inside of the boundary layer and outside of it. Further, an approximation for L can be found from the first problem having chosen the remaining parameters arbitrarily from within their corresponding ranges. Then, with known L the problem can be solved outside of the boundary layer to define α and R_1. Thus calculated, α and R_1 can be used to adjust L and then to find new values for α and R_1, etc. As calculations demonstrate, this algorithm well converges to a solution. However, it is absolutely inapplicable in practice, because analytical expressions for eigenvalues that are necessary for its work can be obtained only in rare instances, and only for elementary problems. Computational investigation of the stability of λ_1 and λ_2 encounters the problem of stiffness and cannot be fulfilled correctly.

In the suggested approach, the sensitivity of *linear connections' coefficients* a_1, a_2 and c_1 are used instead of the sensitivity of eigenvalues λ_1 and λ_2. Analytical expressions for these coefficients can be found for high-order systems (including nonlinear ones), as it requires only multiplication of matrixes. In the problem under consideration, values of matrix **J** elements for coefficients of connections are as follows:

$$\begin{array}{cccc} & a_1 & c_1 & a_2 \\ \mathbf{J} = & \left[\begin{array}{ccc} 0.2 & 1 & 1 \\ 1 & 0.01 & 1 \\ 0.1 & 0.3 & 0.25 \end{array}\right] & \begin{array}{c} \xi_1 \\ \xi_2 \\ \xi_3 \end{array} \end{array}.$$

Obviously, the coefficient c_1 depends weakly on the variable ξ_2 (the normalized inductance L). Therefore, the solution of Eq. (3.36) can be divided into two stages:

- in the first stage, we shall find L from any of the equations of (3.35) an arbitrary α and R_1, from their respective ranges of definition;
- in the second stage, we find α and R_1 using data about the process outside of the boundary layer.

Further, by means of calculated values α and R_1, we adjust the value L and find new values for α and R_1, etc. This process converges well, but requires (during realization of the second stage) integration of a system of stiff differential equations on each iteration. Therefore we shall use only linear connections (3.35) for determination of α, L, R_1.

For this purpose, we can use the following iterative process:

- find the inductance L from Eq. (3.35) for a_1: $a_1\left(\alpha^{(k)}, L, R_1^{(k)}\right) \to L^{(k+1)}$;
- find the coefficient α from the expression for a_2: $a_2\left(\alpha, L^{(k+1)}, R_1^{(k)}\right) \to \alpha^{(k+1)}$;
- find the resistance R_1 from the expression for c_1: $c_1\left(\alpha^{(k+1)}, L^{(k+1)}, R_1\right) \to R_1^{(k+1)}$.

As our analysis shows, this process converges well at any initial values $\alpha^{(0)}$ and $R_1^{(0)}$ from their respective ranges of definition. The relative error of parameters' calculation versus error of experimental data Δ_{max} is shown in Fig. 3.13. For estimation of the quality of this, results recall that the error of α, L, R_1 calculation by the traditional method (3.10) at $\Delta_{max}=10^{-4}$ was equal to hundreds of percents.

Summing up this section, let's note that the property of stiffness of an inverse problem can be used for simplification of its solution. Use of linear connections between the problem's variables discussed above is one of the effective methods of simplification of its solution.

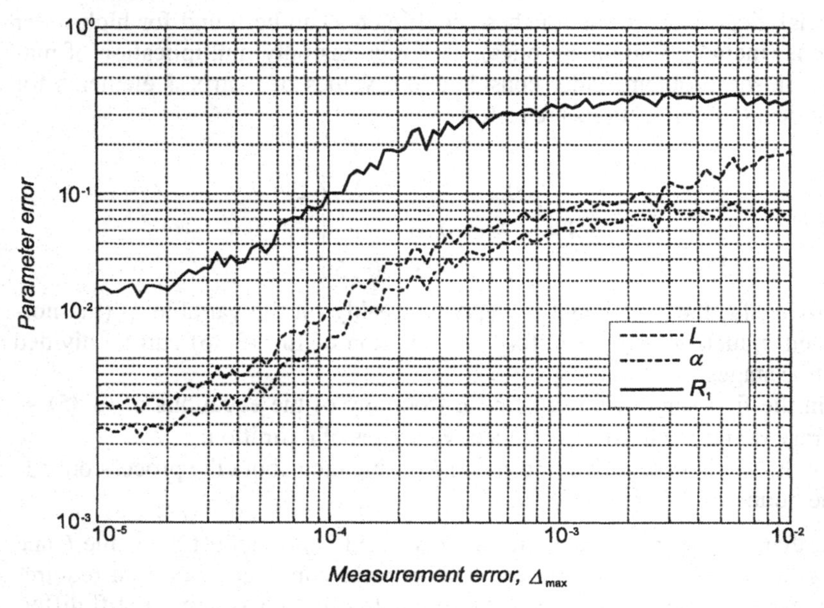

Fig. 3.13. Dependence of parameters L, α, R_1 definition error from the measurement error Δ_{max}

3.4 The problems of diagnostics and the identification of inverse problems in circuit theory

This section is concerned with furthering the discussion of problems arising from solution of inverse problems in circuit theory, and those resulting from the stiffness of equations of the circuits under consideration.

As it was noted earlier in Section 1.3, the problem of electric circuit diagnostics involves definition of parameters of the elements of the circuit by its measured responses to known actions. Each problem of diagnostics consists of two stages: the experimental stage, at which the physical device characteristics are measured, and the calculation stage, at which parameters of the device are calculated by its measured characteristics. As a rule, at solution of a diagnostics problem, we have extensive information on the circuit's topology, and rated values of parameters of its elements. The purpose of diagnostics is in detecting deviations of the device parameters from their rated values during manufacture or operation.

A more general type of inverse problem in circuit theory is the identification problem, which aims at constructing a circuit model on the basis of experimental data. The most complicated object of identification is the so-called "black box", – an object with completely unknown internal structure. Lack of information on the device to be identified and simultaneous desire to create an adequate model can lead to excessively complicated models. Occasionally it is necessary to consider the complete model, introducing deliberately or unintentionally to the discussion of weak or insignificant connections between separate elements of the device. Thus, elements with strongly differing parameters may be included into the model. This is the reason of stiffness of the model equations. Further, we shall consider the identification problem as a more common inverse problem of the circuit theory.

Depending on whether the topology of a circuit is known and whether nodes and branches of the circuit are accessible for connecting measuring devices, we shall distinguish the following types of identification problems:

1. Problems with known topological structure, in which all necessary nodes and/or branches are accessible for carrying out experiments.

2. Problems which have known topological structure of the circuit but not all necessary nodes and/or branches are accessible for carrying out experiments.

3. Problems with unknown topology.

When solving problems of the first type it is possible to derive a model completely adequate to the real electric circuit. In this case values of admittances or resistances of branches of the investigated circuit are the unknown parameters of the model.

As a result, in the solution of a problem of the second type finding values of parameters only for some of the circuit branches is possible. Besides, it is possible to also find the values of parameters of the multiterminal network, which is equivalent to the part of the circuit inaccessible for measurements. We shall illustrate this by the example shown in Fig. 3.14. Let nodes 1-6 be accessible for carrying out experiments, and N be the circuit part containing internal (inaccessible) nodes. At solution of the identification problem, the values of admittances g_{12}, g_{34}, g_{56}, g_{13}, g_{35} of the real circuit branches, and values of admittances g_{26}, g_{24}, g_{46} of the multiterminal network equivalent to its part can be found.

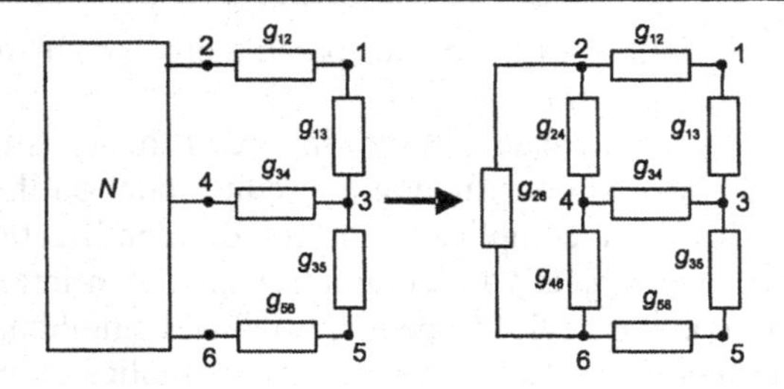

Fig. 3.14. A circuit with only partially known structure

When solving a problem with unknown topology, (problems of the third type) the circuit can be represented in the form of a full equivalent multiterminal network. The number of its nodes is equal to the number of external nodes of the circuit. Values of parameters of multiterminal network elements are found by means of experimentation, for which the general number of nodes of the real circuit is unknown.

We define a circuit observable by current or voltage if the experimental data is sufficient for the determination of currents or voltages for all of its branches. For complete representation of a circuit, it is sufficient to know the currents $\mathbf{I}_c$ of its links, or voltages $\mathbf{U}_c$ of the circuit graph tree branches. Currents and voltages of the remaining branches can be calculated as follows: $\mathbf{I}_t = \mathbf{F}^T \cdot \mathbf{I}_c$, $\mathbf{U}_c = -\mathbf{F} \cdot \mathbf{U}_t$, where $\mathbf{F}$ is a matrix associated with the tree so that $\mathbf{D} = \left[\mathbf{1}, -\mathbf{F}^T\right], \mathbf{C} = \left[\mathbf{F}, \mathbf{1}\right]$, $\mathbf{D}$ is a fundamental cutset matrix, $\mathbf{C}$ is a fundamental loop matrix, $\mathbf{I}_t$ and $\mathbf{U}_t$ are sets of tree currents and voltages, $\mathbf{I}_c$ and $\mathbf{U}_c$ are sets of cotree currents and voltages, and $\mathbf{1}$ is a matrix of unity.

Depending on whether the circuit is observable by current or voltage, different topological bases are used for its identification. If the circuit nodes are accessible, it is expedient to use the basis of cuts; in this case there is a possibility to measure the voltages of tree branches. It is convenient to use the so-called fundamental tree with branches that connect each of the circuit nodes with the basis node. If there is no branch between some node and the basis node, a fictitious branch with zero admittance is introduced between them. Thus, a conversion from the basis of cuts to nodal basis is carried out which simplifies identification of circuits with unknown structure. If circuit branches are accessible for measurements, then it is expedient to solve the problem in the loop basis. In this case, measurements of the link currents is sufficient.

The choice of either topological basis offers two methods of circuit identification: the method based on nodal analysis - the so-called Nodal Impedances Method (or NIM) [1], and the method based on loop analysis – Loop Admittances Method (or LAM).

3.4.1 Methods of identification of linear circuits

Let's consider the statement of the identification problem by means of NIM.

Let all $N+1$ nodes of a passive electric circuit of unknown structure be accessible for experimentation. Let's define the admittances of the equivalent circuit by a circuit presented in the form of a full multiterminal network. We assume that the identification problem is solved if the matrix of circuit nodal admittances $\mathbf{Y}_0=\left(y_{i,j}\right)_{N\times N}$ is found. Indeed, in that case we have the admittance $g_{i,j}$ between nodes i and j as,

$$g_{i,j}=\begin{cases}\left|y_{i,j}\right|, & i,\ j\neq 0;\\ \sum\limits_{k=1}^{N} y_{i,k}, & i\neq j=0\end{cases}.$$

According to NIM, the matrix $\mathbf{Y}_0$ is defined as the inverse of the matrix of nodal resistances $\mathbf{Z}_0$. Elements of the matrix $\mathbf{Z}_0=\mathbf{Y}_0^{-1}$ are defined by experimental results presented in Fig. 3.15, which constitute the experimental stage of the problem's solution. Let's connect a current source of 1 A (generally of one relative unit of current) between nodes 0 and 1. Then the nodal voltages of the circuit will satisfy the following equation:

$$\mathbf{U}_0=\mathbf{Y}_0^{-1}\cdot\mathbf{J}=\mathbf{Z}_0\cdot\mathbf{J},$$

where $\mathbf{U}_0$ is the column of measured nodal voltages and $\mathbf{J}$ is the column of current sources. After designating this experiment as the first one in a series of measurements, we can write the expansion as

$$\mathbf{U}_0^{(1)}=\mathbf{Z}_0\cdot\left(1,\quad 0,\quad \cdots\quad 0\right)^T=\mathbf{Y}_0^{-1}\cdot\mathbf{e}_1^T.$$

Thus, the column $\mathbf{U}_0^{(1)}$ of nodal voltages represents the first column of matrix $\mathbf{Y}_0^{-1}$.

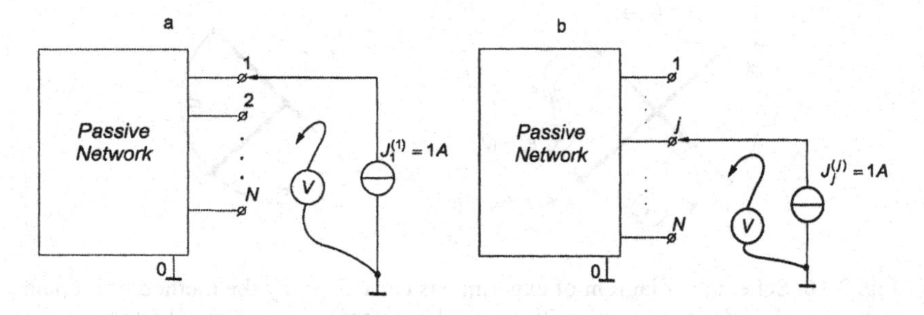

Fig. 3.15. Schematic diagram of experiments by Nodal Impedances Method (NIM)

Similarly, to find the j-th column $\mathbf{U}_0^{(j)}$ of the matrix $\mathbf{Z}_0$, it is necessary to connect a current source between nodes 0 and j of the multiterminal network (Fig. 3.15b) and then measure the nodal voltages $\mathbf{U}_0^{(j)}$.

$$\mathbf{U}_0^{(j)}=\mathbf{Y}_0^{-1}\cdot(0,0,...,1,...,0)^T=\mathbf{Y}_0^{-1}\cdot\mathbf{e}_j^T.$$

By means of connecting the current source to each node and in turn measuring the nodal voltages, we find $\mathbf{Z}_0=\left[\mathbf{U}_0^{(1)},\mathbf{U}_0^{(2)},\cdots,\mathbf{U}_0^{(N)}\right]$. If the measurements are carried out with sufficient accuracy, then the solution of the electric circuit identification problem becomes $\mathbf{Y}_0=\mathbf{Z}_0^{-1}$. Therefore, the calculation stage of NIM involves inversion of the matrix $\mathbf{Z}_0$. An important advantage of NIM is that it requires rather simple experiments, and they can be easily automated. The calculation stage of the method also does not pose any difficulties. However, this method has a disadvantage in that it does not allow determining the admittance of parallel connected branches.

To define parameters of parallel branches that may be included in the diagnosed circuit, the LAM (Loop Analysis) method can be used. Let's consider an identification problem of the first type. According to LAM, the identification problem is solved if the matrix of the circuit loop resistances $\mathbf{Z}_c=\mathbf{CZC}^T$ is found, where N is the number of links of the circuit graph, $\mathbf{C}$ is the fundamental loop matrix, and $\mathbf{Z}$ is a diagonal branch impedance matrix.

The matrix of loop resistances is defined as the inverse matrix to the matrix of loop admittances $\mathbf{Y}_c$, generated by experimental results. To carry out circuit identification it is necessary to connect a 1 V emf source to the circuit links, and in turn, measure the currents of all links in each experiment. An example of such an experiment for a circuit with four links is shown in Fig. 3.16a, where the branches of the circuit network tree are indicated by solid lines.

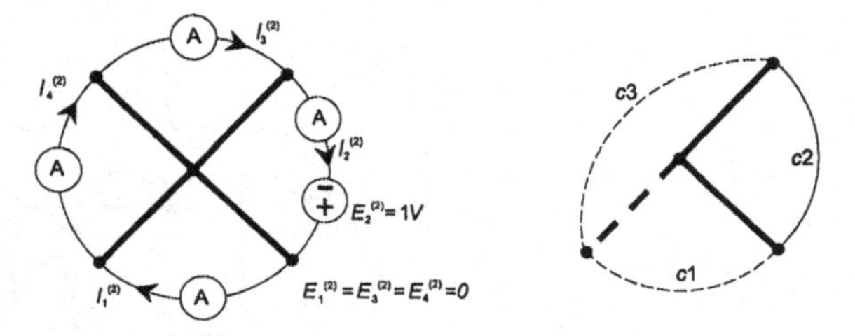

Fig. 3.16. Schematic diagram of experiments carried out by the method of loop admittances (LAM) for a circuit with four independent loops, when solving problems of first **a**) and second **b**) types

Link currents $\mathbf{I}_c^{(j)}$ measured in the *j*-th experiment satisfy the following set of equations:

$$\mathbf{I}_c^{(j)}=\mathbf{Z}_c^{-1}\cdot(0,\, 0,\cdots,\, 1,\cdots,\, 0)^T=\mathbf{Y}_c\cdot\mathbf{e}_j^T,$$

hence, $\mathbf{I}_c^{(j)}$ is the *j*-th column of matrix $\mathbf{Z}_c^{-1}$. Let's compose a matrix $\mathbf{Y}_c=\left[\mathbf{I}_c^{(1)},\, \mathbf{I}_c^{(2)},\cdots,\mathbf{I}_c^{(N)}\right]$ consisting of the measured columns $\mathbf{I}_c^{(j)}$. Then the solution of the electric circuit identification problem becomes $\mathbf{Z}_c=\mathbf{Y}_c^{-1}$, provided experiments were carried out with adequate accuracy. Thus, the calculation stage of LAM, as well as of NIM, involves the inversion of a matrix with factors that have previously been found by measurements.

In cases of incomplete observability of the circuit (problems of the second type), it is possible to find parameters of branches only of those loops which have no common branches with the loop that corresponds to an inaccessible link. For example, if in the circuit shown in Fig. 3.16a, link 4 is inaccessible to measure, it is only possible to define parameters of the equivalent circuit with the graph shown in Fig. 3.16b. The parameters of the branches indicated by solid lines correspond to real ones, and the parameters of the remaining branches represent some combination of real parameters.

3.4.2 Error of identification problem solution

To investigate errors in the solution of an identification problem, we need a model for the measuring devices used in the experimental stage. Assume that these measuring devices allow us to carry out measurements with a maximum relative error $\Delta_{\max}$, in which the error of each real measurement is a random variable uniformly distributed on the interval $\left[-\Delta_{\max}, \Delta_{\max}\right]$.

As has been shown above, the solution of the identification problem either by the LAM or NIM method requires inversion of a matrix (further - matrix **A**) with factors obtained by measurements. The condition number (or Todd's number) of a matrix inversion procedure acts as its characteristic of numerical stability:

$$\Theta(\mathbf{A})=\frac{\max|\lambda(\mathbf{A})|}{\min|\lambda(\mathbf{A})|}=\frac{\lambda_{\max}(\mathbf{A})}{\lambda_{\min}(\mathbf{A})}.$$

Increasing of the number Θ causes deterioration of numerical stability of the matrix inversion procedure. Ill-conditionality of matrix **A** at identification of a circuit may be caused by the following reasons:

- incorrect conceptions about the structure of the diagnosed circuit, leading to occurrence of elements with strongly differing parameters in the equivalent circuit;

- topological singularities of the circuit, such as special cuts (sections) passing only through branches with small admittances and special loops passing only on branches with small resistances.

Let's specify concepts of "small admittance" and "small resistance". Obviously the concept of smallness should be connected with the measurement error Δ_{max}. Let g_{max} and g_{min} be maximum and minimum values of the circuit's branch admittances, accordingly. Assume that some admittance g_k is small, if the inequality $g_k / \sqrt{g_{min} \cdot g_{max}} < \Delta_{max}$ holds. And on the contrary, we assume that an admittance g_k is large, if $g_k / \sqrt{g_{min} \cdot g_{max}} > \Delta_{max}^{-1}$. Accordingly, parameters of branches differ strongly if one of the parameters is small in comparison with others.

In identification of a circuit with unknown structure, in nodal analysis (NIM) fictitious nodes resulting in redundancy of the model, and consequently - in its stiffness, can exist. Let's show this by the example using the circuit in Fig. 3.17a. Here the node 2' is split into two nodes, 2 and 3, one of which is fictitious. We will use nodes 2 and 3 to carry out our experiments, as we have no information that these nodes are connected to one another inside the device. Using NIM results in the equivalent circuit shown in Fig. 3.17b. This circuit corresponds to the real circuit if $G \to \infty$. The equivalent circuit in Fig. 3.17b has an ill-conditioned matrix of nodal admittances Y_0. This matrix has two eigenvalues, close by magnitude to g_k, $k = \overline{1,3}$, and a third one close to $G \to \infty$. A dual situation can arise at identification of circuits by the method of loop admittances. For these cases, the models' ill-conditionality is caused by the lack of information about the identified circuit's structure.

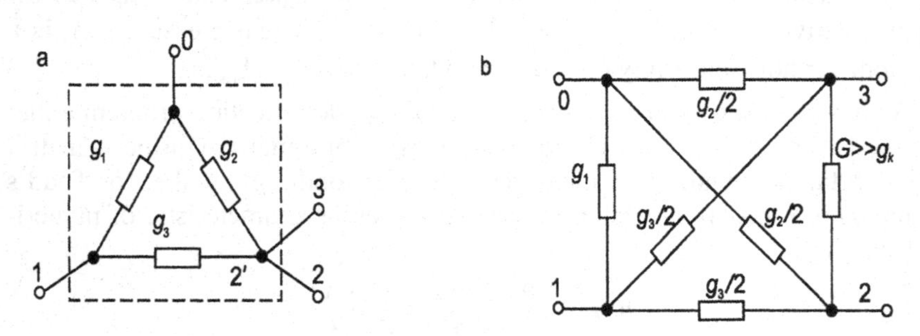

Fig. 3.17. Identification of fictitious nodes by the method of nodal resistances

Let's consider topological singularities. We shall consider a cut as a special one if the sum of admittances of branches intersected by this cut is small in the sense specified above. Accordingly, a loop is considered a special one if the sum of resistances of branches included into it is small. Identification of circuits with special cuts is difficult using the NIM method. In identification of circuits using the loop analysis method (LAM) difficulties arise when the presence of special loops in the circuit exist.

Special cuts and special loops divide the circuit into subcircuits that are poorly connected with each other. In the circuit shown in Fig. 3.18, there are three poorly connected subcircuits, N_1, N_2 and N_3, separated by a special cut s and a special loop l. In the case of a special cut, the currents of branches connecting N_1, and N_2 are small because of admittances of branches in this cut are small. In the case of a special loop, the voltages on its branches are small, and subcircuits N_2 and N_3 are poorly connected as well.

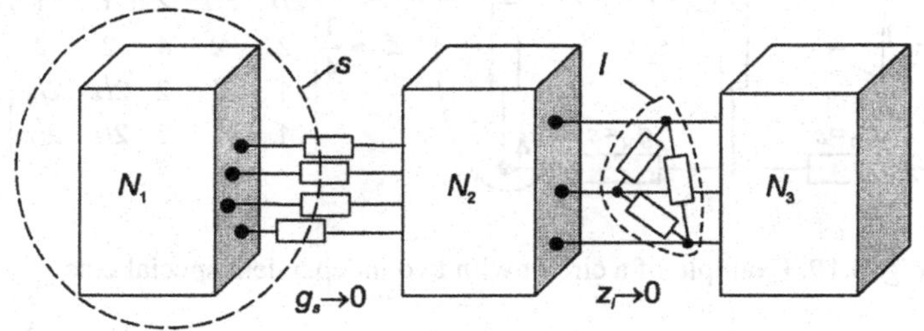

Fig. 3.18. A circuit with topological singularities

Each special cut introduces a small modulo eigenvalue in the spectrum of the nodal admittance matrix **Y**. Indeed, in the extreme case when the admittance of cut branches is zero, the circuit breaks up into two sub-circuits which are not connected. Therefore, while analyzing such a circuit, it would be possible to consider each subcircuit and choose a zero node for each one separately. For this reason the matrix **Y** of the whole circuit will include a single, linearly dependent row (and column) and one of its eigenvalues will be zero. If the admittances of the special cut are small, then by virtue of continuous dependence of eigenvalues on the matrix elements, the whole circuit matrix **Y** will have a small eigenvalue. Accordingly, $\Theta(\mathbf{Y})$ will be large and the matrix **Y** will be ill-conditioned. Similarly, the influence of a special loop on eigenvalues of the loop resistances matrix $\mathbf{Z}_c = \mathbf{CZC}^T$ can be explained.

Topological singularities can be of the "embedded" type. Let there be a special cut s in an identified circuit, inside of which there is a subcircuit N_1. If there is also a special cut s_1 in N_1, and cuts s and s_1 are not intersected, then cut s_1 is embedded in cut s. Similarly, "embedded" special loops can now be introduced. It will further be shown that circuit identification becomes considerably more complex in the presence of "embedded" special cuts in it.

Further, we shall concentrate mainly on the NIM method, focusing attention on circuits with special cuts. According to the duality principle the results will be valid for the LAM method as well.

Let's consider singularities of circuits with special cuts by using the example of the circuit shown in Fig. 3.19a (admittances of its elements are specified in the drawing). At $\varepsilon \ll 1$ the circuit has two special cuts s_1 and s_2. It can be shown for this problem that $\lambda_{min}(\mathbf{Y}_0) \approx \varepsilon$, $\lambda_{max}(\mathbf{Y}_0) \approx 2$ and hence, Todd's number is $\Theta = 2/\varepsilon \gg 1$.

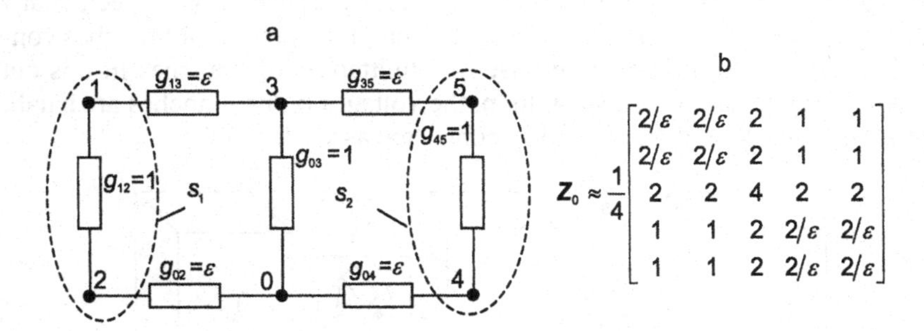

Fig. 3.19. Example of a circuit with two independent special cuts

Let $\varepsilon=10^{-3}$ and measurements carried out have an error $\Delta_{max}=5\cdot10^{-2}$. Then at identification using the method NIM, the matrix $\mathbf{Z}_0$, presented in Fig. 3.19b will be obtained as a result of measurements. We readily verify that matrix $\mathbf{Z}_0$ is a special one, and its eigenvalues are equal to (0, 0, 1, 1000, 1000). Therefore the calculation stage of the method NIM, consisting in an evaluation $\mathbf{Y} = \mathbf{Z}_0^{-1}$, cannot be fulfilled.

The reason for our failure in finding a solution to the identification problem of the circuit with special cuts involves insufficient accuracy of measurements. If elements of matrix $\mathbf{Z}_0$ have been found with exact accuracy, its eigenvalues would be equal to (0.5, 0.5, 1, 1000, 1000). Thus, error of measurements has led to distortion of small eigenvalues of $\mathbf{Z}_0$.

In practice it is unlikely we would derive zero eigenvalues at the realization of measurements, such as in the example above. However, it is safe to say that for ill-conditioned matrixes, small errors in the calculation of their elements generate large errors in small eigenvalues. Elements of the matrix $\mathbf{Y} = \mathbf{Z}_0^{-1}$ that are necessary for identification depend only on small eigenvalues of $\mathbf{Z}_0$, as eigenvalues of these matrixes are connected by the relation $\lambda_{\mathbf{Y}} = \lambda_{\mathbf{Z}}^{-1}$. Therefore, identification errors for circuits with special cuts will always be high.

Dependences of identification error δ from the conditionality Θ of the nodal admittance matrix for a circuit with a single special cut are shown in Fig. 3.20. The mean square error of definition of elements of the matrix, calculated in the following relation, is considered as the error in the solution:

$$\delta=\frac{1}{N}\sqrt{\sum_{i=1}^{N}\sum_{j=1}^{N}\left(\frac{\left|y_{i,j}-\tilde{y}_{i,j}\right|}{\left|y_{i,j}\right|}\right)^{2}}, \tag{3.37}$$

where $\tilde{y}_{i,j}$ is a matrix element determined by some error, and $y_{i,j}$ is the exact matrix element.

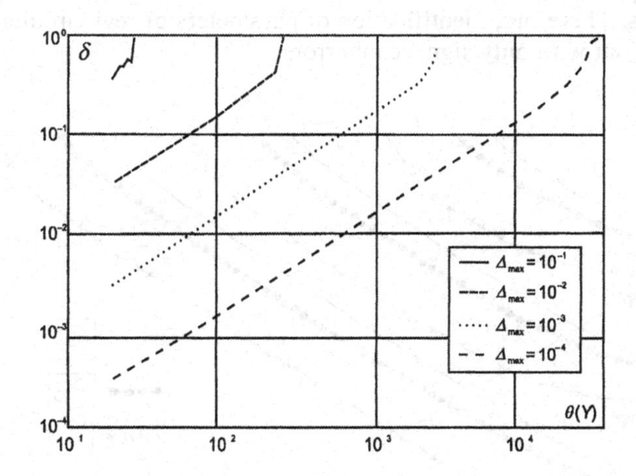

Fig. 3.20. Dependences of identification error δ (3.37) from the conditionality Θ of nodal admittances matrix for a circuit with a single special cut

To derive smooth dependences, calculations for each pair of Δ_{max} and Θ have been carried out 500 times and then averaged.

It is obvious from Fig. 3.20 that the identification error δ raises linearly as Θ increases for all Δ_{max}. For example, if we accept $\delta=10^{-2}$ as the accuracy and carry out measurements with an error $\Delta_{max}=10^{-3}$, then the solution of the identification problem can be obtained only if the conditionality of the circuit matrix **Y** does not exceed ~100. In other words, the method NIM allows performing identification only for circuits with well-conditioned mathematical models.

The dependences of identification error δ from ε^{-1} are shown in Fig. 3.21 (here, as above, ε is the admittance of branches through which the special cut passes) for electric circuits with the following layouts of special cuts:

- circuit with a single special cut (the circuit pictogram - [O]),
- circuit with two independent special cuts (the circuit pictogram - [°o]),

- circuit with two special cuts embedded in one another (the circuit pictogram - ◎).

It is apparent that the identification of the error of parameters for the circuit with embedded cuts considerably exceeds the identification of error for parameters of circuits with another arrangement of special cuts. So, at $\Delta_{max}=10^{-3}$ and $\varepsilon=10^{2}$, the identification error δ is about ~10%. In electric circuits used in practice, admittances of branches can differ considerably, more than 10^2 times. Therefore, identification of parameters of real circuits can often be carried out with only significant errors.

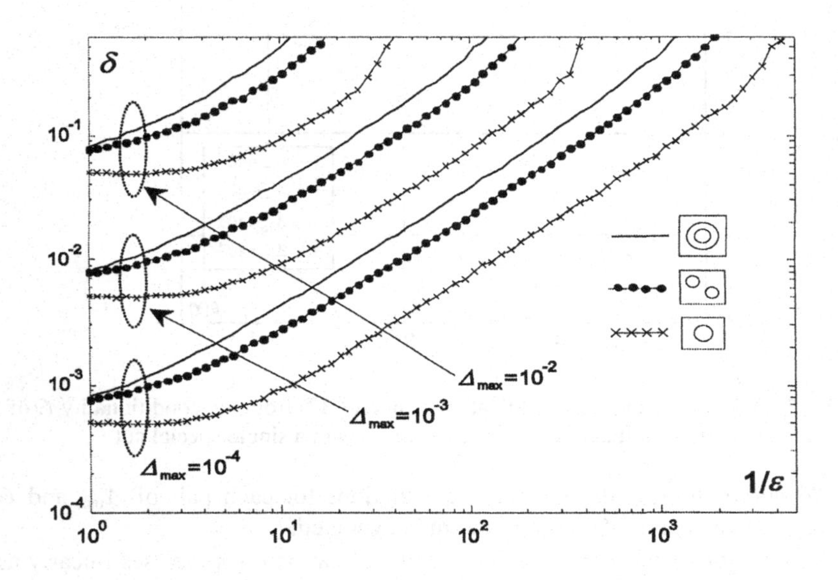

Fig. 3.21. Comparison of identification error δ for various arrangements of special cuts

The above statements are also valid for sinusoidal current circuits. Moreover, in this case additional special cuts caused by resonance phenomena can emerge (Fig. 3.22). Dependence $\Theta=\Theta(f)$ of the circuit matrix $\mathbf{Y}_0$ Todd's numbers from frequency is shown in Fig. 3.23. At f=1 kHz (at $C_{32}=C_{50}$=25.33 μF, $L_{32}=L_{50}$=1 mH, $R_{32}=R_{50}$=1 MOhm) the minimum eigenvalue $\lambda_{min}(\mathbf{Y}_0)\approx 0$. Therefore, there is a special cut s at this frequency. In this circuit special cuts can appear also at other, higher frequencies.

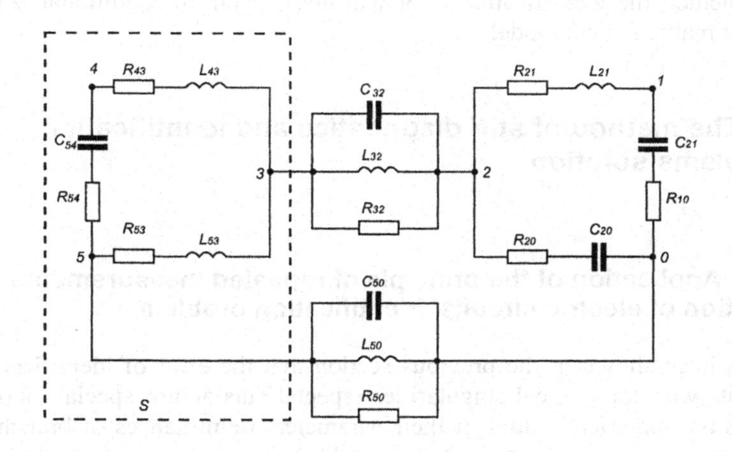

Fig. 3.22. Special cuts in an AC circuit

Therefore, at identification of AC circuits, it is necessary to choose the frequency of experiments in order to obtain sufficiently small conditionality of the matrix of nodal voltages. The above described appearance occurs for the majority of AC circuits. This is especially clear at high frequencies when formation of high-Q resonance contours is possible.

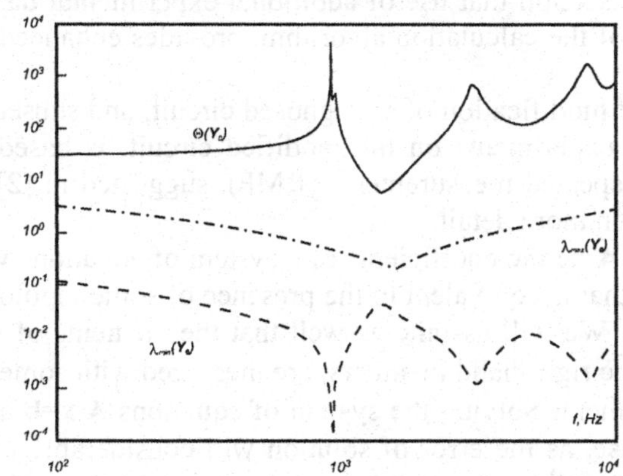

Fig. 3.23. Dependences of conditionality Θ from frequency *f* for the AC circuit

We have demonstrated that the identification error increases quickly with deterioration of conditionality of the circuit mathematical model. It is true for AC as well as for DC circuits. Further, we shall consider a new method of identification based on the principle of quasi-stationarity of derivatives. In this method the identification error will not depend on conditionality of the circuit mathematical model.

3.5 The method of stiff diagnostics and identification problems solution

3.5.1 Application of the principle of repeated measurements for solution of electric circuits' identification problem

It has been shown in the previous section that the error of identification of circuits with topological singularities (special cuts and/or special loops) depends on numerical values of their parameters (admittances of branches intersected by a special cut, etc.). The problem properties are characterized by Todd's number Θ (condition number) for the matrix of coefficients of the circuit's system of equations (or its mathematical model). Larger condition numbers mean less accurate solutions of the identification problem.

The solution of identification problem consists of experimental and calculation stages. Improving only one of these constituents does not lead to substantial gain in accuracy for identification of circuit parameters. It will be shown in this section that use of additional experimental data, together with modification of the calculation algorithm, provides enhanced accuracy in the solution.

The idea of modification of a diagnosed circuit, and subsequent realization of additional experiments on the modified circuit, is based on the general principle of repeated measurements (RMP), suggested in [2]. Let's consider this principle in more detail.

Let matrix **A** be the coefficients of a system of equations which are poorly conditioned, that is equivalent to the presence of some topological singularity in the circuit. We will assume as well that the elements of matrix **A** and of vector **b**, of the right-hand members, are measured with some error δ during a "first" experiment. Solving the system of equations $\mathbf{A}\cdot\mathbf{x}=\mathbf{b}$ at $\Theta(\mathbf{A})>>1$ does not make sense, as the error of solution will considerably exceed δ. Really, its solution $\mathbf{x}^*=\mathbf{A}^{-1}\cdot\mathbf{b}$ essentially depends on small eigenvalues of matrix **A**. However, information concerning small eigenvalues of matrix **A** approximating the representation of its elements is inadequate.

It has been shown above (Sections 3.2 and 3.3) that matrixes **A** and **b** with elements that are determined with an error δ contain certain information on linear connections between elements of the solution vector $\mathbf{x}^*$:

$$(\xi, \mathbf{x}^*) = \xi_0, \tag{3.38}$$

where vector ξ and the number ξ_0 are coefficients of the linear connection, the error of linear connections definition being close to δ. Therefore we use matrix **A** and vector **b**, resulting from the "first" experiment, to determine linear connections between the components of the vector $\mathbf{x}^*$. Then the dimensionality of the problem can be reduced by means of linear connection (3.38). It can be shown [1] that the conditionality of the matrix of the problem with reduced dimensionality (or the reduced problem) does not exceed the conditionality of the initial problem. Then a *repeated experiment* can be carried out to determine parameters of the reduced problem.

It is possible to apply this algorithm recursively, meaning that the results of the repeated experiment can be used for determination of new linear connections. Then a further reduction of the model can be performed and successive, repeated experiments can be carried out. The way to reduce a problem, in order for repeated experiments to be carried out for a considerably less stiff model, will now be considered.

We shall perform the electric circuit identification by the NIM method, comprehensively considered in the previous section. Results will also be valid in the case of application of the LAM method on condition of transfer to the loop basis.

When using RMP for identification of circuits with topological singularities the following should be carried out:

- finding the linear connections (3.38) between variables of the identification problem according to data from initial experiments;
- suggesting the type of repeated experiments that should be carried out for a less stiff problem.

Solution of these particular problems will be considered below.

3.5.2 Definition of linear connections between parameters of circuit mathematical models

Let's find the equation of linear connection between elements of columns of matrix $\mathbf{Y}_0$ in the case of circuit identification by the NIM method. We shall present matrix equation $\mathbf{U}\cdot\mathbf{Y}_0=\mathbf{1}$ in the form of N systems of equations $\mathbf{U}\cdot\mathbf{y}_k=\mathbf{e}_k,\ k=\overline{1,N}$ (N is the number of circuit nodes), where $\mathbf{y}_k$ is the k-th column of the matrix $\mathbf{Y}_0$ of node admittances.

Assume there is a special cut in the circuit. Then the ill-conditionality of matrix **U** is due to the presence of big modulo eigenvalue λ_1 in its spectrum:

$$\lambda_1(\mathbf{U}) >> \lambda_2(\mathbf{U}) \geq \cdots \geq \lambda_N(\mathbf{U}).$$

Let's consider the system of equations $\mathbf{U}\cdot\mathbf{y}_1=\mathbf{e}_1$. Multiplication of its right and left parts s-1 times, by matrix **U** from the left, gives

$$\mathbf{U}^s\cdot\mathbf{y}_1=\mathbf{U}^{s-1}\cdot\mathbf{e}_1. \tag{3.39}$$

Let's represent matrix **U** in the form of $\mathbf{U}=\mathbf{P}\cdot\mathbf{\Lambda}\cdot\mathbf{P}^{-1}$. Then $\mathbf{U}^s=\mathbf{P}\cdot\mathbf{\Lambda}^s\cdot\mathbf{P}^{-1}$, where **P** is the matrix of matrix **U** eigenvectors, and $\mathbf{\Lambda}=\mathrm{diag}(\lambda_1,\lambda_2,\cdots,\lambda_N)$ is the matrix of matrix **U** eigenvalues. When $s\to\infty$, matrix $\mathbf{\Lambda}^s$ degenerates:

$$\mathbf{\Lambda}^s=\lambda_1^s\cdot\mathrm{diag}\left(1,\frac{\lambda_2^s}{\lambda_1^s},\cdots,\frac{\lambda_N^s}{\lambda_1^s}\right)\Bigg|_{s\to\infty}=\lambda_1^s\cdot\mathrm{diag}(1,0,\cdots,0).$$

Hence, all rows of matrix $\mathbf{U}^s$ are pairwise linearly dependent, and any of the equations of (3.39) represents the desired linear connection (3.38) between vector components of $\mathbf{y}_1$. Since $(\mathbf{e}_1,\ldots,\mathbf{e}_N)\equiv\mathbf{1}$, then the set of linear connections for all columns of matrix **U** can be written down as follows:

$$\mathbf{U}^s\cdot\mathbf{Y}_0=\mathbf{U}^{s-1}. \tag{3.40}$$

If coefficients of matrix **U** are obtained as experimental results by some error, then the linear connection will be found with an error along the same order as the measurement error. Really, it is defined only by the maximum modulo eigenvalue, which has a distortion close to the error of matrix **U** coefficients.

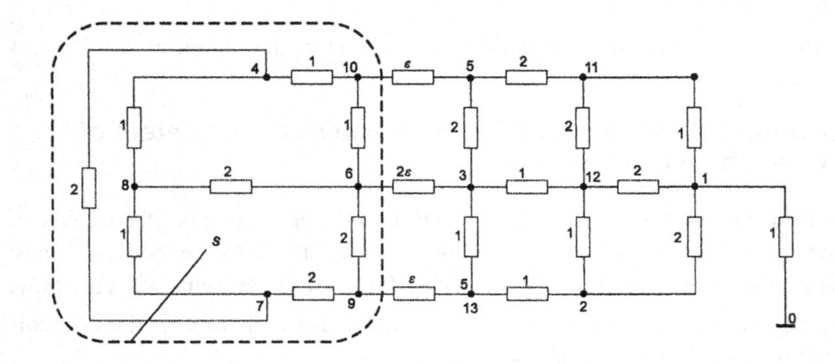

Fig. 3.24. Electric circuit with a special cut s for researching the accuracy of identification problem solution

Let the coefficients of matrix **U** be defined by the NIM method with an error $\Delta_{\max}$. We shall estimate the linear connections' definition error δ_ξ in association with the condition number Θ of matrix $\mathbf{Y}_0$ for the circuit shown in Fig. 3.24. This circuit has a special cut S. Assume δ_ξ is represented by the following quantity:

$$\delta_\xi = \max_k \left|(\xi, \mathbf{y}_k) - \xi_{0,k}\right|,$$

where ξ is the vector of coefficients of linear connection (3.38) and is identical to all columns of matrix $\mathbf{Y}_0$, and $\xi_{0,k}$ is the linear connection's absolute term for k-th column. Figure 3.25 shows dependences of δ_ξ from the condition number of matrix $\mathbf{Y}_0$ for various levels of voltages' measurement error $\Delta_{\max}$. To derive smooth curves, these dependencies were averaged upon a large number of calculations. Apparently, the error of linear connections' definition does not depend on matrix $\mathbf{Y}_0$ condition numbers and is close to $\Delta_{\max}$.

Dependencies shown in Fig. 3.25 are obtained at s=5. Increasing the exponent s leads to diminution of influence on coefficients of linear connections from strongly distorted small eigenvalues of matrix **U**. Therefore, research of dependence of linear connections' definition error from s is of interest. The dependences $\delta_\xi=\delta_\xi(s)$ for various Θ are shown in Fig. 3.26. Obviously, increasing s over 4-5 is not prudent.

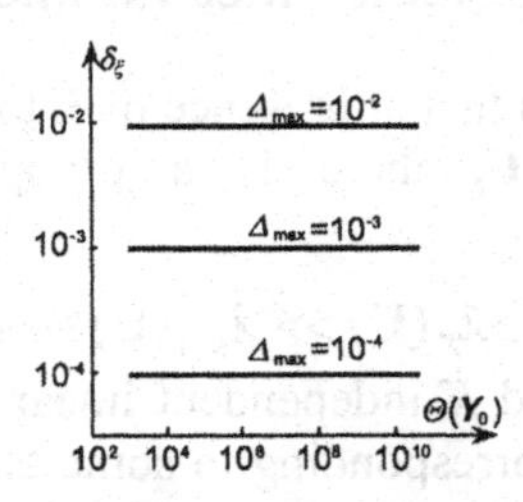

Fig. 3.25. Averaged dependences of linear connections' definition error δ_ξ versus matrix $\mathbf{Y}_0$ condition numbers for various values of measurement error

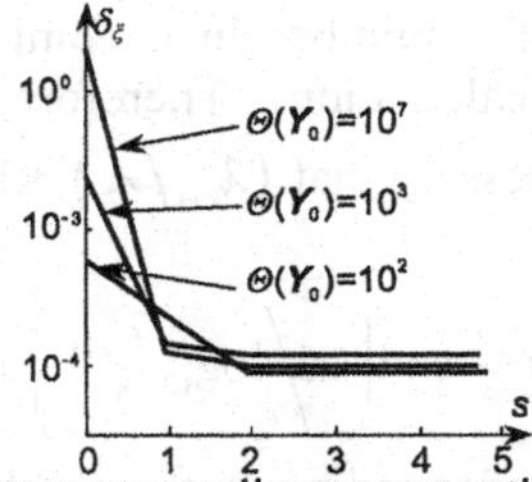

Fig. 3.26. Influence of exponent s on linear connections' definition error δ_ξ for various condition numbers $\Theta(\mathbf{Y}_0)$

In conclusion, we shall note that identification of nonreciprocal circuits (circuits with controlled sources, etc.) in matrix $\mathbf{Y}_0$ is asymmetric. Therefore the linear connections between the elements of columns of matrixes generally do not coincide with linear connections between elements of their rows. However, taking into account that $\mathbf{Y}_0 \cdot \mathbf{U} = \mathbf{U} \cdot \mathbf{Y}_0 = \mathbf{1}$, the following expression for linear connections between elements of rows can be found:

$$\mathbf{Y}_0 \mathbf{U}^s = \mathbf{U}^{s-1}. \tag{3.41}$$

Obviously, properties of linear connections between elements of columns considered above are also valid for connections between elements of rows. Further, we shall assume that if necessary, both groups of connections should be defined: connections between elements of columns by the expression (3.29) and connections between elements of rows by the expression (3.41). It should be noted that in the identification of "black box" type electric circuits, it is always necessary to define both groups of connections.

3.5.3 Algorithm and results of electric circuits' identification problem solution using repeated measurements

It has previously been shown that at presence of K topological singularities in the spectrum of matrix $\mathbf{U}$, there is a group of large eigenvalues $\Lambda_{\max} = \{\lambda_1, \cdots, \lambda_K\}$, such that

$$\lambda_1(\mathbf{U}) \geq \cdots \geq \lambda_K(\mathbf{U}) >> \lambda_{K+1}(\mathbf{U}) \geq \cdots \geq \lambda_N(\mathbf{U}).$$

Thus, it is possible to find K independent linear connections (3.38) from the equations (3.40), each corresponding to some eigenvalue from the group $\Lambda_{\max}$. To numerically calculate all independent connections, one needs the value of exponent s in Eq. (3.40) at which information on all small eigenvalues in matrix $\mathbf{U}^s$ vanishes, whereas information concerning Λ_{max} remains. According to IEEE floating-point arithmetic standards, the error of representation of the stagnant part of a number in a computer is 10^{-16} (at double precision floating-numbers calculation). Therefore, to provide the above-mentioned condition it is necessary that $(\lambda_{k+1}/\lambda_k)^s \leq 10^{-16}$. Hence, we find the optimal exponent

$$s = -\mathrm{ceil}\left(16 \Big/ \log_{10} \frac{\lambda_{k+1}}{\lambda_k}\right),$$

where ceil(x) is the function of rounding up the number x to an integer.

When carrying out measurements with some relative error $\Delta_{\max}$, the latter is the basic quantity that defines s. Therefore,

$$s=\text{ceil}\left(\log_{10}\left(\Delta_{\max}\right)\Big/\log_{10}\frac{\lambda_{k+1}}{\lambda_k}\right). \tag{3.42}$$

The relationship (3.42) shows that acceptance of s=1-2 at $\Theta(\mathbf{U}) >> 1/\Delta_{\max}$ and s=2-5 (in case of smaller stiffness of the problem) is sufficient. Further, we assume s=4, if not specified otherwise.

Thus, we have considered a means to find linear connections between elements of rows and/or columns of matrix **U**. These linear connections can be obtained by results of an initial experiment. The accuracy of their determination does not depend on the stiffness of the mathematical model, and is defined only by the measurement error. Further, we shall consider the way to reduce the problem for carrying out repeated experiments.

The new reduced identification problem should satisfy the following requirements:

- parameters of all elements of the model, with the exception of those which can be restored later by means of linear connections, should be defined from the solution of the reduced problem;

- the new model's coefficients matrix condition number should be less than the condition number of the initial model's matrix.

Obviously, to provide the second requirement it is necessary to delete topological singularities from the model one by one. At the first stage of reduction this will be the topological singularity corresponding to maximum eigenvalue $\lambda_1(\mathbf{U})$. At the next stage it will correspond to the maximum eigenvalue from remaining ones, i.e. to $\lambda_2(\mathbf{U})$, and so on.

Topological singularities divide a circuit into subcircuits that are weakly connected with each other. To remove a topological singularity, one may connect subcircuits isolated by it to the zero-node, by means of an additional link with a large admittance G. If there is a special loop, it requires introducing an additional large resistance R into a link, which is included in this loop.

In practice, short-circuiting ($G\to\infty$) one of the isolated subcircuit nodes to zero, thereby shunting the special cut as well as breaking off ($R\to\infty$) the special loop, is more convenient. At that the matrix of coefficients of the reduced model will be a submatrix of the initial matrix of model. The purpose of repeated experiments involves determination of elements of this submatrix. Missing elements of the initial matrix may be restored later by means of linear connections.

Let's carry out identification of the circuit shown in Fig. 3.27 using the method described above. This circuit contains a single special cut S.

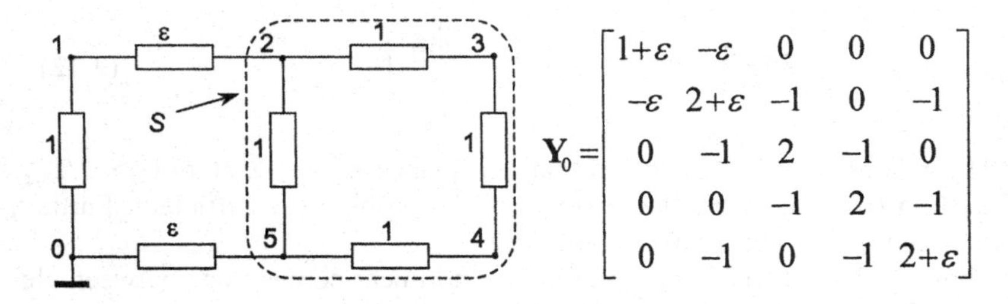

$$\mathbf{Y}_0=\begin{bmatrix} 1+\varepsilon & -\varepsilon & 0 & 0 & 0 \\ -\varepsilon & 2+\varepsilon & -1 & 0 & -1 \\ 0 & -1 & 2 & -1 & 0 \\ 0 & 0 & -1 & 2 & -1 \\ 0 & -1 & 0 & -1 & 2+\varepsilon \end{bmatrix}$$

Fig. 3.27. Resistive circuit with a single special cut S

Assume for definiteness that $\varepsilon=10^{-3}$ Sm. If $\Theta(\mathbf{Y}_0)\approx10^3$, then matrixes $\mathbf{Y}_0$ and $\mathbf{Z}_0=\mathbf{Y}_0^{-1}$ of the circuit are ill-conditioned. We shall carry out experiments according to NIM, using the above-described model of measurement error at $\Delta_{max}=0.01=1\%$. Results of the first series of experiments are as follows:

$$\tilde{\mathbf{U}}=\begin{bmatrix} 1.004 & 0.5005 & 0.4979 & 0.4943 & 0.4989 \\ 0.5005 & 50.86 & 50.31 & 50.38 & 50.11 \\ 0.4979 & 50.31 & 50.57 & 50.39 & 50.24 \\ 0.4943 & 50.38 & 50.56 & 50.94 & 50.31 \\ 0.4989 & 50.06 & 50.19 & 50.31 & 50.43 \end{bmatrix}.$$

This circuit is a reciprocal one. Therefore, the experimental matrix of voltages has been balanced. Hereinafter, we will write down the results with only the significant digits shown, the number of which is defined by the magnitude of Δ_{max}.

Predictably (see Section 3.4), the isolated subcircuit node voltages dominate by their large values. The minimum eigenvalue of the matrix of voltages is $\lambda_{min}(\tilde{\mathbf{U}})=0.162$, which differs from its exact value $\lambda_{min}(\mathbf{U})=0.250$ by 35%. That is, definition error of a small eigenvalue noticeably exceeds the measurement error. At the same time the error of the maximum eigenvalue is – 0.1%, that is, it does not exceed the measurement error. Following NIM, we find the identification problem solution as follows:

$$\tilde{\mathbf{Y}}_0=\tilde{\mathbf{U}}^{-1}=\begin{bmatrix} 1.001 & -5.808\cdot10^{-3} & 0 & 0 & 0 \\ -5.808\cdot10^{-3} & 1.359 & -0.7252 & 0 & -0.3917 \\ 0 & -0.7252 & 3.569 & -0.2732 & 0 \\ 0 & 0 & -0.2732 & 1.099 & -0.5888 \\ 0 & -0.3917 & 0 & -0.5888 & 3.572 \end{bmatrix}.$$

Hereinafter, assuming that the circuit structure is known, the $\tilde{\mathbf{Y}}_0$ elements of the matrix corresponding to objectively nonexistent admittances will be substituted by zeroes. Small eigenvalues of matrix $\tilde{\mathbf{U}}$ are distorted by the measurement error. Therefore, the relative error $\delta\mathbf{Y}_0$ at definition of matrix $\mathbf{Y}_0$ elements essentially exceeds Δ_{max}=1%:

$$\left|\delta\mathbf{Y}_0\right|=\begin{bmatrix} 6\cdot10^{-4} & 480 & - & - & - \\ 480 & 32.1 & 27.5 & - & 60.8 \\ - & 27.5 & 78.4 & 72.7 & - \\ - & - & 72.7 & 45.0 & 41.1 \\ - & 60.8 & - & 41.1 & 78.5 \end{bmatrix}\%.$$

Thus, NIM does not allow identifying the circuit considered above. Let's apply the principle of repeated measurements (RMP) for the identification of this circuit. We shall find linear connections (3.40) between elements of matrix $\mathbf{Y}_0$ from the results of the first experiment:

$$\underbrace{\begin{bmatrix} 9.8438\cdot10^{-3} & 0.99592 & 0.99461 & 1.000 & 0.98757 \\ 9.8435\cdot10^{-3} & 0.99592 & 0.99461 & 1.000 & 0.98757 \\ 9.8435\cdot10^{-3} & 0.99592 & 0.99461 & 1.000 & 0.98757 \\ 9.8435\cdot10^{-3} & 0.99592 & 0.99461 & 1.000 & 0.98757 \\ 9.8435\cdot10^{-3} & 0.99592 & 0.99461 & 1.000 & 0.98757 \end{bmatrix}}_{\mathbf{U}^s}\times\mathbf{Y}_0= \tag{3.43}$$

$$\underbrace{\begin{bmatrix} 4.9071\cdot10^{-2} & 4.9403 & 4.9338 & 4.9606 & 4.8989 \\ 4.9030\cdot10^{-2} & 4.9403 & 4.9338 & 4.9606 & 4.8989 \\ 4.9030\cdot10^{-2} & 4.9403 & 4.9338 & 4.9606 & 4.8989 \\ 4.9030\cdot10^{-2} & 4.9403 & 4.9338 & 4.9606 & 4.8989 \\ 4.9030\cdot10^{-2} & 4.9403 & 4.9338 & 4.9606 & 4.8989 \end{bmatrix}}_{\mathbf{U}^{s-1}}\times10^{-3}.$$

Linear connections (3.43) are normalized, that is, each of the rows in Eq. (3.43) is divided by the maximum element of matrix $\mathbf{U}^s$ in that row. It is apparent that all rows in Eq. (3.43) are identical within an error not exceeding Δ_{max}. As noted above, this follows from the fact that there is only one special cut in the considered circuit. Therefore any row from Eq. (3.43), for example the first one, can be taken as a linear connection. Then we have five equivalent linear connections for columns $\mathbf{y}_i$, $i=\overline{1,5}$ of matrix $\mathbf{Y}_0$, and each of these linear connections can be used for further calculations:

$$\xi\cdot\mathbf{y}_1=\xi_{0,1}=4.9071\cdot10^{-5},\ \xi\cdot\mathbf{y}_2=\xi_{0,2}=4.9403\cdot10^{-3},\ \xi\cdot\mathbf{y}_3=\xi_{0,3}=4.9338\cdot10^{-3},$$

$$\xi\cdot\mathbf{y}_4=\xi_{0,4}=4.9606\cdot10^{-3},\ \xi\cdot\mathbf{y}_5=\xi_{0,5}=4.8989\cdot10^{-3},$$

where $\xi=\left(9.8438\cdot10^{-3},\ \ 0.99592,\ \ 0.99461,\ \ 1.000,\ \ 0.98757\right)$.

Then, we shall remove the special cut by means of modification of the circuit. For this purpose we connect one of nodes of the isolated subcircuit, for example the fourth node, to the zero-node. The matrix $\mathbf{Y}_0^{(-1)}$ of nodal admittances of the new reduced circuit is given by

$$\mathbf{Y}_0^{(-1)} = \begin{bmatrix} 1+\varepsilon & -\varepsilon & 0 & 0 \\ -\varepsilon & 2+\varepsilon & -1 & -1 \\ 0 & -1 & 2 & 0 \\ 0 & -1 & 0 & 2+\varepsilon \end{bmatrix},$$

where the superscript (-1) shows that the matrix size has been reduced by a unit. The matrix $\mathbf{Y}_0^{(-1)}$ has been obtained by removal of the fourth row and the fourth column from the initial matrix $\mathbf{Y}_0$. Then, we shall perform a repeated series of experiments according to NIM for the reduced circuit. The measured nodal voltages matrix is given by

$$\tilde{\mathbf{U}}^{(-1)} = \begin{bmatrix} 1.004 & 9.935\cdot 10^{-4} & 4.974\cdot 10^{-4} & 5.009\cdot 10^{-4} \\ 9.935\cdot 10^{-4} & 1.001 & 0.4963 & 0.4965 \\ 4.974\cdot 10^{-4} & 0.4963 & 0.7524 & 0.2486 \\ 5.009\cdot 10^{-4} & 0.4965 & 0.2486 & 0.7472 \end{bmatrix}.$$

As the special cut has been removed from the reduced circuit, then the matrix $\tilde{\mathbf{U}}^{(-1)}$ is well-conditioned and $\Theta(\tilde{\mathbf{U}}^{(-1)}) \approx 6$. Therefore, the solution of the reduced identification problem can be found with an error of the order Δ_{max}. Indeed,

$$\tilde{\mathbf{Y}}_0^{(-1)} = \begin{bmatrix} 0.9961 & -9.762\cdot 10^{-4} & 0 & 0 \\ -9.762\cdot 10^{-4} & 1.971 & -0.9745 & -0.9854 \\ 0 & -0.9745 & 1.975 & 0 \\ 0 & -0.9854 & 0 & 1.996 \end{bmatrix}, \text{ and}$$

$$\left|\delta \mathbf{Y}_0^{(-1)}\right| = 10^{-2} \begin{bmatrix} 0.5 & 2.3 & - & - \\ 2.3 & 1.5 & 2.5 & 1.5 \\ - & 2.5 & 1.3 & - \\ - & 1.5 & - & 0.2 \end{bmatrix}.$$

Further, it is necessary to solve the initial problem using linear connections and the reduced problem solution, that is, to find all elements of matrix $\mathbf{Y}_0$. For this purpose we first restore, by linear connections, the fourth column, and then the fourth row:

$$\tilde{\tilde{\mathbf{Y}}}_0 = \begin{bmatrix} 0.9961 & -9.762\cdot 10^{-4} & 0 & 0\ (\mathbf{-8.784\cdot 10^{-3}}) & 0 \\ -9.762\cdot 10^{-4} & 1.971 & -0.9745 & 0\ (\mathbf{-1.560\cdot 10^{-2}}) & -0.9854 \\ 0 & -0.9745 & 1.975 & \mathbf{-0.9889} & 0 \\ 0\ (\mathbf{-8.784\cdot 10^{-3}}) & 0\ (\mathbf{-1.560\cdot 10^{-2}}) & \mathbf{-0.9889} & \mathbf{1.977} & \mathbf{-0.9849} \\ 0 & -0.9854 & 0 & \mathbf{-0.9849} & 1.996 \end{bmatrix}.$$

Here, calculated values which are equal to zero in the exact matrix are shown in brackets. Apparently, all of them have the same order of error, equal to the order of measurements error. Thus, the relative error of the initial identification problem solution is of the order of magnitude Δ_{max}:

$$\left|\delta\tilde{\bar{\mathbf{Y}}}_0\right| = 10^{-2}\begin{bmatrix} 0.5 & 2.3 & - & - & - \\ 2.3 & 1.5 & 2.5 & - & 1.5 \\ - & 2.5 & 1.3 & 1.1 & - \\ - & - & 1.1 & 1.2 & 1.5 \\ - & 1.5 & - & 1.5 & 0.2 \end{bmatrix}.$$

In this example, the special cut has been excluded by means of connecting the fourth node to zero. Let's ascertain how to find the node number p_{opt} that gives maximum drop in the condition number of the reduced problem when connected to zero. One of the trivial though rather laborious methods of solution of this problem is searching through all the nodes. In that, for each reduced problem it is necessary to perform experimental definition of matrix **U** and estimation of $\Theta(\mathbf{U})$.

Searching through all nodes can be done virtually, that is without realization of numerous experiments with the reduced problems. This requires only matrix **U** to be obtained as a result of the first series of experiments. We shall describe the k-th step ($k=\overline{1,N}$, where N is the number of circuit nodes) of the algorithm for searching the node number as follows:

1. Transpose the k-th row and k-th column of matrix **U** to the first place.
2. Perform one step of direct Gauss elimination that gives matrix $\mathbf{U}^{(-1)}$, with size on a unit less than **U**.
3. Calculate $\Theta(\mathbf{U}^{(-1)})$. As proved in [7], $\Theta(\mathbf{U}^{(-1)}) \le \Theta(\mathbf{U})$.
4. Enter the value $\Theta(\mathbf{U}^{(-1)})$ into $\theta(k)$.

After realization of N steps of this algorithm, p_{opt} is defined as follows: $p_{opt}: \theta(p_{opt}) = \min_k \theta(k)$.

The problem of searching of p_{opt} can have a non-unique solution. So in the above example, nodes 2, 3, 4 and 5 are practically equivalent to be chosen as the node p_{opt}, that is $\theta(2) \approx \theta(3) \approx \theta(4) \approx \theta(5)$. Indeed, choosing any of these nodes as p_{opt} results in removal of a large eigenvalue from the spectrum. In this case any of these nodes can be chosen as p_{opt}.

Further, we consider application of RPM for identification of a circuit with several topological singularities. For each topological singularity, there is a corresponding eigenvalue from the group Λ_{max} and a number p_{opt} of the best element for reduction of the problem (for elimination of this topological singularity). Let's designate the set of these numbers as $\Pi = \{p_{1,opt}, \cdots, p_{K,opt}\}$.

For identification of a circuit with several topological singularities we shall find all elements of Π by means of experimental matrix **U**, using above-described algorithm. Further, we can find k independent linear connections by application of Eq. (3.41). Then we shall simultaneously exclude all topological singularities and perform repeated experiments. Finally, we can identify circuit parameters by independent linear connections and results of repeated experiments. This method is convenient for practice as it requires only two series of experiments. However, its use ensures the desired accuracy for the problem solution only when special cuts do not encompass each other. In the general case of nested special cuts, the circuit identification should be performed by use of the following algorithm:

1. Perform the first series of experiments according to NIM to determine the matrix of voltages **U** as a result.
2. Using matrix **U**, define the first linear connection corresponding to maximum eigenvalue of matrix **U**.
3. Analyze matrix **U** and define the node number $p_{1,opt}$, then perform reduction of the problem by means of connecting the node of number $p_{1,opt}$ to the zero-node.
4. Perform a series of repeated experiments on the reduced circuit to determine a new experimental matrix $\mathbf{U}^{(-1)}$ as a result.
5. If the condition number $\Theta(\mathbf{U}^{(-1)})>1/\Delta_{max}$, then a second linear connection should be defined by repetition of steps 3-5 of the algorithm to determine $p_{2,opt}$, matrix $\mathbf{U}^{(-2)}$, etc. When the condition number becomes comprehensible at the specified level of measurements error Δ_{max}, then matrix $\mathbf{U}^{(-k)}$ should be converted to determine parameters of the reduced circuit branches. Remaining parameters of the initial circuit are defined from k independent linear connections.

Results of application of this algorithm for identification of the circuit with a single special cut are shown in Fig. 3.27. On ordinate axis lays the error of the identification problem solution, calculated by the formula (3.37). Comparing dependences $\delta=\delta(\Theta)$, (shown in Fig. 3.20 and Fig. 3.28), obtained for identical circuits, one can arrive at the following important conclusions:

- use of RPM has allowed reducing the error of definition of matrix $\mathbf{Y}_0$ elements to the level of measurements error; and
- the identification error at use of RPM does not depend on the conditionality of the circuit's mathematical model.

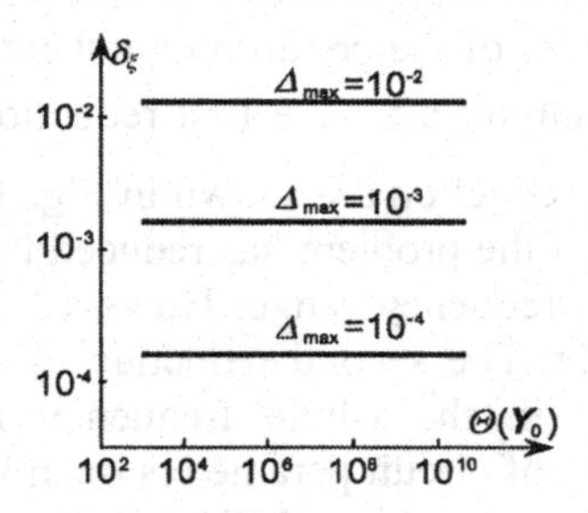

Fig. 3.28. Averaged mean square error of definition of matrix $\mathbf{Y}_0$ elements by use of RPM for a circuit with a single special cut

In this section we have considered the identification of resistive circuits by means of NIM. However these results are also valid for the LAM method and for identifications of AC circuits. Special cuts in this case may arise on frequencies close to resonance. An alternative to application of RPM in this case is realization of experiments at various frequencies. However, presence of a large number of inductance-capacitance connections in complex devices may be an impediment for choice of frequencies to carry out diagnostic experiments. Moreover, the properties of devices' mathematical models can vary considerably at change of frequency, especially for high-Q devices. Thus, arbitrary choice of frequencies for realization of experiments as a rule is impossible in practice.

Let's consider an example of identification for the AC circuit shown in Fig. 3.29a. This circuit has two nested special loops, l_1 and l_2, for a frequency close to 1 kHz. Therefore, the circuit should be reduced twice for identification. As has been shown in Section 3.4, identification of parameters for circuits with nested singularities is of particular complexity.

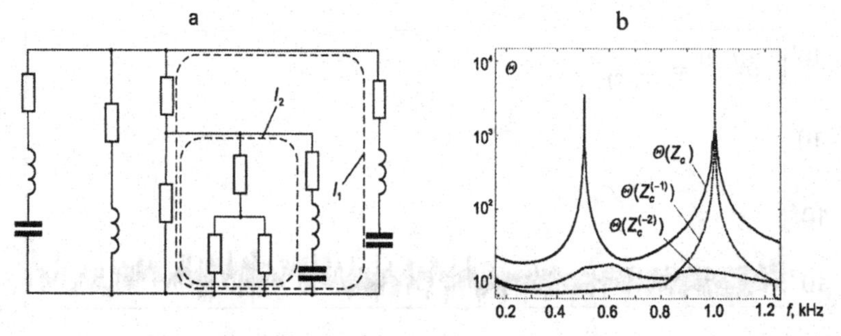

Fig. 3.29. An AC circuit **a)** with two special loops and **b)** - dependences of the condition numbers $\Theta(\mathbf{Z}_c)$ for the initial circuit, $\Theta\left(\mathbf{Z}_c^{(-1)}\right)$ for the circuit after the first reduction, and $\Theta\left(\mathbf{Z}_c^{(-2)}\right)$ for the circuit after the second reduction from the frequency

Frequency dependences of the condition numbers $\Theta(\mathbf{Z}_c)$ for the initial circuit, $\Theta\left(\mathbf{Z}_c^{(-1)}\right)$ for the circuit after the first reduction, and $\Theta\left(\mathbf{Z}_c^{(-2)}\right)$ for the circuit after the second reduction are shown in Fig. 3.29b. One can easily see that the first reduction of the problem has reduced its condition number practically over the whole frequency range. However, it is not enough for frequencies close to 1 kHz. The second reduction allows obtaining a comprehensible conditionality in the whole frequency range, including 1 kHz. Therefore, identification of circuit parameters on any frequency after the second reduction does not represent any difficulties.

According to our calculations, the error of construction of linear connections was of the same order of magnitude as the measurements error at each iteration step for all frequencies. In the above considered problem it was – $\Delta_{\max}$=1%, therefore definition of the initial circuit parameters by linear connections also did not involve any difficulties.

Comparison of errors of circuits' identification after the first and the second reductions, (shown in Fig. 3.30) is of interest.

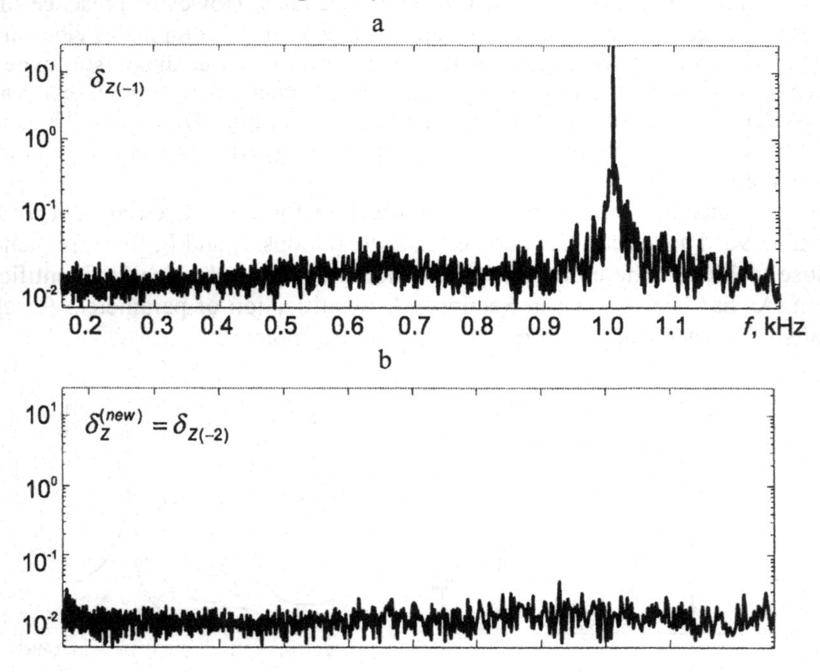

Fig. 3.30. Results of identification of an AC circuit with nested special loops by RPM. Identification errors after the first **a**) and the second **b**) reductions of the circuit

One can easily see that the circuit identification after the first reduction is successfully fulfilled for all frequencies, with the exception of frequencies close to 1 kHz. This fact well agrees with the frequency dependence of $\Theta\left(\mathbf{Z}_c^{(-1)}\right)$ shown in Fig. 3.29. The circuit identification error after the second reduction (Fig. 3.29b) does not depend on frequency and is close to the measurement error Δ_{max}=1%.

Thus, application of RPM essentially allows raising the accuracy of identification problems solution for DC and AC circuits at use of both NIM and LAM methods.

3.6 Inverse problems of localization of disturbance sources in electrical circuits by measurement of voltages in the circuit's nodes

In previous sections of this chapter we have solved several problems to determine parameters of passive elements in electric circuits. In this section, we shall consider the problem of searching for location of disturbance sources in electric circuits. At that we assume that only current sources, i.e. active elements, should be acting as disturbance sources. This problem can also be referred to as an inverse problem in electric circuits' theory, considered as a particular case of the identification problem. Its urgency for practical purposes is obvious.

Let's consider this problem in the following statement:

- we shall find the location of a disturbance source in a linear electric AC circuit with operating frequency ω=ω_0, assuming that parameters of its elements are known quantities;
- we assume that the electric circuit has some nodes that are accessible for measurements of their voltages to earth (test nodes);
- we have reliable information that disturbances can occur in the circuit nodes carrying numbers E_1, E_2, ... E_n (excited nodes);
- we assume that in neither of the test nodes, occurrence of a disturbance is possible;
- the disturbance source is similar to a source of non-stationary noise J_d connected between the circuit node N_x and the ground.

The problem involves definition of the node number N_x to which the disturbance J_d is connected, on the condition that voltages measured in the circuit's test nodes are known. This problem will be referred to as a problem of localization of a non-stationary disturbance source. It should be noted that assumption about non-stationarity of disturbances is important for the furthering of this concept, and well agrees with practice.

For the electric circuit in which we seek the location of disturbance source, the measuring system and the disturbance source are shown in Fig. 3.31. The measuring system is connected to the electric circuit through digital filters. These filters exclude the operating frequency ω_0 from the measurements. Therefore, the input signal on the measuring system does not contain harmonics with frequency ω_0. We assume that the measuring system measures effective values of voltages.

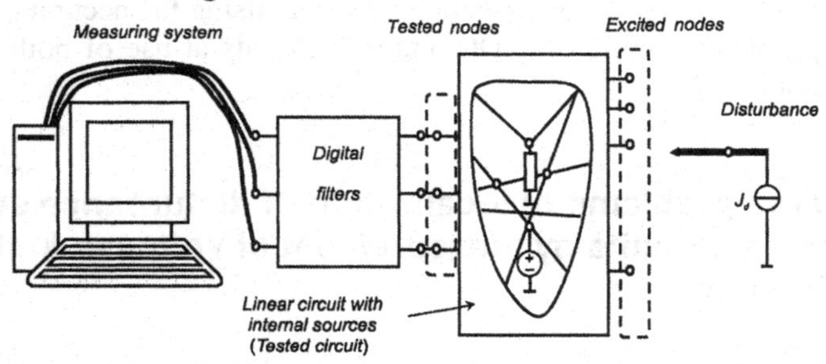

Fig. 3.31. The measuring system, filters, the disturbance source J_d ,and the tested electric circuit

Voltages measured by the measuring system will not change by representing the circuit shown in Fig. 3.31 to the circuit shown in Fig. 3.32. Thus all voltage sources of operating frequency will be short-circuited, whereas current sources of operating frequency will be open. In addition, harmonics with frequency ω_0 should be excluded from the disturbance spectrum. Inasmuch as the disturbance source is a source of non-stationary noise, removal of any harmonics from its spectrum will not essentially change its properties and will not have any impact upon the accuracy of its localization procedure.

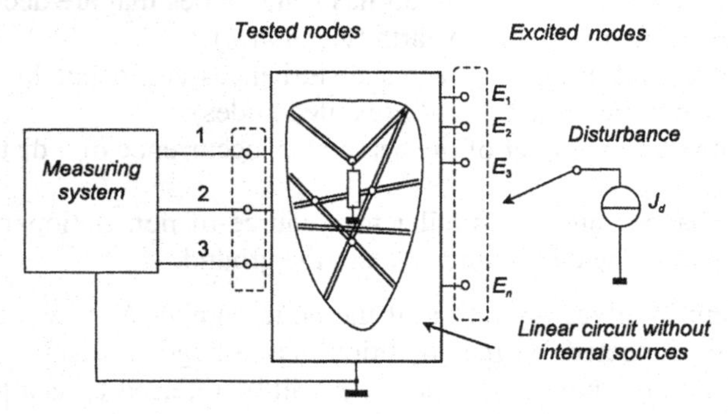

Fig. 3.32. The measuring system and the tested electric circuit

Thus, electric circuits shown in Fig. 3.31 and Fig. 3.32 are equivalent from the point of view of searching for the location of the disturbance source. Apparently, the circuit in Fig. 3.32 is simpler compared to the circuit in Fig. 3.31. Furthermore, we shall consider the problem of localizing the disturbance in the circuit shown in Fig. 3.32.

Solving the problem of localization of the source of disturbance, we begin from the particular case of sinusoidal disturbance localization when $j_d = J_{m,d} \sin \omega t, \quad \omega \neq \omega_0$. For its solution, voltages V_1 and V_2 measured in two test nodes (Fig. 3.33) are sufficient. We solve this problem in two stages. At the first (calculation) stage we shall connect a source of disturbance $\underline{J}_d$ to all exited nodes in turn, and then calculate factors $\alpha_p = V_1/V_2$, named identifying factors. Here, index "*p*" means that at calculation of the factor α_p, the disturbance source is connected to the node with number "*p*" as shown in Fig. 3.33.

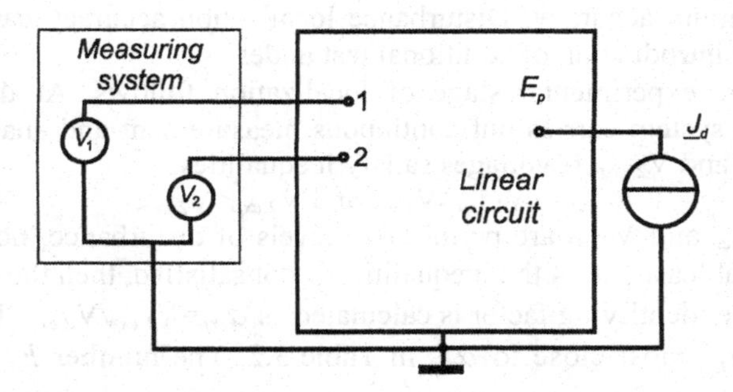

Fig. 3.33 Circuit diagram for the calculation stage of localization of sinusoidal disturbance $\underline{J}_d$

Voltages in test nodes V_1 and V_2 are proportional to J_d:

$$V_1 = k_1 J_d, \; V_2 = k_2 J_d, \quad \alpha = \frac{k_1 J_d}{k_2 J_d} = \frac{k_1}{k_2}.$$

Therefore for any "*p*", the identifying factor α does not depend on the magnitude J_d of the disturbance source and is defined only by the node to which the disturbance source is connected. This property of identifying factors will be used further for localization of the disturbance source. It should be noted that all factors α_p are real.

Results of the calculation stage of the localization problem are summarized in Table 3.2.

Table 3.2. Identifying factors α_{E_k} and numbers of corresponding excited nodes E_k, $k=\overline{1,n}$, determined during the calculation stage of localization procedure

Node number	E_1	E_2	...	E_n
Identifying factor	α_{E_1}	α_{E_2}	...	α_{E_n}

The way of choosing test nodes is of great importance, as it essentially affects localization accuracy. The best is such an arrangement of test nodes when the disturbance signal comes to test nodes from excited nodes through various circuit branches. Further, we shall elucidate particularities of choice of test nodes and the influence of their arrangement upon the accuracy of the localization process by means of appropriate examples. In practice, numbers of nodes in which disturbance occurrence is possible are often known. Therefore test nodes can be chosen from the condition of disturbance localization with maximum accuracy. Disturbance localization accuracy can also be increased by introduction of additional test nodes.

Then the experimental stage of localization follows. At this stage the measuring system carries out continuous measurement and analysis of voltages $V_{1,exp}$ and $V_{2,exp}$. If voltages satisfy inequalities:

$$V_{1,exp}<V_{1,per} \text{ and } V_{2,exp}<V_{2,per},$$

where $V_{1,per}$ and $V_{2,per}$ are permissible levels of disturbance, no actions will follow. If at least one of the inequalities is not satisfied, then the experimental value of the identifying factor is calculated as $\alpha_{exp}=V_{1,exp}/V_{2,exp}$. Then, we find a factor α_{E_k} most close to α_{exp} in Table 3.2. The number E_k in the upper cell of the table corresponds to the number of the node in which the disturbance occurs. Thus, the problem of localization of a sinusoidal source is solved.

In practice, the above-described algorithm of localization shows low selectivity because of the following circumstances:

- values of identifying factors obtained at the first, calculation stage can be equal (or rather close) for various excited nodes.

- the experimental value of an identifying factor α_{exp} can be equally close to its several calculated values owing to measurement errors.

The actual disturbance is not sinusoidal and often has a wide frequency spectrum. As noted above, a good model of disturbance is its representation as a source of non-stationary noise. Let's show that this circumstance allows a considerable increase in the accuracy of disturbance localization.

Assume further that there is an infinite number of harmonics in the disturbance spectrum. Let, for the sake of definiteness, them be harmonics with frequencies $2\omega_0$, $3\omega_0$, ... $M\omega_0$. We can find identifying factors $\alpha_p^{(k)}$ for each harmonic at the calculation stage of localization. Here, the subscript specifies the excited node to which the disturbance source is connected, and the superscript - the harmonic number in the disturbance spectrum. Since factors $\alpha_p^{(k)}$ do not depend on the amplitude of the disturbance source, then when calculating, the amplitude of the disturbance source can be assumed equal to 1 for all frequencies. Results of calculations are summarized in Table 3.3.

Table 3.3. Identifying factors $\alpha_{E_k}^{(m)}$, $m=\overline{2,M}$ and corresponding numbers of excited nodes E_k, $k=\overline{1,n}$ for harmonics with frequencies $2\omega_0$, $3\omega_0$, ... $M\omega_0$

	E_1	E_2	...	E_n
$2\omega_0$	$\alpha_{E_1}^{(2)}$	$\alpha_{E_2}^{(2)}$	...	$\alpha_{E_n}^{(2)}$
$3\omega_0$	$\alpha_{E_1}^{(3)}$	$\alpha_{E_2}^{(3)}$	...	$\alpha_{E_n}^{(3)}$
...	...	...	...	...
$M\omega_0$	$\alpha_{E_1}^{(M)}$	$\alpha_{E_2}^{(M)}$	...	$\alpha_{E_n}^{(M)}$

Data shown in Table 3.3 can be presented in the form of a two-dimensional graph shown in Fig. 3.34. Functions $\alpha_{E_k}(\omega)$, $k=\overline{1,n}$ in Fig. 3.34 can be found, for example, by spline-interpolation of the columns in Table 3.3. Thus, the calculation stage involves determination of a set of functions $\alpha_p(\omega)$, similar to those presented in Fig. 3.34. Functions $\alpha_p(\omega)$ can also be called identifying functions. It should be noted that the calculation stage is run only once.

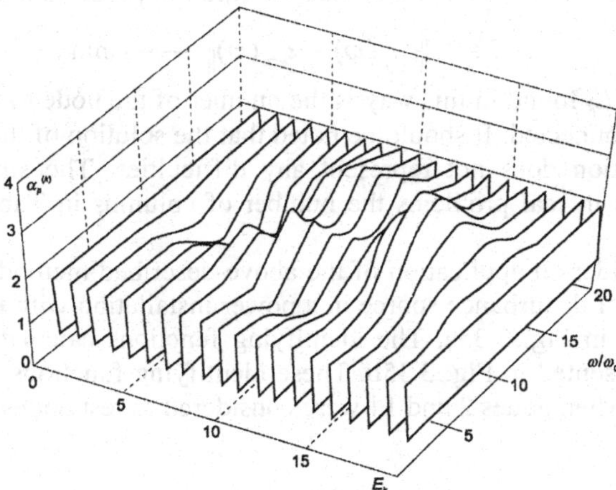

Fig. 3.34. Graphic representation of identifying functions $\alpha_p(\omega)$, corresponding to Table 3.3.

The experimental stage involves the following:
- the measuring system carries out continuous measurement and analysis of voltages $V_{1,exp}$ and $V_{2,exp}$. When the disturbance level exceeds a permissible level ($V_{1,exp}>V_{1,per}$ or $V_{2,exp}>V_{2,per}$), then $2M$ values of voltages $V_{1,exp}(t)$ and $V_{2,exp}(t)$ are recorded for instants of time $t_k = \dfrac{\pi}{\omega_0 M}k,\ k=\overline{1,2M}$;

- we carry out Fourier-series expansion of functions $V_{1,exp}(t)$ and $V_{2,exp}(t)$ (for example, by means of fast Fourier transform):

$$\upsilon_{1,exp}(k) = \sum_{k=1}^{k=2M} V_{1,exp}(t_k)e^{-j\frac{\pi}{M}k},\quad \upsilon_{2,exp}(k) = \sum_{k=1}^{k=2M} V_{2,exp}(t_k)e^{-j\frac{\pi}{M}k}; \tag{3.44}$$

- we calculate experimental identifying function for frequencies $k\omega_0,\ k=\overline{1,M}$:

$$\alpha_{exp}^{(k)} = \frac{\upsilon_{1,exp}(k)}{\upsilon_{2,exp}(k)},\quad k=\overline{1,M} \tag{3.45}$$

and then we construct the function $\alpha_{exp}(\omega)$ by means of, for example, spline-interpolation;
- we find a function closest to $\alpha_{exp}(\omega)$ in Table 3.3 (or among functions $\alpha_{E_k}(\omega),\ k=\overline{1,n}$ in Fig. 3.34) and thus we find the problem solution:

$$E_k:\quad \left\|\alpha_{E_k}(\omega)-\alpha_{exp}(\omega)\right\| \xrightarrow[E_k]{} \min .$$

The value E_k found in this way is the number of the node to which the disturbance is connected. It should be noted that the solution of the last problem of minimization does not represent any difficulties. The simplest way is searching, as in real problems the number of columns in Table 3.3 seldom exceeds 100.

Let's consider an application of the above-described method for searching the location of disturbance source in a power installation with the circuit diagram shown in Fig. 3.35a. The identifying functions calculated at the first stage are presented in Fig. 3.35b. These identifying functions were obtained for the case when nodes 2 and 14 were considered as test nodes.

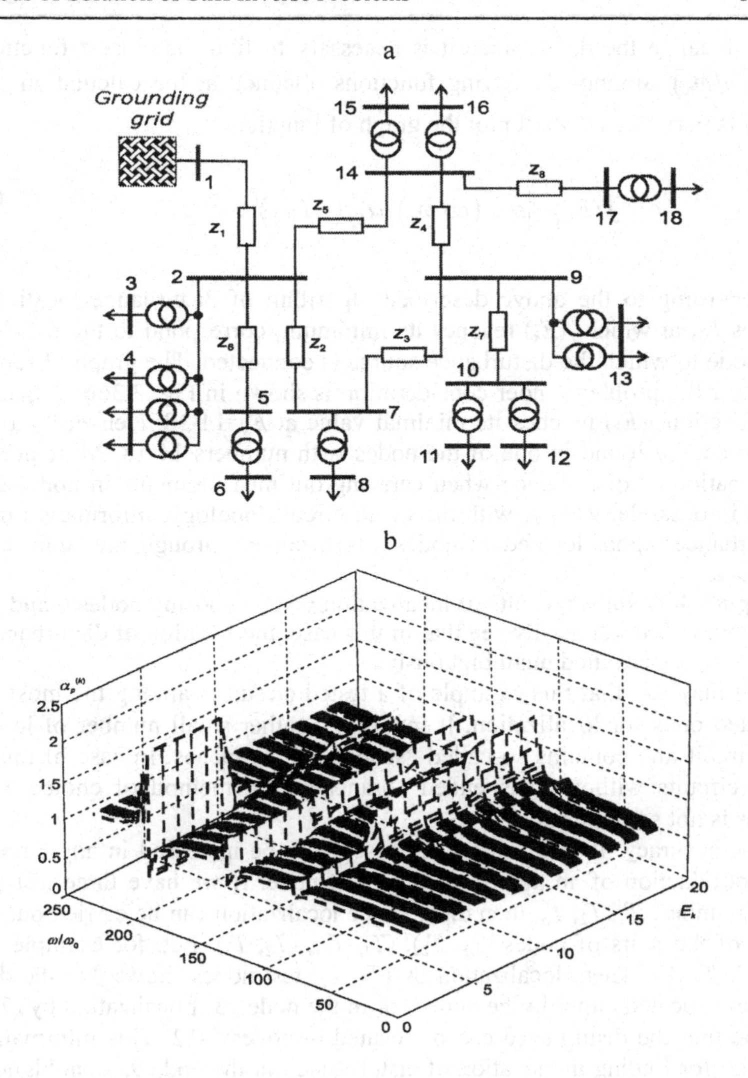

Fig. 3.35. The circuit diagram of a power installation a), and graphic representation of Table 3.3 for this circuit b)

As a result of the experimental stage, the function $\alpha_{exp}(\omega/\omega_0)$ shown in Fig. 3.36a is determined. The error of $\alpha_{exp}(\omega/\omega_0)$ definition has been roughly 10%.

To localize the disturbance it is necessary to find the closest function to $\alpha_{exp}(\omega/\omega_0)$ among identifying functions obtained at the calculation stage. For this purpose, we shall plot the graph of function

$$f(E_k)=\left\|\alpha_{exp}(\omega/\omega_0)-\alpha_{E_k}(\omega/\omega_0)\right\|. \tag{3.46}$$

According to the above described algorithm of disturbance localization, values E_k, at which $f(E_k)$ reaches its minimum, correspond to the number of the node to which the disturbance source is connected. The graph of function $f(E_k)$ for the problem under consideration is shown in Fig. 3.36b. Apparently the function $f(E_k)$ reaches its minimal value at E_k=14-18. Hence, the disturbance can be found in one of the nodes with numbers 14-18. More accurate localization of disturbance when carrying out measurements in nodes 2 and 14 is impossible. Really, with the given circuit topology, information on the disturbance signal located in nodes 14-18 comes through the same circuit branches.

Figure 3.37 shows results of localization when choosing nodes 6 and 14 as test nodes. One can easily see that in this case, the location of disturbance (at node 5) is determined unambiguously.

We may say that this example of a tested circuit is among the most complicated ones for localization. It involves a rather small number of loops in the circuit and contains so-called “suspended” branches. In case of multiple loop circuits without “suspended” branches, the method of choice of test nodes is not so important.

The accuracy of localization can essentially be increased in any circuit by the introduction of an additional third test node. If we have three test nodes with numbers T_1, T_2, T_3, then disturbance localization can be carried out using each of the pairs of nodes (T_1, T_2), (T_1, T_3), (T_2, T_3). Let, for example T_1=2, T_2=12, T_3=13. Then, localization by (T_1, T_2) test nodes shows that the disturbance is located either in the node 9 or in the node 13. Localization by (T_1, T_3) shows that the disturbance can be located in nodes 9-12. This information is enough for finding the location of disturbance - at the node 9, unambiguously. Thus use of localization data by (T_2, T_3) is not required.

There may be several sources of disturbances in a tested circuit. The above-described algorithm allows defining the number of these sources and localizing them. Let’s consider briefly the problem of localization of two disturbance sources arranged in various nodes. At first, we find from measurements of $V_{1,exp}$ and $V_{2,exp}$ that there are more than one disturbance sources in the circuit. With regard to the second source of disturbance, let’s assume that it is a source of non-stationary noise similar to the first one. We shall assume as well that characteristics of both sources are not correlated.

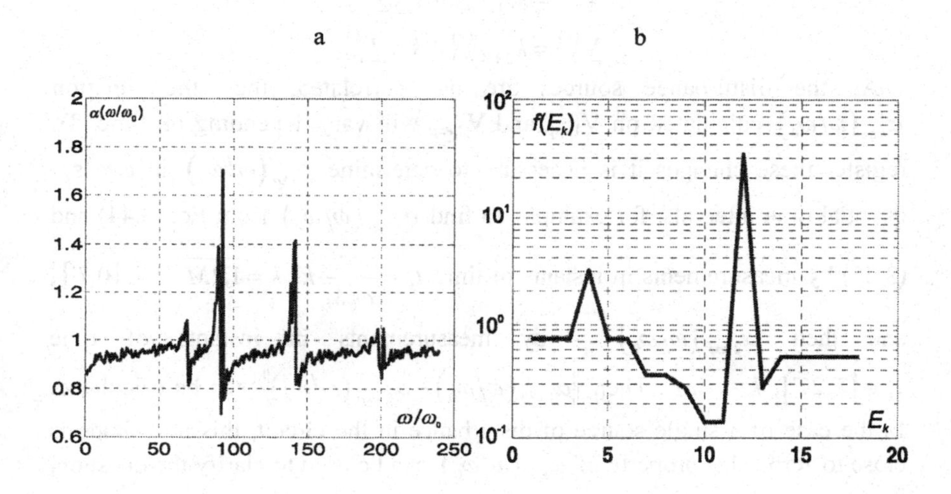

Fig. 3.36. Experimental dependence $\alpha_{exp}\left(\omega/\omega_0\right)$ **a)** and the residual function **b)**

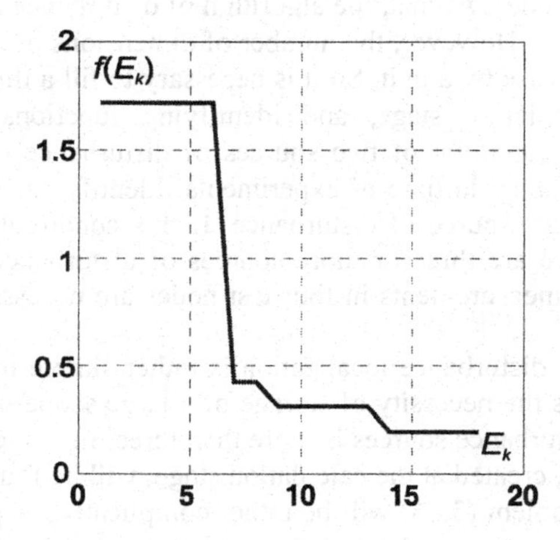

Fig.3.37. The residual function at choice of nodes 6 and 14 as test nodes

If two sources of disturbance $J_{d,1}$ and $J_{d,2}$ operate in a circuit, then we can write down for the k-th harmonic:

$$\underline{V}_1^{(k)} = k_{1,1}\underline{J}_{d,1}^{(k)} + k_{1,2}\underline{J}_{d,2}^{(k)},$$
$$\underline{V}_2^{(k)} = k_{2,1}\underline{J}_{d,1}^{(k)} + k_{2,2}\underline{J}_{d,2}^{(k)}.$$

As the disturbance sources are not correlated, then the function $\alpha_{exp}(\omega/\omega_0)$ calculated on $V_{1,exp}$ and $V_{2,exp}$ will vary depending on time. To register these changes it is necessary to determine $\alpha_{exp}(\omega/\omega_0)$ in two sequential time intervals; for example, to find $\alpha_{1,exp}(\omega/\omega_0)$ from Eq. (3.44) and (3.45) by measurements in instants of time $t_k = \dfrac{\pi}{\omega_0 M}k,\ k = \overline{1,2M},\ t_k \in [0,T]$, and then $\alpha_{2,exp}(\omega/\omega_0)$ - by measurements in instants of time $t_k \in [T,2T],\ k=\overline{1,2M}$. Then $\left\|\alpha_{1,exp}(\omega/\omega_0) - \alpha_{2,exp}(\omega/\omega_0)\right\|$ can be calculated. In the case of a single source of disturbance in the circuit, this magnitude is close to zero. This property of $\alpha_{exp}(\omega/\omega_0)$ can be used to clarify the question, if indeed there is only a single source of disturbance in the circuit.

Assume we have two sources of disturbance operating in the circuit. It can be shown that their identification requires measurements to be carried out in at least three test nodes. At that, the algorithm of disturbance source localization will not change. However, the number of dimensions of all matrixes and functions will increase by a unit. So it is necessary to fill a three-dimensional table at the calculation stage, and identifying functions will be two-dimensional. The condition of two sources of disturbance operating in the circuit is the constancy in time of experimental identifying function, similar to the case of single-source of disturbance. If this condition fails, one may conclude that there are three or more sources of disturbance in the circuit. This implies that measurements in four test nodes are necessary for their localization, etc.

This method of disturbance localization is rather simple in its realization. Its disadvantage is the necessity of storage of a large scope of information if the number of disturbance sources is more than three. In that case, the dimensionality of tables, created at the calculation stage, will be four and more, and the solution of problem (3.46) will be rather complicated. A possible as well as perspective alternative to the storage and handling of large tables can become a combination of this method with the method of neural networks. In this case, the training of the neural network will become the most long-run stage of calculations.

This method is also valid for DC circuits with essential simplifications. It can be applied for localization of disturbance sources in nonlinear circuits and, in particular, in circuits with semiconductor elements. This, however, requires fulfillment of the condition that the disturbance should represent a "small signal" in comparison with the operating parameters of the device.

References

1. Demirchian, K.S., and P.A. Butyrin (1988). *Modeling and machine computation of electric circuits* (In Russian). Moscow: High school Publ House.
2. Rakitski, Yu.V., S.N.Ustinov, and I.G.Chernorutski (1979). *Numerical methods for the solution of stiff systems* (In Russian). Noscow:Nauka.
3. Nelder, J.A., and R.Mead (1965). A simplex method for function minimization. *The Comp. Journal*, no7:308-313.
4. Wilkinson, J.H. (1965). *The Algebraic Eigenvalue Problem*. Oxford:Clarendon Press.
5. Parlett, B.N. (1980). *The Symmetric Eigenvalue Problem*. Englewood Cliffs, NJ:Prentice-Hall.
6. Faddeev, D.K., and V.N.Faddeeva (1963). *Computational methods of linear algebra*. San Francisco-London: H. Freeman and Co.
7. Kosteliansky, D.M. (1955). About influence of transformation of Gauss on spectra of matrixes (In Russian). *Successes of math sciences*,10, 1,117-12.

Chapter 4. Solving Inverse Electromagnetic Problems by the Lagrange Method

In this chapter we shall consider an effective method for solving inverse electromagnetic problems by applying Lagrange multipliers. In this chapter we shall also explore the properties and features of this method for practical use. In Section 4.1, we shall examine the application of Lagrange multipliers as continuous functions for electromagnetic optimization problems. When derived, the equations for field potentials and auxiliary adjoining functions can be used to show how to construct the boundary conditions for these functions and the algorithm for the numerical solution of optimization problems. Furthermore, in Section 4.2 we illustrate, through a number of examples, the procedure of finding field sources of the adjoining function, including the appropriate equations. The search for optimum distribution of a substance in a space can be carried out in various classes of media such as homogeneous, non-uniform, isotropic, nonlinear, etc. In Section 4.3, we shall also consider an algorithm for variations of the medium properties, allowing one to achieve local minima of the objective functional. Specific features of the method and its numerical realization will be considered by means of practical examples and application of benchmark problems in Section 4.4. Section 4.5 deals with the problems of computing values. In Section 4.6 some issues of the Lagrange method application for solution of optimization problems in non-stationary electromagnetic fields will be discussed.

4.1 Reduction of an optimization problem in a stationary field to boundary-value problems

The problem of seeking an optimum shape for bodies embedded in the electromagnetic field should be stated generally for a non-stationary three-dimensional electromagnetic field with consideration of the medium's nonlinear and anisotropic properties. Such problems are rather complex, therefore we shall first reduce optimization problems to boundary-value problems, taking into consideration the stationary three-dimensional and two-dimensional fields. We shall then take into account the media for which the solution is sought, for example, in a class of homogeneous or non-uniform media. By using simple examples, simplification of the statement will enable us to obtain the essence of the method and its properties.

One of the peculiarities of inverse electromagnetic problems is the fact that the number of parameters, i.e. the number of degrees of freedom determined by geometrical and physical characteristics of the media and sources is indefinitely large. In fact, for a numerical solution the degrees of freedom can be measured to be in the tens of thousands, leading to significant computing time. When we use gradient methods, they are associated with the necessity to repeat the many calculations of derivatives of the objective function with respect to the parameters sought [1]. As is shown below, application of the Lagrange method to the solution of these problems helps to alleviate this situation.

When using the Lagrange method, field equations included in the objective functional may be considered as constraints in composing the augmented functionals and multiplying them by the function $\lambda_i(x,y,z)$, which is dependent on the spatial coordinates. Thus, the augmented functionals in the general case will include, as unknowns, numbers λ (Lagrange multipliers that are discrete constraints) and functions $\lambda_i(x,y,z)$, which can be found using the procedure shown in the previous chapters.

We shall describe stationary fields with the help of potentials $\varphi(x,y,z)$ and $A(x,y,z)$. In the case of three-dimensional fields, the potential $A(x,y,z)$ is a vector function generally having several components in the given coordinate system, and the potential $\varphi(x,y,z)$ is a scalar function. In the case of two-dimensional fields, the potentials $A(x,y)$ and $\varphi(x,y)$ are scalar functions.

The shape and the structure of bodies to be sought are determined by the spatial distributions of functions $\mu(x,y,z)$, $\varepsilon(x,y,z)$ and $\sigma(x,y,z)$ which describes the medium properties. For simplicity we shall designate any of these functions - $\mu(x,y,z)$, $\varepsilon(x,y,z)$, and $\sigma(x,y,z)$ - as $\xi(x,y,z)$. It is obvious that in order for the spatial distribution of the $\xi(x,y,z)$ function to be found, both the shape and the structure of bodies must be known. The $\xi(x,y,z)$ function will be regarded as the optimization parameter forming a vector **p** in the numerical solution with a finite number of components, namely $p_1, p_2, \ldots p_n$. The forms of the objective functionals and constraints usually imposed on the field intensity and other characteristics, and on the unknown function $\xi(x,y,z)$, have been considered earlier.

It is expedient to look for an algorithm of optimum search using variational methods. This allows one to solve the problem analytically and to realize the numerical optimization algorithm only at the final stage. The peculiarity of problems found in seeking the shape and structure of bodies positioned in an electromagnetic field is that the quantity $\xi(x,y,z)$ may vary in a limited range $(\xi_{\min} \le \xi \le \xi_{\max})$. Therefore, they are regarded as non-classical problems of calculation of variations.

Let us consider the optimization problem in an electrostatic or stationary magnetic field with the following conditions. Assume that the field satisfies the equation $\operatorname{div}(\xi \operatorname{grad}\varphi) = -\rho$, where the function $\xi(x,y,z)$ describes the distribution of dielectric permeability in the case of an electrostatic field or the distribution of magnetic permeability in the case of a stationary magnetic field. The quantity $\rho = \rho(x,y,z)$ determines the density of given electric charges for an electrostatic field or the density of magnetic charges equivalent to the given electric currents [2].

In the domain V limited by the surface S_{bd}, the field sources of density ρ occupy the domain V_{sr} (Fig. 4.1). It is therefore necessary to find a function $\xi(x,y,z)$ in the domain V_{ξ} of admissible material position for which the objective functional, given in the domain V_{ob}, assumes its minimal value (or achieves the lowest boundary).

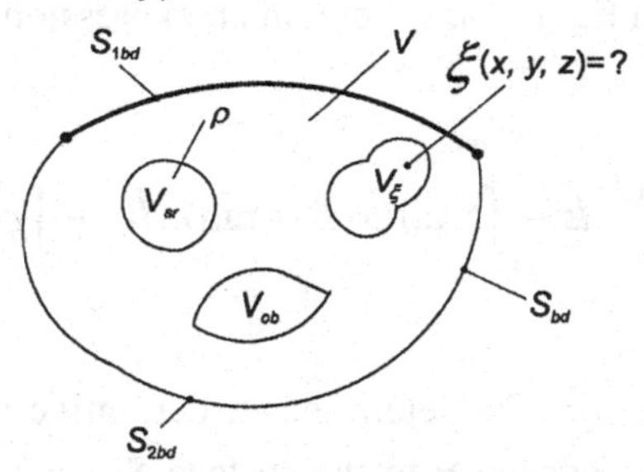

Fig. 4.1. Calculation domain in the problem of searching for optimal distribution of field sources and the medium in an electromagnetic field

The boundary condition of the first kind for the potential $\varphi(x,y,z) = h_1(x,y,z)$ is given at the boundary S_{1bd}, and the boundary condition of the second kind $\frac{\partial \varphi}{\partial n}(x,y,z) = h_2(x,y,z)$ is imposed at the S_{2bd} section.

In most problems, the objective functional I depends not only on the coordinates x,y,z and on the potential $\varphi(x,y,z)$, but also on the field intensity, i.e. on the potential derivatives $\frac{\partial \varphi}{\partial x} = \varphi'_x$, $\frac{\partial \varphi}{\partial y} = \varphi'_y$, $\frac{\partial \varphi}{\partial z} = \varphi'_z$:

$$I = \int_{V_{ob}} f(x,y,z,\varphi,\varphi'_x,\varphi'_y,\varphi'_z)dV . \tag{4.1}$$

For example, in solving some problems a magnetic pole tip of unknown shape is considered where it provides the least deviation from a constant of the normal to some surface component of field intensity. In this case, the function $\varphi_n' = \frac{\partial \varphi}{\partial n}(x,y,z)$ enters the functional I.

Regarding the equation $\mathrm{div}(\xi \mathrm{grad}\varphi) + \rho = 0$ as a constraint, the augmented functional according to the Lagrange method can be composed as:

$$L(\varphi,\xi,\lambda) = I + \int_V (\mathrm{div}\xi \mathrm{grad}\varphi + \rho)\lambda dV\,, \tag{4.2}$$

in which the quantity λ is a function of coordinates $\lambda(x,y,z)$, being distinct from the discrete Lagrange multipliers usually used [1,3].

Applying Green's theorem to the transformation of the integral $\int_V (\mathrm{div}\xi \mathrm{grad}\varphi)\lambda dV$ in Eq. (4.2), we obtain an expression [4]

$$L = I + \oint_{S_{bd}} \lambda\xi \frac{\partial \varphi}{\partial n} ds - \int_V \mathrm{grad}\varphi \cdot \xi \cdot \mathrm{grad}\lambda dV + \int_V \rho\lambda dV\,, \tag{4.3}$$

in which the quantity $\partial\varphi/\partial n$ determines a derivative of the potential φ directed along the outer normal $\boldsymbol{n}$ to the surface S_{bd} (Fig. 4.1). Application of Green's theorem allows us to write down the augmented functional L as follows:

$$L = I - \oint_{S_{bd}} \varphi\xi \frac{\partial \lambda}{\partial n} ds + \oint_{S_{bd}} \lambda\xi \frac{\partial \varphi}{\partial n} ds + \int_V \varphi(\mathrm{div}\xi \mathrm{grad}\lambda) dV + \int_V \rho\lambda dV. \tag{4.4}$$

We will seek the equation for the function $\lambda(x,y,z)$ as well as the medium parameter $\xi(x,y,z)$ based on the necessary condition of the extremum of the augmented functional L. First calculating the variation δL (see Appendix B) of the augmented functional with respect to the potential φ, and using the condition $\delta L = 0$, we get the equation for the function $\lambda(x,y,z)$. Calculating the variation δL of the augmented functional with respect to the medium characteristics ξ and using the condition $\delta L = 0$, we get the equation for the function $\xi(x,y,z)$.

In the calculation of the variation of both parts of Eq. (4.4) with respect to the potential φ inside the volume V, we obtain the expression

$$\int_V (\mathrm{div}\xi\,\mathrm{grad}\lambda)\ \delta\varphi dV = -\delta_\varphi I = -\int_{V_{ob}} \delta_\varphi f(x,y,z,\varphi,\varphi'_x,\varphi'_y,\varphi'_z)dV .$$

By virtue of randomness of a variation $\delta\varphi$, it yields that the function $\lambda(x,y,z)$ satisfies the equation $\mathrm{div}(\xi \mathrm{grad}\lambda) = 0$ for $\delta_\varphi I = 0$, and particularly, where $I = 0$, i.e. at the points beyond V_{ob}.

If the quantity $\delta_\varphi f(x,y,z,\varphi,\varphi'_x,\ \varphi'_y,\ \varphi'_z)$ is represented in the form $\delta_\varphi f(x,y,z,\varphi,\varphi'_x,\varphi'_y,\varphi'_z) = f_1\delta\varphi$, the equation satisfied by the function $\lambda(x,y,z)$ in the domain V_{ob} may be written as $\mathrm{div}(\xi \mathrm{grad}\lambda) = -f_1$.

The specific kind of the right-hand part f_1 of this equation, determining the field sources of function $\lambda(x,y,z)$ named adjoint to the potential $\varphi(x,y,z)$, depends on the objective functional I. In the following Section (4.2), we show how to search for the function f_1. There will also be examples of its calculation for various objective functionals I, as these are frequently used in practice.

Let us determine the boundary conditions on the surface S_{bd} for the variable λ. To this end, we equate to zero the variation of the augmented functional on the surface S_{bd}, with respect to the potential and to its normal surface derivative. We take into account that at the part S_{1bd} (see Fig. 4.1), the boundary condition of the first kind $\varphi(x,y,z)=h_1$ is set, and the boundary condition of the second kind $\frac{\partial\varphi}{\partial n}(x,y,z) = h_2$ is imposed at the part S_{2bd}.

If the objective function is set in the domain V_{ob}, having no common points with S_{bd}, then we have $\delta I = 0$ on the surface S_{bd}, and from the condition $\delta L = 0$ it follows that

$$\oint_{S_{bd}} \xi \frac{\partial\lambda}{\partial n}(\delta\varphi)ds = 0, \quad \oint_{S_{bd}} \xi\lambda\delta\left(\frac{\partial\varphi}{\partial n}\right)ds = 0 .$$

Since the potential on the part S_{1bd} of the surface is fixed, it should not be varied on that part of the surface. It may be varied only on that part of the boundary where its derivative normal to S_{bd} is set, i.e. on S_{2bd}. Therefore, by virtue of randomness of a variation $\delta\varphi$, we obtain the condition $\frac{\partial\lambda}{\partial n} = 0$ on S_{2bd} from the integral $\int_{S_{2bd}} \xi \frac{\partial\lambda}{\partial n}(\delta\varphi)ds = 0$ (Fig. 4.2). Consequently, as the normal derivative of the potential is given on the part S_{2bd}, it may be varied only on the part of the boundary where the potential is fixed, i.e. on the part S_{1bd} as is shown below. Therefore, on the part S_{1bd} of the domain V boundary we obtain the boundary condition $\lambda = 0$ for the adjoint variable from the integral $\int\limits_{S_{1bd}} \xi \lambda \delta \left(\frac{\partial \varphi}{\partial n} \right) ds = 0$.

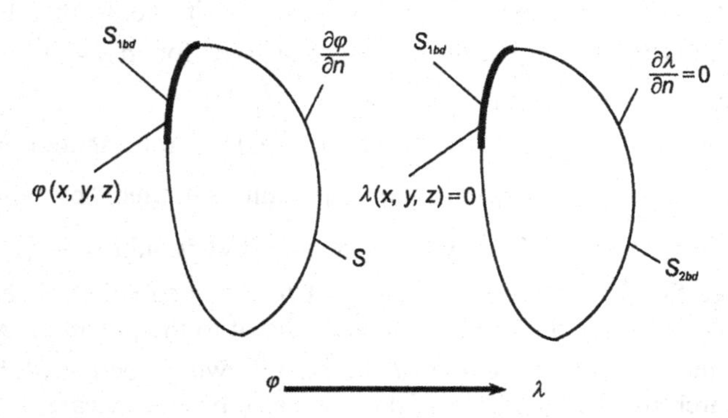

Fig. 4.2. Boundary conditions for the potential and the adjoint variable

Thus, the boundary condition of the first kind for potential φ on the part S_{1bd} is transformed on that part of the boundary into a homogeneous boundary condition $\lambda = 0$ for the adjoint variable.

The boundary condition of the second kind for the potential on the part S_{2bd} is transformed on this part of the boundary into a homogeneous boundary condition of the second kind $\frac{\partial \lambda}{\partial n} = 0$ for the adjoint variable.

When the objective function is defined on the domain boundary S_{bd} or on some part of it, the form of boundary conditions for the adjoint variable will vary on this part of the boundary, and it may become non-uniform. Actually, in this case the variation of the objective functional on the domain boundary, with respect to the potential or its normal derivative, may differ from zero.

Let, for example, the objective functional be specified on the surface S_{1bd} as $I|_{S_{1bd}} = \int\limits_{S_{1bd}} F\left(\frac{\partial \varphi}{\partial n} \right) ds$. We shall present its variation with respect to the potential normal derivative to the boundary S_{1bd} as $\delta I|_{S1bd} = \int\limits_{S_{1bd}} F_1 \delta \left(\frac{\partial \varphi}{\partial n} \right) ds$.

From the condition $\delta L=0$, follows that equation $\int\limits_{S_{1bd}} \xi\lambda\delta\left(\frac{\partial\varphi}{\partial n}\right)ds =$ $= -\int\limits_{S_{1bd}} F_1\delta\left(\frac{\partial\varphi}{\partial n}\right)ds$ should hold on the surface S_{1bd}. This gives the boundary condition $\lambda|_{S_{1bd}} = -F_1/\xi$.

Similarly, the boundary condition for variable λ, in the case where the objective functional is specified on the surface S_{2bd} as $I|_{S_{2bd}} = \int\limits_{S_{2bd}} F(\varphi)ds$, can be obtained. Its variation on the surface S_{2bd}, with respect to potential φ, can be written down as $\delta I|_{S_{2bd}} = \int\limits_{S_{2bd}} F_2\delta\varphi ds$. From the condition $\delta L = 0$ follows that equation $\int\limits_{S_{2bd}} \xi\frac{\partial\lambda}{\partial n}\delta\varphi ds = \int\limits_{S_{2bd}} F_2\delta\varphi ds$ should hold on the surface S_{2bd}. This gives the boundary condition $\left.\frac{\partial\lambda}{\partial n}\right|_{S_{2bd}} = F_2/\xi$.

The function $\lambda(x,y,z)$ is described by the equation $\mathrm{div}(\xi\mathrm{grad}\lambda) = -f_1$, which is similar to that for the potential φ: $\mathrm{div}(\xi\mathrm{grad}\lambda) = -\rho$. Distinction of these two problems (for variables φ and λ) consists due to the fact that the variables φ and λ have different sources and different (in the general case) boundary conditions. At the same time, the function $\xi(x,y,z)$ determining the substance distribution has the same form in both problems. Field sources of the variable λ, characterized by the function f_1, depend on the objective functional and are positioned either in the domain V_{ob} (including its boundary S_{ob}) or on the boundary S_{bd} of the domain V, if the objective function is given also on the boundary S_{bd}.

Analogy of the equations for the potential φ and for the function λ allows us to consider the calculated adjoint function λ as a potential of an auxiliary electric or a magnetic field "conjugate" to a real field with the potential φ.

Hence, with analysis and calculation it is possible to use such concepts accepted for static and stationary fields, such as a charge, a current, field intensity, a flux, energy, a field map, etc.

Note that a problem of the function λ calculation is adjoint to the direct problem of calculating the potential φ.

Let us determine the conditions for finding the medium characteristics $\xi(x,y,z)$ on the assumption that a minimum of the objective functional is provided. To do this we calculate a variation of both parts of the relationship in Eq. (4.3) with respect to ξ and use the condition $\delta_\xi L = 0$. In seeking the optimum in a class of non-uniform media, when the medium with the characteristic $\xi = \xi(x,y,z)$ may be present in the domain V_ξ, we consider the variation of Eq. (4.3) with respect to ξ:

$$\int_{V_\xi} \text{grad}\varphi \cdot \text{grad}\lambda (\delta\xi) dV = \delta_\xi I + \oint_{S_{bd}} \lambda \frac{\partial \varphi}{\partial n} (\delta\xi) ds .$$

If the domain V_ξ has no common points with the domain V_{ob} where the objective functional is given, and also has no common points with the boundary S_{bd}, the right-hand part of the last expression is equal to zero, and by virtue of randomness of a variation $\delta\xi$ we obtain the necessary condition of the stationarity of the functional L in the form $\text{grad}\varphi \cdot \text{grad}\lambda = 0$.

This condition means that in the solution of the optimization problem in a class of non-uniform media the vectors $\text{grad}\varphi$ and $\text{grad}\lambda$ should be mutually orthogonal at the points of the domain V_ξ. Generally, when the objective functional I is given in the domain V_{ob} intersecting with the domain V_ξ, the stationarity condition $\delta_\xi L = 0$ allowing one to calculate $\xi(x,y,z)$, is obtained as $\text{grad}\varphi \cdot \text{grad}\lambda = f_2$, where f_2 is a function whose form depends on the objective functional. The necessary condition found of an optimality linking the quantities $\text{grad}\varphi$ and $\text{grad}\lambda$ is essential for the construction of an effective algorithm of the problem solution.

Depending on the form of constraints superimposed on medium properties that are characterized by the quantity $\xi(x,y,z)$, problems of searching for an optimum structure and shape of body can be distinguished. If the function $\xi(x,y,z)$ in the body volume is specified, for example, as a constant value, then the problem of searching for the body shape should be solved, inasmuch as its structure will be known.

In some cases the quantity $\xi(x,y,z)$ is not specified beforehand. It can be a tensor quantity, and simultaneously discontinuous and nonlinear as well. Such problems require searching for the body structure. At that, the body topology, i.e. the medium distribution in the body volume is unknown beforehand, and is defined as a result of the solution. Such problems are referred to as problems of topological optimization [5,6].

Thus, optimization is possible in various classes of media: homogeneous, piecewise homogeneous, inhomogeneous, linear or nonlinear, isotropic or non-isotropic. Usually the class of medium in optimization problems is defined by the technical possibilities of realization of obtained solutions. In many cases, available technical possibilities are very limited and allow realizing only solid bodies made of homogeneous substances.

Then, the problem is reduced to searching for the body shape. However even at such constraints, a search for a solution in a wider class of media, for example, in the class of inhomogeneous non-isotropic media, is expedient at the first stage. Indeed, the objective function will have a smaller value in this class of media. Transition to homogeneous continuous medium, easily realizable in practice (that worsens the solution as the value of the objective function at such transition generally increases), can be fulfilled by various methods [7,8]. This problem will be illustrated in more detail in Section 4.4 by using several examples.

Thus, the search for the medium distribution, bringing a minimum to the objective functional, requires solving the boundary-value problems for the variables φ and λ. Let us construct an algorithm of search for the function $\xi(x,y,z)$.

At first we find the boundary conditions for the variable λ on the surface S_{bd} and determine the distribution of the density f_1 of field sources λ.

We will further set a particular initial distribution of substance $\xi^0(x,y,z)$ and solve the equation $\operatorname{div}\left(\xi^0 \operatorname{grad}\varphi\right) = -\rho$ for the potential φ, as given on the S_{bd} boundary conditions.

Next, we solve the equation $\operatorname{div}\left(\xi^0 \operatorname{grad}\lambda\right) = -f_1$ for the adjoint variable $\lambda(x,y,z)$ under the boundary conditions found from the first step for $\lambda(x,y,z)$ and its sources.

Further, we check the implementation of the condition $\operatorname{grad}\varphi \cdot \operatorname{grad}\lambda = f_2$. The solution is considered to be found if the accepted criterion is satisfied. The criterion will be satisfied if the condition $\operatorname{grad}\varphi \cdot \operatorname{grad}\lambda = f_2$ is carried out with the given error or if the rate of change of the objective functional does not exceed the required value.

If the criterion at the termination of the process is not met, we find the corrected distribution $\xi^1(x,y,z)$, repeat the solution of the equations for the functions φ, λ, and again check the implementation of this condition. Correction of the function $\xi(x,y,z)$ may be effected using various approaches. The effective algorithm of correction of substance distribution in the iterative process of the solution will be considered in Section 4.3.

Since functional L variation $\delta_\xi L$ represents the main linear part of its increment, then function $\operatorname{grad}\varphi \cdot \operatorname{grad}\lambda$ defines the approximate value of the functional "material" derivative $\partial L/\partial \xi$. Its calculation enables the application of known algorithms for the search of an augmented functional minimum.

Thus, the Lagrange method allows us to construct such an optimization algorithm that does not require the calculation of "material" derivatives of the objective function with respect to the optimization parameters. Applying the Lagrange method, it is sufficient to calculate gradients of potentials of the initial and auxiliary adjoint problem formulated with respect to Lagrange multipliers.

For search of optimum spatial distribution of field sources density $\rho(x,y,z)$, the variation of augmented functional (see Eq. (4.3)) with respect to desired density should be put to zero: $\delta_\rho L = 0$. If the domain of admissible position of sources lies outside the definitional domain of the functional we obtain the condition $\int\limits_{V_\rho} \lambda(x,y,z)(\delta\rho)dV = 0$. It means that for an optimum distribution of sources, the condition $\lambda(x,y,z) = 0$ should be valid. Just as in the case of search of medium distribution, calculation of the functional derivative allows applying known optimization algorithms for finding the source's density distribution $\rho(x,y,z)$.

In the following section, we consider the sort and character of the spatial distribution of sources of the adjoint variable λ in the solution of the elementary problems of search for the optimum shape of bodies in an electromagnetic field. We will derive the sources using the scalar and vector magnetic potentials.

4.2 Calculation of adjoint variable sources

As indicated in the previous section, finding the adjoint variable λ source's density, which determines the right part of the equation $\operatorname{div}(\xi \operatorname{grad}\lambda) = -f_1$, requires calculation of the objective functional variation with respect to the potential φ: $\delta I = \int\limits_V f_1 \delta\varphi dV$.

In this section, we shall find variations δI of objective functionals to be minimized when solving inverse problems in magnetic and electric fields, as well as adjoint variables' field sources. The definition of functional variation and examples of its calculation can be found in Appendix *B*.

Such quantities as inductance, mutual inductance or emf are expressed via magnetic flux. Therefore, finding their extremum is associated with the search for extremum for magnetic flux [9].

In a plane-parallel field the magnetic flux Φ through a segment connecting points a and b relates to the vector magnetic potential by the expression $\Phi = A_a - A_b$. Assume, as *a priori* known fact, that during a search for conditions of magnetic flux extremum the sign on the difference between A_a–A_b does not change. Then the functional $I = \Phi$ can be accepted as the objective functional. Its variation in the point a with respect to the potential is $\delta I = \delta A_a$. Besides, $\delta I = -\delta A_b$ is in the point b and $\delta I = 0$ in all other points of the domain. To write down the functional as $I = \int_V f dV$, we introduce a two-dimensional delta function $\eta(x,y)$ which is equal to zero everywhere except for points a and b: $I = \int_V (\eta_a - \eta_b) A dV$. Thus, the variation of the objective functional can be given by $\delta I = \int_V (\eta_a - \eta_b) \delta A dV$. Now we can write down an equation with regard to the variable λ adjoining to the vector potential.

Let the vector magnetic potential satisfy the equation $\operatorname{div}(\nu \operatorname{grad} A) = 0$ (where $\nu = 1/\mu$). Assume $\varphi = A$, $\xi = \nu$, $\rho = 0$ in Eq. (4.4). Then, from the relationship $\int_V (\operatorname{div} \nu \operatorname{grad} \lambda) \delta A dV = -\int_V (\eta_a - \eta_b) \delta A dV$ (see Eq. (4.4)), we find the following equations: $\operatorname{div}(\nu \operatorname{grad} \lambda) = -\eta_a$ in the point a, $\operatorname{div}(\nu \operatorname{grad} \lambda) = \eta_b$ in the point b and $\operatorname{div}(\nu \operatorname{grad} \lambda) = 0$ in all other points of the domain.

The field of variable λ can be considered as a magnetic field adjoining to A. Its sources are linear electric currents $i_a = 1$ and $i_b = -1$. They are located at the ends a and b of the segment through which the magnetic flux flows. The density of current in points a and b is infinite (Fig. 4.3).

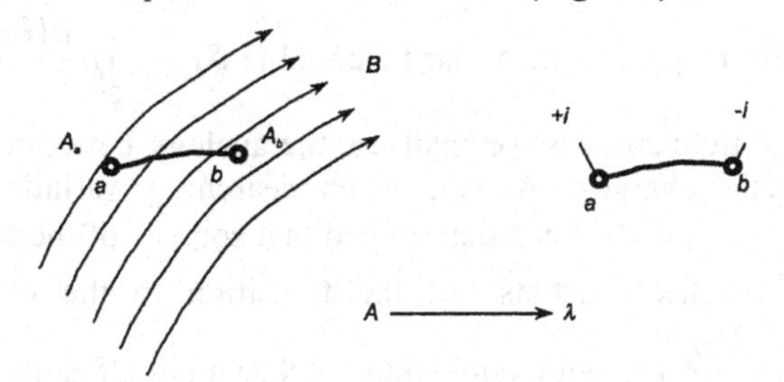

Fig. 4.3. Magnetic flux through the segment *ab* (left) and sources of the adjoint variable λ in points *a,b* (right)

If the quantity $A_a - A_b$ can change its sign during search of the magnetic flux extremum, then $I = 0.5(A_a - A_b)^2$ can be accepted as the objective functional. By analogy with the previous example, it can be written down as

$I = 0.5\left(\int\limits_V \eta_a A dV - \int\limits_V \eta_b A dV\right)^2$. Here, as above, η_a and η_b are delta functions differing from zero in points a and b, accordingly. Variation of this functional with respect to the potential A is $\delta I =$ $= (A_a - A_b)\left(\int\limits_V \eta_a \delta A dV - \int\limits_V \eta_b \delta A dV\right)$. The equation for λ becomes $\mathrm{div}(\nu \mathrm{grad}\lambda) = -\eta(A_a - A_b)$ in the point a, $\mathrm{div}(\nu \mathrm{grad}\lambda) = \eta(A_a - A_b)$ in the point b and $\mathrm{div}(\nu \mathrm{grad}\lambda) = 0$ in all other points of the domain.

Apparently, sources of λ field are linear electric currents numerically equal to $A_a - A_b$ in the point a and $-(A_a - A_b)$ in the point b, correspondingly.

The scalar magnetic potential can be applied for the analysis of both two-dimensional and three-dimensional magnetic fields. Therefore, it is of interest to find expressions for the variation of functionals by use of the scalar magnetic potential.

When seeking the conditions of magnetic flux extremum and using scalar magnetic potential for a plane-parallel field calculation in a direct problem, the objective function can be written down as $I = \Phi = -\int\limits_a^b \mu \frac{\partial \varphi_m}{\partial n} dl$. As stated above, it can be used if the function $\partial \varphi_m / \partial n$ does not change its sign during the search of magnetic flux extremum. Then the variation of the objective function with respect to the scalar potential is $\delta I = -\int\limits_a^b \mu \frac{\partial(\delta \varphi_m)}{\partial n} dl$.

To find λ field sources we shall use the analogy between electric currents and magnetic charges. Above, when searching variation of functional $I = \Phi = A_a - A_b$, it has been determined that sources of the adjoint variable λ are linear electric currents. At the transition to the objective function $I = \Phi = -\int\limits_a^b \mu \frac{\partial \varphi_m}{\partial n} dl$ with equivalent replacement of sources by magnetic charges in a direct problem, it is necessary to replace sources λ by magnetic charges equivalent to them. As is well-known, linear electric currents $+i$ and $-i$ located in points a, b are equivalent to a double layer of magnetic charges with a constant moment. This layer of charges rests upon points a, b. Therefore adjoint variable λ field sources, which in this case can be considered as a scalar magnetic potential, will form a double layer of magnetic charges with a moment μi. At its crossing the quantity λ has a discontinuous jump equal to $i=1$. Thus, the integrand expression $\mu \frac{\partial(\delta\varphi_m)}{\partial n}$ in the variation of functional $\delta I = \delta\Phi = -\int_a^b \mu \frac{\partial\delta(\varphi_m)}{\partial n} dl$ determines a double layer of magnetic charges with a constant moment.

The functional $I = 0.5\left(\int_a^b \mu \frac{\partial\varphi_m}{\partial n} dl\right)^2$, similar to the above-considered functional $I = 0.5\left(A_a - A_b\right)^2$, defines the square of the magnetic flux crossing a segment *ab* in a plane-parallel field. Using the equivalence of electric currents and magnetic charges we can find sources of the field λ when the direct problem is solved by means of the scalar magnetic potential. Double layers of magnetic charges with constant moment numerically equal $\mu\left(A_a - A_b\right)$ are equivalent to electric currents found above, numerically equal $\left(A_a - A_b\right)$ in the point *a* and $-\left(A_a - A_b\right)$ in the point *b*. Therefore in this case, adjoint variable λ field sources are magnetic charges forming a double layer. At crossing it, the scalar magnetic potential has a discontinuous jump $\int_a^b \mu \frac{\partial\varphi_m}{\partial n} dl$.

In the general case of three-dimensional magnetic field, the magnetic flux through a surface bounded by a contour l is related to the vector magnetic potential by expression $\Phi = \oint_l \boldsymbol{A}\, d\boldsymbol{l}$. When searching for extremum for the functional $I = \Phi$ and solving the direct problem with the help of vector magnetic potential, the source of field λ can easily be shown to be the linear electric current flowing along the contour l. Starting at the solution of the direct problem by means of the scalar magnetic potential we shall present the objective functional as $I = \Phi = -\int_S \mu \frac{\partial\varphi_m}{\partial n} ds$. In this case magnetic charges forming a double layer become sources of field λ. This layer has a constant moment and rests on the contour l.

Thus, in problems of search of extremum conditions for magnetic flux expressed by vector magnetic potential, sources of adjoint variable λ are linear electric currents. If the magnetic flux is expressed by a scalar magnetic potential, then double layers of magnetic charges equivalent to electric currents become the sources of adjoint variable λ.

The problem of finding conditions of extremum for magnetomotive force $F_{ab} = \int_a^b \boldsymbol{H}d\boldsymbol{l} = \varphi_{ma} - \varphi_{mb}$ between points *a*, *b* has a solution similar to one obtained above for the search for conditions of extremum for magnetic flux expressed by vector magnetic potential. Variation of the objective functional $I = F$ with respect to potential φ_m is $\delta I = \delta\varphi_{ma}$ in the point *a* and $\delta I = \delta\varphi_{mb}$ in the point *b*. In case of three-dimensional field the sources of field λ are magnetic charges $m_a = 1$ and $m_b = -1$ located in points *a*, *b*. In the case of a plane-parallel magnetic field they are distributed along lines. In both cases their cubic density is infinite.

Obviously, similar solution can be derived when searching for conditions of extremum for electric voltage between points $\varphi_{eab} = \varphi_{ea} - \varphi_{eb}$, equal to the difference of electric potentials. In the case of a three-dimensional electric field, when the objective functional is specified as $I = \varphi_{ab}$, field λ sources are point electric charges $q = 1$ and $q = -1$. They are located in points *a*, *b*, correspondingly.

Let's find the variation of the objective functional for the problem of finding the pole 1 shape of a ferromagnetic core that provides an extreme value for the electromagnetic force component F_k in the direction of the ***k*** axis, acting upon conductor 2 with current density J_z (Fig. 4.4).

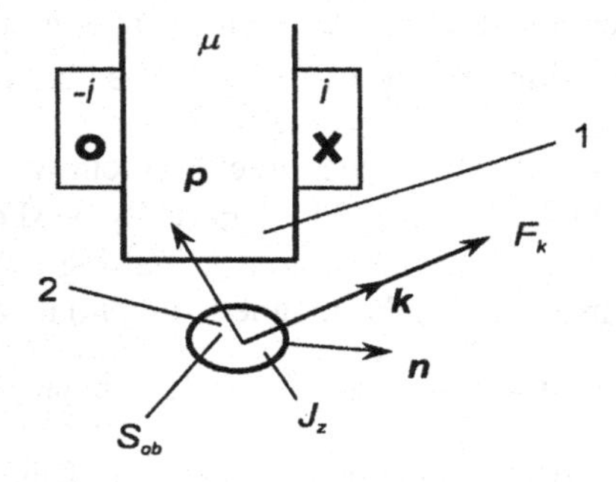

Fig. 4.4. Searching for the pole 1 shape to find the extremum of the electromagnetic force

The objective function

$$I = F_k = -\int_{S_{ob}} J_z B_p ds = \int_{S_{ob}} J_z \left(\frac{\partial A}{\partial k}\right) ds$$

defines the component of force in the plane-parallel magnetic field, acting upon a conductor with a current in the direction of the *k* axis (here B_p is the projection of magnetic induction vector on the axis *p*). It can be written down as $I = F_k = \oint_l J_z A\cos(\boldsymbol{n},\boldsymbol{k})dl$, where $\boldsymbol{n}$ is the unit vector normal to a point on conductor cross-section contour l.

Variation of the objective function $\delta I = \int_l J_z \cos(\boldsymbol{n},\boldsymbol{k})(\delta A)dl$ shows that sources of the adjoint variable λ form a simple layer of current on the contour of the conductor cross section. The linear density of this current is numerically equal to $J_z \cos(\boldsymbol{n},\boldsymbol{k})$.

Above-discussed elementary objective functions will find application when searching for extremum conditions for flux, voltage, and integral quantities associated with them, such as electric capacitance, inductance or mutual inductance. There are other problems that require the value of a certain quantity F to be as close as possible to a specified one. In other words, the deviation of F from the specified F_w should be minimal.

In particular, when the required value is zero $(F_w = 0)$, F will take on its extremal value.

Let's find the variation of the functional $I = 0.5 \int_{V_{ob}} (F - F_w)^2 dV$ and the density of sources of adjoint variable λ for selected quantities $F(x,y,z)$ used in practice.

Taking into account that $\delta F_w = 0$, we have

$$\delta I = \int_{V_{ob}} (F - F_w)\delta F dV .$$

Let F be the scalar potential of a field with required distribution of potential specified by the function $F_w = \varphi_w(x,y,z)$ in domain V_{ob}. The objective function becomes $I = 0.5 \int_{V_{ob}} (\varphi - \varphi_w)^2 dV$. Then,

$$\delta I = \int_{V_{ob}} (\varphi - \varphi_w)(\delta\varphi)dV$$

and the equation for λ is $\text{div}(\xi \text{grad}\lambda) = -(\varphi - \varphi_w)$ within the domain V_{ob}, and $\text{div}(\xi \text{grad}\lambda) = 0$ outside of it.

Apparently, the cubic density of field sources λ numerically equal to $\varphi - \varphi_w$ depends on the direct problem potential φ. Changes of potential φ during the search of optimum lead to change of adjoint problem source's density. Therefore, direct and adjoint problems appear to be interconnected.

In practice, it requires solving the problem of creating a specified distribution of a certain component of magnetic field intensity on a surface or along a line. Frequently, the problem concerns the component of magnetic field intensity or magnetic induction that is normal to this surface or line. Let $F = B_n$ and $F_w = B_{wn}$. The minimum of functional $I = 0.5\int_S (B_n - B_{wn})^2 ds$ provides the least deviation of B_n - the normal component of magnetic induction to the surface S, from desired B_{wn}. Its variation with respect to the scalar magnetic potential is

$$\delta I = -\int_S \mu(B_n - B_{wn})\frac{\partial(\delta\varphi_m)}{\partial n} ds .$$

As follows from stated above, the integrand expression here defines a double layer of magnetic charges on the surface S. Its moment numerically equals $\mu(B_n - B_{nw})$, and the jump of potential λ on the surface S generally are functions of coordinates.

To obtain a required distribution of the magnetic induction component B_n normal to line ab in two-dimensional statement, both vector and scalar magnetic potentials can be used. For simplicity, we set the segment ab along the axis x. In the objective function $I = 0.5\int_a^b (B_n - B_{nw})^2 dl$ we assume $dl_{ob} = dx$, $B_n = B_y$. Expressing magnetic induction through the vector magnetic potential A and calculating the functional variation with respect to A, we have

$$\delta I = \int_a^b \left(\frac{\partial A}{\partial x} + B_{yw}\right)\delta\left(\frac{\partial A}{\partial x}\right)dx = \int_a^b \left(\frac{\partial A}{\partial x} + B_{yw}\right)\frac{\partial}{\partial x}(\delta A)dx .$$

Integrating by parts, we find

$$\delta I = \left(\frac{\partial A}{\partial x} + B_{yw}\right)\delta A\bigg|_{x=b} - \left(\frac{\partial A}{\partial x} + B_{yw}\right)\delta A\bigg|_{x=a} - \int_a^b \frac{\partial}{\partial x}\left(\frac{\partial A}{\partial x} + B_{yw}\right)\delta A dx .$$

Adjoint variable field sources form a simple layer of current distributed on line ab with density $j = -\frac{\partial}{\partial x}\left(\frac{\partial A}{\partial x} + B_{yw}\right)$. Linear currents $i_a = -\partial A/\partial x - B_{yw}$, $i_b = \partial A/\partial x + B_{yw}$ are located in points a and b. As one can easily verify, the sum of adjoint variable field sources equals zero.

When using the scalar magnetic potential to solve this problem, variation of the objective function $I = 0.5\int_a^b (B_y - B_{yw})^2 dx = 0.5\int_a^b \left(-\mu\frac{\partial\varphi_m}{\partial y} - B_{yw}\right)^2 dx$ with respect to the potential φ_m results in the following expression:

$$\delta I = \int_a^b \left(\mu\frac{\partial\varphi_m}{\partial y} + B_{yw}\right)\delta\left(\mu\frac{\partial\varphi_m}{\partial y}\right)dx .$$

It shows that adjoint variable field sources form a double layer of magnetic charges with a moment numerically equal to $\mu \dfrac{\partial \varphi_m}{\partial y} + B_{yw}$. The same result can be derived by integrating the current distributed along the line *ab* obtained above at solution of the direct problem by means of the vector magnetic potential.

Evidently, adjoint variable sources can be both magnetic charges and electric currents. The type of sources in an adjoint problem is determined by the kind of potential in the direct problem. Since they should be equivalent respective of the created magnetic field, then it is possible to calculate equivalent magnetic charges by the currents found and vice versa (see Appendix *A*).

The above-considered problem of obtaining a specified distribution of magnetic induction normal to a line in two-dimensional statement can be put differently by use of the expression $B_y = -\partial A/\partial x$. The objective function $I = 0.5\int_a^b (A(x) - A_w(x))^2\, dx$ becomes its minimum at $A(x) = A_w(x)$. At that, we have $B_y = B_{yw}$.

Variation of this objective function with respect to potential A

$$\delta I = \int_a^b [A(x) - A_w(x)](\delta A)\, dx$$

shows that in this case the adjoint variable field source is an electric current distributed along the line *ab* with linear density $j(x) = A(x) - A_w(x)$. Apparently, sources in this case have more simple form in comparison with the case of the above-considered objective function $I = 0.5\int_a^b (B_y - B_{yw})^2\, dx$. However, the required accuracy of calculation in this case should be higher, as the closeness of the potential A to the specified one A_w does not mean that functions $\dfrac{\partial A}{\partial x}, \left.\dfrac{\partial A}{\partial x}\right|_w$ will be as close to each other as potentials A and A_w.

Let's find the objective function variation and adjoint variable field sources in the problem of obtaining a magnetic field of specified intensity $\boldsymbol{H} = \boldsymbol{H}_w$ within the domain V_{ob}. Assume that $\boldsymbol{F} = \boldsymbol{H} = \text{-grad}\varphi_m$, $\boldsymbol{F}_w = \boldsymbol{H}_w$; then variation of the objective function $I = 0.5\int_{V_{ob}} (\text{-grad}\varphi_m - \boldsymbol{H}_w)^2 dV$ can be written down as:

$$\delta I = \int_{V_{ob}} (\text{-grad}\varphi_m - \boldsymbol{H}_w)\delta(\text{-grad}\varphi_m) dV = \int_{V_{ob}} (\text{grad}\varphi_m + \boldsymbol{H}_w)(\text{grad}\delta\varphi_m) dV\,.$$

Transforming the integral with the use of Green's theorem, we find

$$\delta I = \int\limits_{S_{ob}} \left(\frac{\partial \varphi_m}{\partial n} + \boldsymbol{H}_{wn} \right) (\delta \varphi_m) ds - \int\limits_{V_{ob}} (\text{divgrad} \varphi_m + \text{div} \boldsymbol{H}_w)(\delta \varphi_m) dV .$$

Here, n defines the external normal to the domain V_{ob} bounding surface.

The variable λ field sources within the volume V_{ob} are characterized by a cubic density numerically equal to $-\text{div}(\text{grad}\varphi_m) - \text{div}\boldsymbol{H}_w$, and on the surface S_{ob} - by surface density $\partial \varphi_m / \partial n + H_{wn}$.

Sources located on the surface S_{ob} form a simple layer of magnetic charges, causing a discontinuous jump of the normal derivative $\partial \lambda / \partial n$.

A known problem of this type involves finding of the optimum shape for a ferromagnetic screen providing minimal field within the volume V_{ob}. At its solution, the source's density and equation for λ are found from the above expressions, assuming that $\boldsymbol{H}_w = 0$.

Another similar problem involves producing a homogeneous magnetic field within the volume V_{ob}.

Let's find the adjoint variable sources by search of conditions at which the functional $I = 0.5 \int\limits_{S_{ob}} \left(\boldsymbol{H} - j H_{wy} \right)^2 ds$, specified inside the rectangular area S_{ob} is its minimal value.

Let assume that the x,y axis of rectangular coordinates are parallel to the sides of region S_{ob}. Assume that the direct problem is solved on the basis of scalar magnetic potential φ_m in plane-parallel statement $(J = J_z, B_z = 0)$. The source's desired density can be found using the general expression derived above for variation of the functional $I = 0.5 \int\limits_{V_{ob}} (\text{-grad}\varphi_m - \boldsymbol{H}_w)^2 dV$:

$$\delta I = \int\limits_{l_{ob}} \left(\frac{\partial \varphi_m}{\partial n} + H_{wn} \right) (\delta \varphi_m) dl - \int\limits_{S_{ob}} (\text{divgrad} \varphi_m + \text{div}\, \boldsymbol{H}_w)(\delta \varphi_m) ds .$$

Surface sources form simple layers of magnetic charges on the sides of the area S with densities $\sigma_1 = \frac{\partial \varphi_m}{\partial x}$, $\sigma_2 = \frac{\partial \varphi_m}{\partial y} + H_{wy}$, $\sigma_3 = -\frac{\partial \varphi_m}{\partial x}$, and $\sigma_4 = -\frac{\partial \varphi_m}{\partial y} - H_{wy}$. Charges with cubic density $\rho = -\frac{\partial^2 \varphi_m}{\partial x^2} - \frac{\partial^2 \varphi_m}{\partial y^2} - \frac{\partial H_{wy}}{\partial y}$ are distributed inside the whole area S_{ob}.

When solving the direct problem by means of the vector magnetic potential, the adjoint variable field sources are electric currents distributed on sides as well as inside the whole area S_{ob}.

Solving of some problems requires the use of objective function $I = 0.5 \int_{V_{ob}} (B - B_w)^2 dV$, containing the magnetic induction absolute value B. Minimization of this function allows finding conditions at which the module B of magnetic induction is close to a specified value B_w within the domain V_{ob}. Need of consideration of such functionals arises in problems of moving particle beams focusing by means of quadrupole devices. At assumption of a plane-parallel magnetic field, the domain V_{ob} can be a circle S_{ob} with a given radius R. Under condition $B_w = kr$ of linearity of the required field induction along the radial coordinate r, the objective function becomes

$$I = 0.5 \int_{S_{ob}} (B - kr)^2 ds .$$

Here k is a constant. The coordinate r is counted from the circle center and the magnetic permeability inside the circle is assumed to be constant.

If the magnetic induction is expressed through the vector magnetic potential, then variation of this functional with respect to vector potential A can be written down as $\delta I = \int_{S_{ob}} \frac{B - kr}{B} \text{grad}A \cdot \text{grad}(\delta A) ds$. It can be transformed so as to have variation δA under the integral sign as a multiplier. Taking into account that $dS_{ob} = rdrd\alpha$, where α is the angular coordinate, we find

$$\delta I = \oint_l \frac{B(R,\alpha) - kR}{B(R,\alpha)} \cdot \frac{\partial A}{\partial r} (\delta A) R d\alpha - \int_{S_{ob}} \text{div}\left(\frac{B - kr}{B} \text{grad}A \right) (\delta A) ds.$$

The integrand expression of the second integral can be presented as $\text{div}\left(\frac{B - kr}{B} \text{grad}A \right) \delta A = \left(\frac{B - kr}{B} \text{divgrad}A + \text{grad}A \cdot \text{grad} \frac{B - kr}{B} \right) \delta A$. Under the condition $\text{divgrad}A = 0$ inside the circle we finally have:

$$\delta I = \oint_l \frac{B(R,\alpha) - kR}{B(R,\alpha)} \cdot \frac{\partial A}{\partial r} (\delta A) ds - \int_S \text{grad}A \cdot \text{grad} \frac{B - kr}{B} (\delta A) ds .$$

This expression shows that the adjoint variable field sources are distributed inside the circle with a cubic density $J = -\text{grad}A \cdot \text{grad} \frac{B - kr}{B}$, and on a circle of radius R with surface density $j = \frac{B(R,\alpha) - kR}{B(R,\alpha)} \cdot \frac{\partial A}{\partial r}$.

It is easy to find adjoint variable field sources in the above-considered problem for the case when for direct problem solution the scalar magnetic potential is applied instead of the vector magnetic potential. It can be done in two ways: by expressing the magnetic induction included in the objective functional through scalar potential, and executing transformations similar to shown above. Also it is possible to take advantage of equivalence of electric currents and magnetic charges and to transform the above found currents into equivalent magnetic charges.

In conclusion, we shall calculate adjoint variable field sources for the problem with objective function $I = 0.5\oint_{S_{ob}} (B - B_w)^2 ds$. Here, S_{ob} is a closed surface in a three-dimensional domain and B is the magnetic induction module in points belonging to the surface S_{ob}. We shall find the variation of the functional with respect to the potential provided that the direct problem is solved by means of the scalar magnetic potential φ_m: $\delta I = \oint_{S_{ob}} (B - B_w)(\delta B) ds$.

The module of magnetic induction can be written down as $B = \sqrt{(\partial\varphi_m/\partial\tau_1)^2 + (\partial\varphi_m/\partial\tau_2)^2 + (\partial\varphi_m/\partial n)^2}$, where τ_1, τ_2, n define tangent τ_1, τ_2 and normal n orts of an orthogonal coordinate system on the surface S_{ob}. For simplification, Lame's constants are omitted in this expression. Taking into account that variation of the magnetic induction module with respect to the potential φ_m is $\delta B = \frac{1}{B}\left(\frac{\partial\varphi_m}{\partial\tau_1}\cdot\frac{\partial\delta\varphi_m}{\partial\tau_1} + \frac{\partial\varphi_m}{\partial\tau_2}\cdot\frac{\partial\delta\varphi_m}{\partial\tau_2} + \frac{\partial\varphi_m}{\partial n}\cdot\frac{\partial\delta\varphi_m}{\partial n}\right)$, variation δI can be written down as

$$\delta I = \int_{\tau_1}\int_{\tau_2} \tilde{B}\cdot\frac{\partial\varphi_m}{\partial\tau_1}\cdot\frac{\partial\delta\varphi_m}{\partial\tau_1} d\tau_1 d\tau_2 + \int_{\tau_1}\int_{\tau_2} \tilde{B}\cdot\frac{\partial\varphi_m}{\partial\tau_2}\cdot\frac{\partial\delta\varphi_m}{\partial\tau_2} d\tau_1 d\tau_2 + \oint_{S_{ob}} \tilde{B}\cdot\frac{\partial\varphi_m}{\partial n}\cdot\frac{\partial\delta\varphi_m}{\partial n} ds.$$

Here, designation $\tilde{B} = (B - B_w)/B$ has been used.

After transformation, the expression for δI becomes

$$\delta I = -\oint_{S_{ob}} \left[\frac{\partial}{\partial\tau_1}\left(\tilde{B}\frac{\partial\varphi_m}{\partial\tau_1}\right) + \frac{\partial}{\partial\tau_2}\left(\tilde{B}\frac{\partial\varphi_m}{\partial\tau_2}\right)\right](\delta\varphi_m) ds + \oint_{S_{ob}} \tilde{B}\frac{\partial\varphi_m}{\partial n}\cdot\frac{\partial\delta\varphi_m}{\partial n} ds.$$

Thus, we have a simple layer of magnetic charges on the surface S_{ob} with density $\sigma_m = -\text{div}_\tau\left(\frac{B - B_w}{B}\cdot\text{grad}_\tau\varphi_m\right)$. Here the index τ means that operations div and grad should be calculated on the surface S_{ob}. The second integral in the expression of δI shows that besides the simple layer, there is also a double layer of magnetic charges on the surface S_{ob} with a moment that is numerically equal to $\frac{B - B_w}{B}\cdot\frac{\partial\varphi_m}{\partial n}$. Therefore not only does the adjoint variable λ have a discontinuity on the surface S_{ob}, but so does $\partial\lambda/\partial n$ - its normal derivative to the surface.

Similar to the above-considered problem, it is possible to search in other problems for adjoint variable λ sources, having forms and spatial distributions determined by given objective functions. It should be noted that the objective function depends on the potential chosen for field calculation. Therefore, both the form and character of spatial distribution of adjoint variable λ field sources also depend on the type of potential accepted at the direct problem solution. However, the approach for calculating variation of the objective function can be the same as the one used in this section.

As appears from above, when performing optimization in a magnetic field, transition from vector magnetic potential to scalar at solution of the direct problem implies equivalent replacement of sources in the adjoint problem: electric currents should be replaced by equivalent magnetic charges.

The procedure and results of search of optimum media distribution based on application of the Lagrange method depend on properties of used materials. Optimization in the class of materials with homogeneous electromagnetic properties allows finding only the shapes of bodies.

However, at use of materials with inhomogeneous electromagnetic properties, not only shapes but also the structure of bodies can be determined as well. Aspects of optimization procedure development for various classes of materials will be considered in the following section.

4.3 Optimization of the shape and structure of bodies in various classes of media

When solving problems to find optimum shapes and structures of bodies embedded in the electromagnetic field, the constraints on properties of used materials are of great importance. In some cases, only application of substances with homogeneous structure and constants of electromagnetic characteristics independent of spatial coordinates are allowed. In other problems, use of not only non-uniform, but also anisotropic materials is possible. Solutions sought after for various classes of media can differ considerably.

The Lagrange method allows finding optimum shapes and structure of bodies in various medium classes: homogeneous, non-uniform, isotropic and anisotropic, linear and nonlinear.

The class of media used for seeking an optimum solution should be defined at the problem statement. When defining the type of medium the possibility of practical realization of the device, acquired as a result of solution on the basis of available materials, is of great importance. Most often homogeneous and isotropic materials are used. Therefore, if a solution is found in the class of homogeneous isotropic media, its practical realization most probably will not pose any difficulties.

Practical realization of solutions found in the class of non-uniform or anisotropic media is much more difficult. Moreover, in many cases they are practically unfeasible. Then the distribution of such medium resulting from calculations should be replaced, for example, by the distribution of a homogeneous medium.

It however does not mean that solutions acquired in the class of non-uniform or anisotropic media do not represent any practical interest. They are interesting because, as a rule, they result in smaller values of an objective function in comparison with solutions in the class of homogeneous isotropic media. Having the optimization problem solution in the class of non-uniform isotropic or anisotropic media enables the estimation of errors arising at its practical realization. It is also possible to determine whether it is expediently to apply, for example, anisotropic instead of isotropic materials for reduction of the objective function to its limit value.

Let's consider features of the Lagrange method application when seeking a substance's optimal distribution in order to find solutions in various classes of media: homogeneous isotropic, non-uniform isotropic and non-uniform anisotropic. At realization of the method and comparison of optimization results in various classes of media, it is possible to estimate advantages of each statement of the problem, to compare solutions by their laboriousness and their degree of approximation to the best results.

In the beginning let's consider the elementary problem of seeking the optimum shape for a body in the class of homogeneous magnetizable media. In this case the body's structure is specified, as the substance magnetic permeability positioned in the area V_μ, has a constant value. So the procedure of search of the body shape can then be considered as finding of function $\mu(x,y,z)$, which can have only two values: μ_{min} and μ_{max}. In the domain of admissible material position it can take values μ_{min} or μ_{max}. Outside of the domain V_μ, it is equal to μ_{min} [9,10,11].

As noted in Section 4.1, the search algorithm of a body shape and structure, i.e. of finding the function $\xi(x,y,z)$, can be constructed by use of equation $\delta_\xi L = 0$, i.e. equality to zero of the augmented functional variation with respect to medium characteristic ξ. This equation has enabled us to find the following conditions.

If the domain V_μ of admissible, material position has no common points with the area boundary S_{bd}, then the following conditions are valid:

$$\text{grad}\varphi_m \cdot \text{grad}\lambda = 0 \text{ and } \text{grad}\varphi_m \cdot \text{grad}\lambda = f_2(I).$$

The first condition $(\text{grad}\varphi_m \cdot \text{grad}\lambda = 0)$ is the necessary condition of the augmented functional extremum if the objective functional I is given outside the domain V_μ of admissible material position. In this case, variation I with respect to the medium characteristic μ $\delta_\mu I = \int\limits_{V_{ob}} \delta_\mu f(x,y,z,\varphi_m,\varphi'_{mx},\varphi'_{my},\varphi'_{mz})dV$ is equal to zero.

In the general case, when the objective functional domain of definition intersects with the domain V_μ, we have $\delta_\mu I \neq 0$. After calculating the variation of the integrand function f of the objective functional, it can be written down as $\delta_\mu f = f_2 \delta\mu$, thus giving the second condition $\text{grad}\varphi_m \cdot \text{grad}\lambda = f_2(I)$.

The peculiarity of search algorithm for body shape discussed in Section 4.1 at its given structure and under the above-mentioned constraints on values of magnetic permeability, involves the way of changing the shape of body surface. As stated in Section 4.1, in the first step we solve a direct problem of the calculation of the potential φ_m at a given, generally arbitrary, distribution of a medium in domain V_μ, i.e. at some initial shape of a solid body. This solution allows finding field intensity everywhere in the whole space, including $\text{grad}\varphi_m(x,y,z)$ on the surface. Since during search for the body shape only the shape of its surface varies, then $\text{grad}\varphi_m(x,y,z)$ can be calculated only on the body surface. Furthermore, we calculate field sources of the function $\lambda(x,y,z)$ and its boundary conditions.

For the second step, by solving the adjoint problem we find Lagrange multipliers $\lambda(x,y,z)$ and then $\text{grad}\lambda(x,y,z)$ on the body surface.

Further, we check fulfillment of the condition $f_2\text{-grad}\varphi_m \cdot \text{grad}\lambda = 0$ at body surface points. If it holds at a point, then this point remains fixed during this step.

If $f_2\text{-grad}\varphi_m \cdot \text{grad}\lambda \neq 0$ at a point on the surface, then it should be moved along the normal to the body surface. When moving this point into an area where $\mu = \mu_{\min}$, i.e. outside the body, we, in addition, position substance near the surface. Accordingly, when moving a point inside the body we remove some volume of substance with magnetic permeability $\mu_{\max}$ in the vicinity of this point, replacing it by medium with magnetic permeability $\mu = \mu_{\min}$.

This algorithm of solution in the class of homogeneous media does not determine the sizes of deformation of the body surface portions on each step of the process. They should be set based upon the experience of calculations. Usually at initial steps when the surface shape is still far from optimum, rather large movements of surface points are accepted. As approaching the optimum shape, they should be reduced.

Interestingly, if searching for a numerical solution in the class of piecewise homogeneous isotropic media, it can result (as shows experience of calculations) to a layered medium. It can be characterized as structural anisotropic.

In that case the functions gradφ_m and gradλ should be calculated in the whole volume of the body, not just on its surface.

When seeking an optimum solution in the class of non-uniform isotropic media, the distribution of substance in the admissible domain V_μ is characterized by an unknown function $\mu_{\min} < \mu(x,y,z) < \mu_{\max}$ beforehand. Thus, it is possible to not only find the body shape, but also its structure.

Just as in the case of seeking the distribution of a piecewise homogeneous medium, it is necessary to calculate functions gradφ_m and gradλ not only on the body surface, but also in all points of domain V_μ at each step of the solution. By the value $\delta_\mu L$ found at a point of domain V_μ, the required character of change of magnetic permeability in that point can be determined. Simultaneously, the increment $\Delta\mu$, proportional to the difference f_2-grad$\varphi_m \cdot$gradλ, can be found.

Difficulties of finding solutions in the class of non-uniform isotropic media are caused by the impossibility of calculating exact values of $\Delta\mu$ in points on each step of the process. They should be determined empirically, proceeding from experience that complicates finding the optimum.

Let's consider the search algorithm for the shape and structure of bodies in a magnetic field in the class of anisotropic media [11]. Its efficiency has been verified by optimization in plane-parallel magnetic fields [9,12].

We shall consider the medium as a layered composite formed by substances with absolute magnetic permeabilities $\mu_0 = \mu_{\min}$ and $\mu = \mu_{\max}$. Properties of the anisotropic non-uniform medium in a volume element are defined by concentrations m_1 and m_2 of these substances, and also by direction of anisotropy axes [11].

The medium can be characterized by the magnetic permeability matrix:

$$\hat{\boldsymbol{\mu}} = \begin{bmatrix} \mu_{xx1}, & \mu_{xy} \\ \mu_{yx}, & \mu_{yy} \end{bmatrix}.$$

The eigenvalues μ_1, μ_2 of this matrix are determined from the equation:

$$\det \begin{bmatrix} \mu_{xx} - \mu_1, & \mu_{xy} \\ \mu_{yx}, & \mu_{yy} - \mu_2 \end{bmatrix} = 0.$$

The angle α between the ort $\boldsymbol{i}_1$ of anisotropy axis and ort $\boldsymbol{i}$ of coordinate x (Fig. 4.5) is bound to matrix elements μ by the following relationships:

$$\mu_{xx} = 0.5\left[\mu_1 + \mu_2 + (\mu_1 - \mu_2)\cos 2\alpha\right],\ \mu_{yy} = 0.5\left[\mu_1 + \mu_2 - (\mu_1 - \mu_2)\cos 2\alpha\right],$$

$$\mu_{xy} = \mu_{yx} = 0.5\left[(\mu_1 - \mu_2)\sin 2\alpha\right].$$

We shall characterize each point in the anisotropic medium by two parameters, namely by the angle α determining the main anisotropy axes direction, and by eigenvalue μ_2 of the magnetic permeability matrix $\hat{\boldsymbol{\mu}}$.

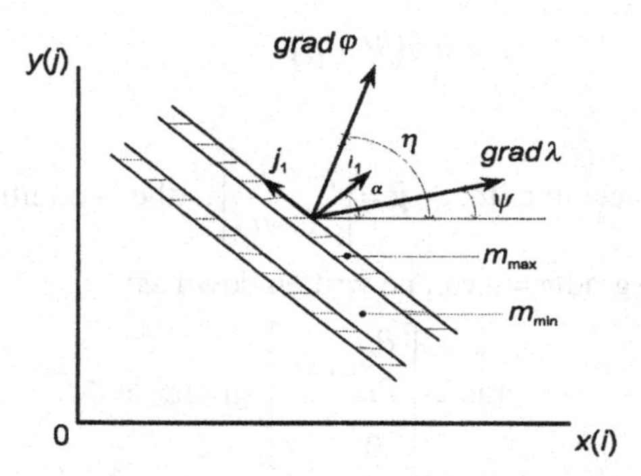

Fig. 4.5. Orientation of substance layers and directions of gradients of main and adjoint variables

Other variants of choice of two parameters determining the medium properties are also possible, for example, μ_1 and m_1 or μ_1 and m_2.

The search algorithm of medium distribution can be found from the stationary condition of the augmented functional L with respect to parameters α and μ_2: $\delta_\alpha L = 0$, $\delta_{\mu_2} L = 0$. These conditions are equivalent (see (4.3)) to conditions $\mathrm{grad}\lambda \cdot (\partial\hat{\boldsymbol{\mu}}/\partial\alpha) \cdot \mathrm{grad}\varphi = 0$, $\mathrm{grad}\lambda \cdot (\partial\hat{\boldsymbol{\mu}}/\partial\mu_2) \cdot \mathrm{grad}\varphi = 0$, if the objective function I is specified outside the domain V_μ of admissible position of the body.

Substitution of the relationship $\dfrac{\partial\hat{\boldsymbol{\mu}}}{\partial\alpha} = (\mu_1 - \mu_2) \cdot \begin{bmatrix} -\sin 2\alpha, & \cos 2\alpha \\ \cos 2\alpha, & \sin 2\alpha \end{bmatrix}$ into the optimality condition $\mathrm{grad}\lambda \cdot (\partial\hat{\boldsymbol{\mu}}/\partial\alpha) \cdot \mathrm{grad}\varphi = 0$ yields the equation:

$$|\mathrm{grad}\lambda| \cdot |\mathrm{grad}\varphi_m| [\cos\eta, \sin\eta] \cdot \begin{bmatrix} -\sin 2\alpha, & \cos 2\alpha \\ \cos 2\alpha, & \sin 2\alpha \end{bmatrix} \cdot \begin{bmatrix} \cos\Psi \\ \sin\Psi \end{bmatrix} \cdot (\mu_1 - \mu_2) = 0.$$

Here, η is the angle between the ort $\boldsymbol{i}$ of coordinate x (Fig. 4.5) and the vector $\mathrm{grad}\varphi_m$.

After matrix multiplication, we find the condition

$$|\mathrm{grad}\lambda| \cdot |\mathrm{grad}\varphi_m| \cdot \sin(\Psi + \eta) \cdot (\mu_1 - \mu_2) = 0.$$

Here, Ψ is the angle between the ort $\boldsymbol{i}$ of coordinate x (Fig. 4.5) and the vector $\mathrm{grad}\lambda$. It follows from this condition that the equality $\Psi = -\eta$ (at $\mu_1 \neq \mu_2$) should hold at optimum.

Thus, at medium optimum distribution the anisotropy axis $\boldsymbol{i}_1$ should be the bisector of the angle between vectors $\mathrm{grad}\varphi_m$ and $\mathrm{grad}\lambda$. Then, angle α should be calculated by the formula:

$$\alpha = 0.5(\Psi + \eta). \tag{4.5}$$

Taking into account that $\hat{\boldsymbol{\mu}} = \begin{bmatrix} \mu_1, & 0 \\ 0, & \mu_2 \end{bmatrix}$, the condition of optimum $\mathrm{grad}\lambda \cdot (\partial\hat{\boldsymbol{\mu}}/\partial\mu_2) \cdot \mathrm{grad}\varphi = 0$ can be written down as:

$$\mathrm{grad}\lambda \cdot \begin{bmatrix} \dfrac{\partial\mu_1}{\partial\mu_2}, & 0 \\ 0, & 1 \end{bmatrix} \cdot \mathrm{grad}\varphi_m = 0.$$

To calculate the derivative $\partial\mu_1/\partial\mu_2$, we shall at first express μ_2 by means of μ_1. For this, the following expressions are used:

$$\mu_2 = m_1\mu_{\min} + m_2\mu_{\max},\ \mu_1 = \mu_{\min} \cdot \mu_{\max} / (m_1\mu_{\max} + m_2\mu_{\min}).$$

These expressions result from the condition of equality of magnetic resistances of layered medium in directions of main axes i_1 and i_2 to their corresponding equivalent values. Thus, we find the concentrations:

$$m_1 = (\mu_2 - \mu_{\max})/(\mu_{\min} - \mu_{\max}),\ m_2 = (\mu_2 - \mu_{\min})/(\mu_{\max} - \mu_{\min}). \tag{4.6}$$

After simple transformations, this gives:

$$\mu_2 = \mu_{\min} + \mu_{\max} - \mu_{\min} \cdot \mu_{\max} \cdot \mu_1^{-1}, \tag{4.7}$$

from which we have $\partial\mu_1/\partial\mu_2 = \mu_1^2 (\mu_{\min} \cdot \mu_{\max})^{-1}$.

Now, the condition $\mathrm{grad}\lambda \cdot (\partial\hat{\boldsymbol{\mu}}/\partial\mu_2) \cdot \mathrm{grad}\varphi = 0$ allows writing down the relationship:

$$|\mathrm{grad}\varphi_m| \cdot |\mathrm{grad}\lambda| \cdot [\cos\eta, \sin\eta] \cdot \begin{bmatrix} \mu_1^2 (\mu_{\min} \cdot \mu_{\max})^{-1}, & 0 \\ 0 & , 1 \end{bmatrix} \cdot \begin{bmatrix} \cos\Psi \\ \sin\Psi \end{bmatrix} = 0.$$

Hence, taking into account that $\Psi = -\eta$, we have $\dfrac{\mu_1^2 \cdot \cos^2\Psi}{\mu_{\min} \cdot \mu_{\max}} - \sin^2\Psi = 0$

and

$$\mu_1^2 = \mu_{\min} \cdot \mu_{\max} \cdot \mathrm{tg}^2\Psi \,. \tag{4.8}$$

We can find the value of μ_2 with the help of (4.7).

Thus, the condition of optimality $\delta_\mu L = 0$ allows calculating material concentrations m_1, m_2 in a layered (composite) medium from Eq. (4.6) by use of ψ, μ_1 and μ_2 values preliminarily found from Eqs.(4.8) and (4.7). The expression (4.8) can be used for the calculation of μ_1 value for such angles ψ that bring the calculated value μ_1 into the limits $\mu_{\min} \le \mu \le \mu_{\max}$, i.e. if $\mu_{\min} \cdot \mu_{\max}^{-1} \le \mathrm{tg}^2\Psi \le \mu_{\max} \cdot \mu_{\min}^{-1}$.

In case of violation of latter inequalities μ_1, μ_2 become their limit values, namely $\mu_{\min}$ or $\mu_{\max}$.

Above-mentioned expressions for calculation of anisotropic axis $\boldsymbol{i}_1$'s direction (see Eq. (4.5)) and substance's concentration m_1, m_2 (see Eq. (4.6)) allow uniquely correcting the medium parameters within volume V_μ on each step of the search process. At that, as distinct from search algorithms for optimum in the class of isotropic media, it does not require the use of empirical relationships. This is the essential difference of this algorithm from the above-mentioned ones. At the same time the laboriousness of realization of all considered algorithms is approximately identical.

The realization to convergence of the solution of any of these algorithms can be controlled in various ways. The most natural and reliable method involves calculating the objective function during each iteration or after several iterations. Use of other criteria for finding a solution can lead to premature termination of calculations or to their excessive prolongation. So, for example, if changes of shape of the optimized surface are accepted as criterion, then fluctuations of the surface shape may be significant even at values of the objective function which are close to the minimum. This is a subject of not only errors of domain digitization, but probably also of the objective function low sensitivity to changes of optimization parameters. This circumstance, being rather essential for design purposes, should be taken into account when performing calculations.

When solving the optimization problem in the class of non-uniform anisotropic media distribution, of medium characteristics in the form of complicated functions, of coordinates are frequently obtained. As a rule, practical realization of such distributions by means of available materials is inconvenient. At the same time, as it will be shown in Section 4.4, in many cases application of anisotropic materials results in a minimal objective function.

One of the methods for the creation of non-uniform continuous distribution of a substance (for example, a ferromagnetic one) is the application of powdered materials. Change of powder local density leads to changes of the elementary volume averages of magnetic permeability. The subsequent process of powder baking allows fixing the shape and structure of the body.

Another way involves using layered materials. With their help, realization of structurally anisotropic substances is possible. However, means of making of thin anisotropic non-uniform structures, as against the powder method, are rather limited.

Transition from theoretically best non-uniform anisotropic medium to its practical realization can be made in various ways. The choice of method depends on available materials, their properties, and cost. When using a homogeneous isotropic material, the found distribution of $\mu(x,y,z)$ can be equivalently replaced by estimation of material concentration value m_1 (or m_2). In the elementary case at $m_1 < 0.5$ the value $\mu = \mu_{min}$ can be assumed in a volume element, and in the case of $m_1 > 0.5$ the value $\mu = \mu_{max}$, correspondingly.

However, this approach can give no optimal shapes of optimized body surface even in the class of homogeneous isotropic media. Therefore with such limited means of practical realization this equivalent replacement should be considered as a rather difficult one. Experience for calculations shows (see Section 4.4) that for the best unknown criterion of replacement of a non-uniform anisotropic medium by a homogeneous isotropic one, it is expedient to search for the optimum shape of a body all at once in the class of homogeneous isotropic media. Then, as noted above, alongside with a homogeneous body it is possible to obtain a structurally anisotropic body as a result of optimization. In some cases practical realization of the latter appears to be feasible (as against practical realization of the solution derived in the class of anisotropic non-uniform media).

4.4. Properties and numerical examples of the Lagrange method

In this section we shall discuss the properties of numerical optimization by the Lagrange method and compare the results of optimization for various classes of media. To this end, we shall consider several problems of search of optimum shape and structure of bodies in a magnetic field. For their numerical solution the finite difference method and the finite element method have been used.

4.4.1 Focusing of magnetic flux

The problem of focusing (concentration) a magnetic flux involves searching for the distribution of the substance in the admissible domain of its position providing concentration of the given magnetic flux on a certain part of a boundary. We will assume the magnetic field to be plane-parallel.

The area where the field is present is limited by a square contour *abcda* (Fig. 4.6) with a side dimension of 1 in relative units. There is a specified magnetic flux uniformly distributed along the upper side *dc*.

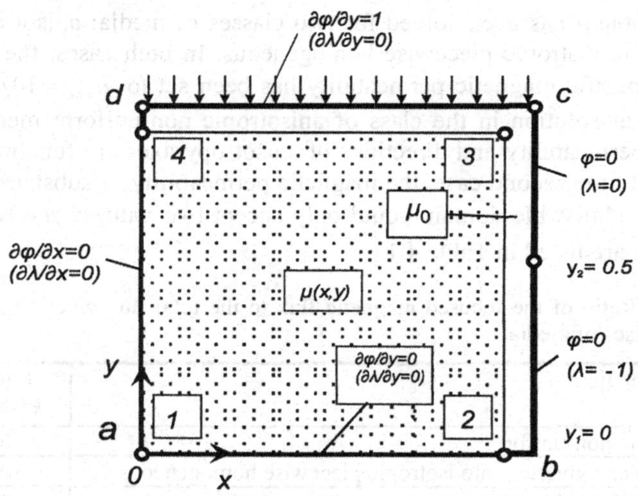

Fig. 4.6. Calculation domain and boundary conditions for the potential and the adjoint variable (in parenthesis)

The magnetic field intensity on this boundary has a single component $H_n = H_y = -1$. On the right side *bc* the scalar magnetic potential $\varphi = 0$. On the lower side *ab* it satisfies the condition $\partial\varphi/\partial y = 0$, and on the left side *ad* we have the condition $\partial\varphi/\partial x = 0$.

Let's find such a distribution of substance with magnetic permeability μ, which provides the largest value of magnetic flux by its module on a certain part of the right side $(x = 1, y_1 < y < y_2, y_1 = 0, y_2 = 0.5)$, i.e. on the segment $y_1 y_2$. The admissible domain of substance position is limited by the contour *1234*. In all points outside of this area the magnetic permeability is μ_0.

We shall write down the objective functional as: $I = \int_{y_1}^{y_2} \mu_0 \frac{\partial\varphi}{\partial x} dy$. The following boundary conditions for the adjoint variable correspond to the given boundary conditions for the potential, as follows from results obtained in sections 4.1 and 4.2: $\lambda = 0$ on the segment y_2c, $\partial\lambda/\partial y = 0$ on the side ab and $\partial\lambda/\partial x = 0$ on the side ad. Taking into account that $\delta I = \int_{y_1}^{y_2} \mu_0 \delta\left(\frac{\partial\varphi}{\partial x}\right) dy$, we have $\lambda = -1$ on the boundary y_1y_2.

The quality of focusing is determined by the ratio of the magnetic flux crossing the segment y_1y_2, and the full magnetic flux crossing the side bc: $k = \Phi_{y_1y_2}/\Phi_{bc}$. Ideal focusing means equality to 1 of this ratio: $k = 1$.

This problem has been solved for two classes of media: anisotropic non-uniform, and isotropic piecewise homogeneous. In both cases, the maximal value of specific magnetic permeability has been set to $\mu_{\max} = 10\mu_0$. In the first case (at solution in the class of anisotropic non-uniform media), both magnetic permeability and directions of anisotropy axes are functions of coordinates. In the second case, the magnetic permeability of substance located within the admissible domain could only accept two values: $\mu = 10\mu_0$ or μ_0. The results are listed in Table 4.1.

Table 4.1. Ratio of the focused magnetic flux to the total flux when optimizing in various classes of media

Class of medium	Flux ratio $k=\Phi_{y1y2}/\Phi_{bc}$
Anisotropic non-uniform	0.713
Anisotropic, converted into isotropic piecewise homogeneous	0.672
Isotropic piecewise homogeneous	0.687

In the second row of the table, the ratio of magnetic fluxes $\Phi_{y_1y_2}/\Phi_{bc}$ $(k = 0.672)$ obtained when replacing the optimal distribution of anisotropic non-uniform medium by an isotropic piecewise homogeneous one is shown. The replacement has been carried out depending on the value m_2, the concentration of substance with maximum magnetic permeability. In areas where it was equal or higher than 0.5 a changeover to a homogeneous medium with the maximum magnetic permeability $10\mu_0$ was performed. In areas with concentrations $m_2 < 0.5$, minimal magnetic permeability μ_0 was accepted. The ratio k of magnetic fluxes has decreased for this elementary (and not the best) method of conversion from the optimum anisotropic non-uniform medium to an isotropic piecewise homogeneous one. It became even less than $k = 0.687$ when seeking the solution in the class of isotropic homogeneous media.

As may be seen from the table, the best solution is obtained when using an anisotropic non-uniform material. The ratio k of magnetic fluxes in this case is maximal.

The body structure and shape found are shown in Fig. 4.7.

Continuous lines in the area of substance location divide zones with various concentrations of the ferromagnetic material m_2.

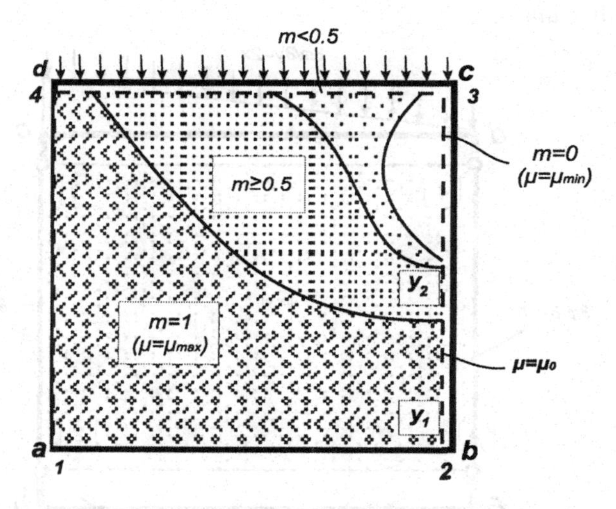

Fig. 4.7. Optimal distribution of substance in the problem of magnetic flux focusing

When $m_2 < 0.5$ prevails, the substance with magnetic permeability $\mu = \mu_{\min}$, and at $m_2 \geq 0.5$ the substance with magnetic permeability $\mu = \mu_{\max}$. Directions of anisotropy axes at points of substance location are not shown.

4.4.2 Redistribution of magnetic flux

The problem of redistribution of magnetic flux involves searching for such a function $\mu(x, y)$ in the admissible domain of substance position, which provides distribution of a specified magnetic flux in a certain part of the boundary according to a prescribed law. The magnetic field is considered to be plane-parallel.

On the the upper side dc of the square area $abcd$ (Fig. 4.8) with dimensions 1×1, the magnetic field intensity changes according to linear law $H_y = -2x$. It is necessary to find such structure and shape of a body that will provide change of magnetic field intensity according to the law $H_y = -2 + 2x$ on the lower side *ab*. The admissible domain of substance position is the area *1234* where $\mu = \mu(x,y)$. In areas *ab21* and *cd43*, the medium properties are fixed and $\mu = \mu_0$.

Boundary conditions for the scalar magnetic potential φ are shown in Fig. 4.8. By virtue of definition of homogeneous conditions of the 2nd kind for the sides *ad* and *bc*, all the magnetic flux that passes through the area *1234* falls on line *ab*.

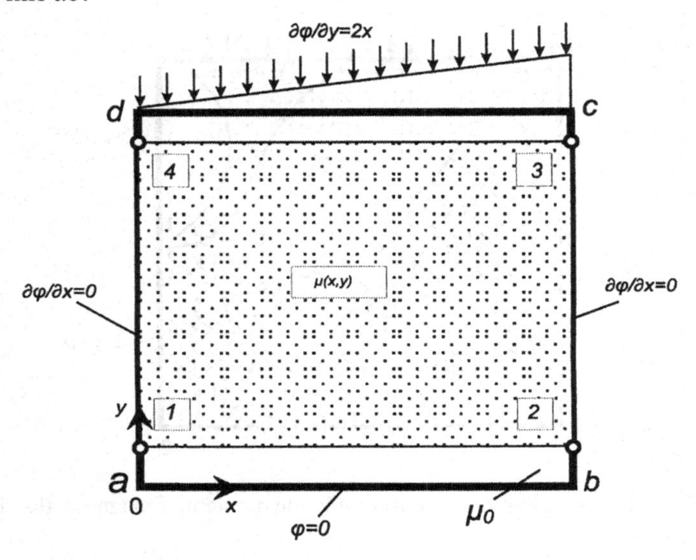

Fig. 4.8. Boundary conditions for the scalar magnetic potential in the problem of obtaining linear distribution of magnetic induction on the line *ab*

The objective functional can be written down as $I = 0.5 \int_a^b \mu_0 \left(\frac{\partial \varphi}{\partial y} + H_{yw} \right)^2 dx$. Boundary conditions for the adjoint variable λ are shown in Fig. 4.9.

The boundary conditions definition for the sides *bc*, *cd* and *da* becomes clear from Fig. 4.2 in Section 4.1. When writing down the boundary condition on the side *ab*, it has been taken into account that the objective functional *I* is defined on it. As the variation of the objective functional is equal to $\delta I = \int_a^b \mu_0 \left(\frac{\partial \varphi}{\partial y} + H_{yw} \right) \delta \left(\frac{\partial \varphi}{\partial y} \right) dx$, the adjoint variable is subject to the condition $\lambda = -\partial\varphi/\partial y - H_{yw} = -\partial\varphi/\partial y + 2 - 2x$ on the side *ab*.

Similar to the above-considered problem of magnetic flux focusing, we now search for the solution in classes of anisotropic non-uniform and isotropic piecewise homogeneous media.

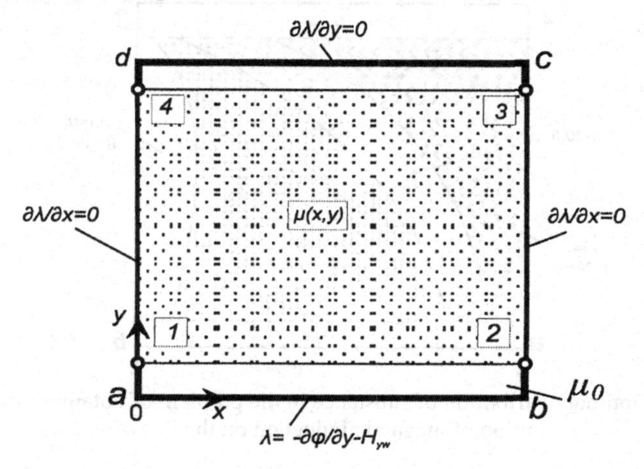

Fig. 4.9. Boundary conditions for the adjoint variable in the problem of obtaining linear distribution of magnetic induction on the line *ab*

Results (see Table 4.2) were compared according to the value of root-mean-square deviation of derived function $H_y(x)$ distribution from the specified linear distribution.

Table 4.2. Root-mean-square (RMS) deviation of magnetic field strength from the linear field when optimizing in various classes of media

Class of medium	RMS deviation σ
Anisotropic non-uniform	0.150
Anisotropic, converted into isotropic piece-wise homogeneous	0.236
Isotropic homogeneous	0.178

RMS deviation $\sigma = \sqrt{\int_a^b (\frac{\partial \varphi}{\partial y} + H_{yw})^2 dx}$

The conversion of an anisotropic non-uniform medium into an isotropic piecewise homogeneous one has been the same, as in the above considered problem of magnetic flux focusing. It is evident from the table that the best solution is obtained, just as in the previous problem, through the use of an anisotropic non-uniform medium. For this case the substance distribution is shown in Fig. 4.10 ($\mu_{\max} = 10\mu_0$ and $\mu_{\min} = \mu_0$).

Directions of anisotropy axes in the substance location points are not shown in the figure.

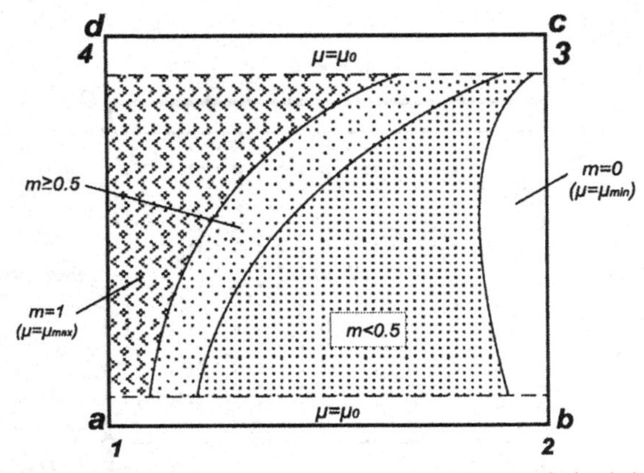

Fig. 4.10. Optimal distribution of substance in the problem of obtaining linear distribution of magnetic induction on the line *ab*

The substance is divided into zones, where values of its concentration m with magnetic permeability $\mu_{\max} = 10\mu_0$ are assumed to be equal $m = 0, m < 0.5$, $0.5 \le m < 1$ and $m = 1$.

Dependence of the field intensity module H_y, from coordinate x along the lower side *ab* at substance optimum distribution, is shown in Fig. 4.11 (curve 2).

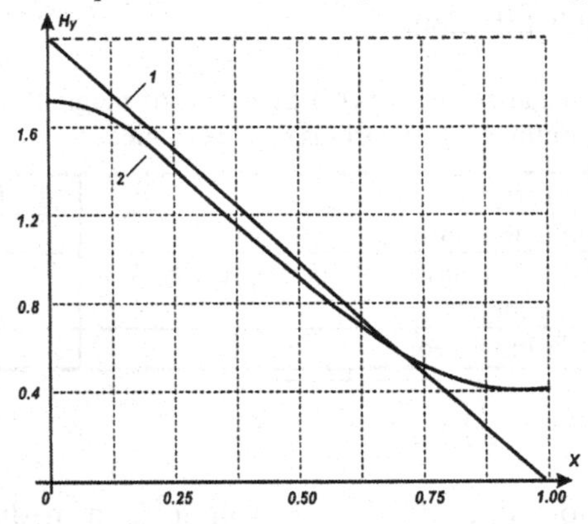

Fig. 4.11. Required (line 1) and obtained (curve 2) distributions of magnetic field strength on the line *ab*

As may be seen from the drawing, the largest deviation from the given distribution of field intensity occurs at the area's edges. In the middle of the area at $0.1 < x < 0.8$ field intensity, H_y changes close to the linear law.

Similar conclusions can be made when considering the problem of redistribution of a uniform flux with field intensity $H_y = -1$ on the upper side *dc* of a square area to a cosine flux $H_{yw} = -\frac{\pi}{2}\cos\frac{\pi}{2}x$ on its lower side *ab*. The admissible domain of substance position is the area *1234*. In areas *ab21* and *cd43* the medium properties are fixed and $\mu = \mu_0$ (Fig. 4.12).

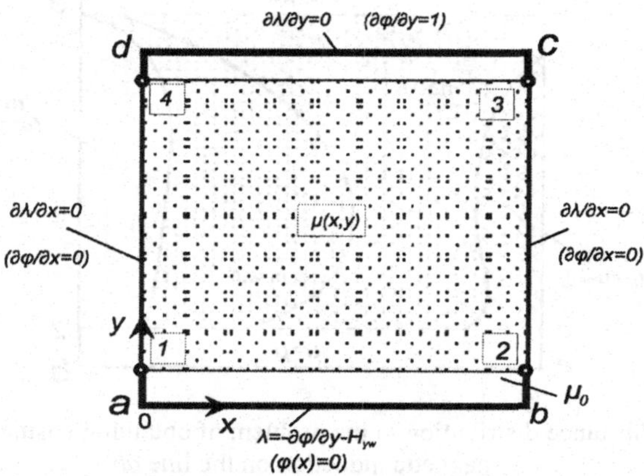

Fig. 4.12. Boundary conditions for the potential (in parenthesis) and adjoint variable in the problem of obtaining cosine distribution of magnetic induction on the line *ab*

The minimized functional is of the same form as in the previous problem: $I = \int_a^b \mu_0 \left(\frac{\partial \varphi}{\partial y} + H_{yw} \right)^2 dx$. Boundary conditions for potentials of direct (in parenthesis) and adjoint problems are shown in Fig. 4.12.

This problem has been solved at the same media classes as the previous one. Results are shown in Table 4.3.

Table 4.3. Root-mean-square (RMS) deviation of magnetic field strength from the cosine field when optimizing in various classes of media

Class of medium	RMS deviation σ
Anisotropic non-uniform	0.099
Anisotropic, converted into isotropic piecewise homogeneous	0.132
Isotropic homogeneous	0.125

RMS deviation is $\sigma = \sqrt{\int_a^b (\frac{\partial\varphi}{\partial y} + H_{yw})^2 dx}$.

The distribution of substance (with $\mu_{\max} = 10\mu_0$ and $\mu_{\min} = \mu_0$) is shown in Fig. 4.13.

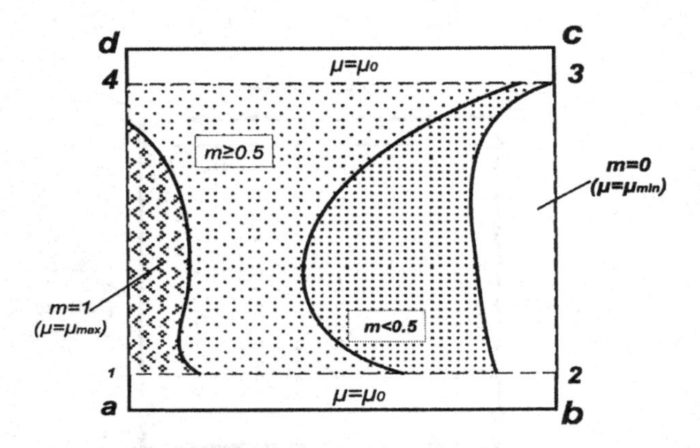

Fig. 4.13. Substance distribution in the problem of obtaining cosine distribution of magnetic induction on the line *ab*

The dependence of the field intensity module H_y, from the coordinate x along the lower side *ab* at the substance optimum distribution, is shown in Fig. 4.14 (curve 2). Curve 1 defines the function $|H_w(x)| = \frac{\pi}{2}\cos\frac{\pi}{2}x$.

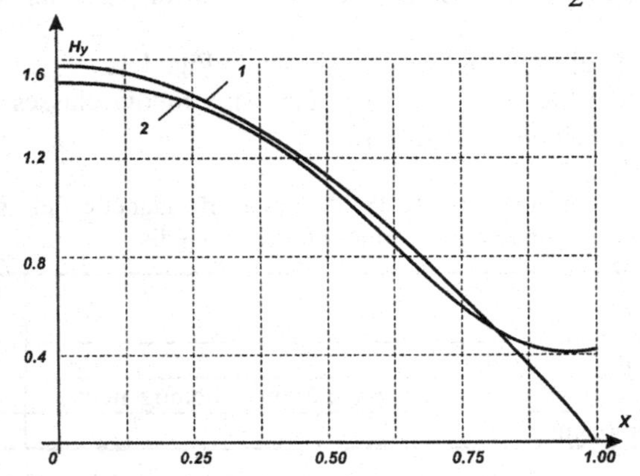

Fig. 4.14. Required (curve 1) and obtained (curve 2) distributions of magnetic induction on the line *ab*

4.4.3 The extremum of electromagnetic force

Let's consider the problem of search of such a shape of a ferromagnetic pole, which provides the maximum value of the vertical electromagnetic force F acting upon a conductor of rectangular cross-section with a current i (Fig. 4.15).

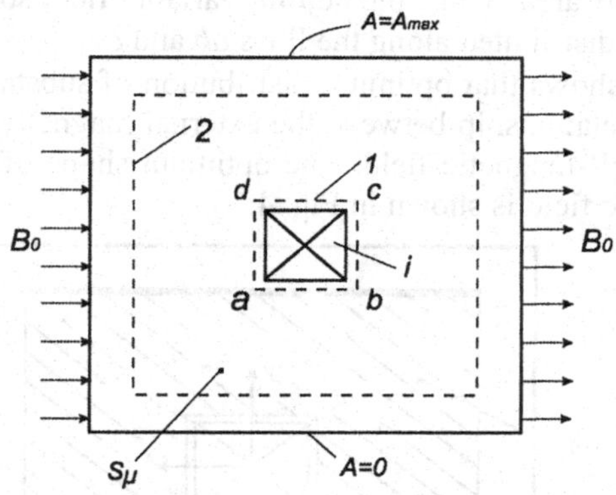

Fig. 4.15. Boundary conditions for the potential and domain S_μ of substance distribution in the problem of searching for extremum of electromagnetic force

The magnetic field is assumed to be plane-parallel. The conductor current density has a component J_z.

The admissible domain of substance position S_μ with a given magnetic permeability $\mu = \text{const}$ is located between two rectangular contours 1 and 2 marked by dashed lines. Additionally, there is an external uniform magnetic field with induction B_0. On the lower and upper sides of the calculation's area (in Fig. 4.16 they are shown by continuous lines) the vector magnetic potential A is assumed to be constant and equal to 0 and $A_{\max}$, correspondingly. On the left and right sides of the area, the potential changes by linear law from 0 to $A_{\max}$.

We shall transform the objective function $I = -F =$ $= -\int\limits_{S_{ob}} J_z B_x dS = -\int\limits_{S_{ob}} J_z \frac{dA}{dy} dS$ to the form $I = -\oint\limits_{l_{ob}} J_z A\cos(\boldsymbol{n}, y) dl$. Here, $\boldsymbol{n}$ is the external normal to the contour l of the wire cross-section.

As $\cos(\boldsymbol{n}, y) = 0$ on the sides bc and da of the conductor, then the objective function can be written down as:

$$I = -\int\limits_a^b J_z A\cos(\boldsymbol{n}, y) dl - \int\limits_c^d J_z A\cos(\boldsymbol{n}, y) dl = -J_z \int\limits_a^b A dl + J_z \int\limits_c^d A dl .$$

Taking into account that $\delta I = -J_z \int_a^b (\delta A) dl + J_z \int_c^d (\delta A) dl$, we have the equation for the adjoint variable $\operatorname{div} \nu \operatorname{grad} \lambda = +J\eta$ on the line *ab*, $\operatorname{div} \nu \operatorname{grad} \lambda = -J\eta$ on the line *cd*, and $\operatorname{div} \nu \operatorname{grad} \lambda = 0$ in other internal points of the calculations area. Thus, the adjoint variable field sources form simple layers of current distributed along the lines *ab* and *cd*.

This solution shows that optimum distribution of substance in the area S_μ, depends on the relationship between the external magnetic field and the conductor current self-magnetic field. The optimum shape of a body in a weak external magnetic field is shown in Fig. 4.16.

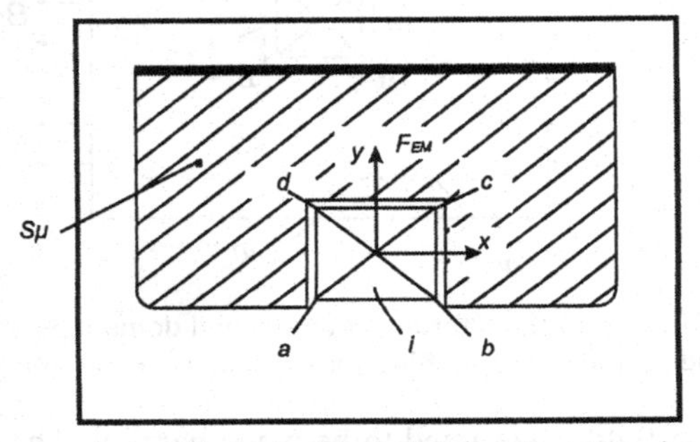

Fig. 4.16. Optimum shape of a body (shaded) in weak external magnetic field

If the external field considerably exceeds the current self-magnetic field, then the optimally shaped body focuses the magnetic flux in the area of the conductor location (Fig. 4.17).

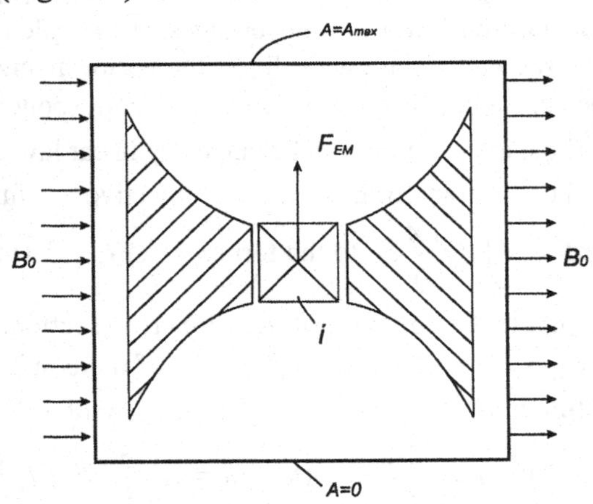

Fig. 4.17. Optimum shape of a body (shaded) in strong external magnetic field

Similar to previous examples, this problem also has been solved by optimization in classes anisotropic and isotropic media. Calculation has confirmed the conclusion that the application of anisotropic substances gives a lesser value for the functional compared with using the cases of homogeneous or non-uniform isotropic substances.

4.4.4 Identification of substance distribution

Let's consider features of the identification problem solution in a magnetic field by the Lagrange method. Such problems arise, in particular, when searching for the shape of buried magnetic bodies causing distortions of ground uniformity.

Surface distributions of potential or field intensity are known, for example, as a result of measurements. Therefore, the identification problem involves seeking such a shape of a body located in the ground which will satisfy the given boundary conditions on the ground surface.

The identification problem was divided into two stages with the purpose of estimation of its solution accuracy. At the first auxiliary stage, the magnetic induction distribution on the ground surface has been found at a given potential of its surface and *at a given distribution of substance*. At the second stage, the actual identification problem was solved by search of such distribution of substance, which furnished minimal difference between the obtained and given distributions of magnetic induction on the ground surface.

At the first stage the problem of magnetic field calculation in the square area *abcd* with dimensions 1×1 (Fig. 4.18) was solved.

In the central part of this area limited by lines *1,2,3,4*, a body with dimensions of its cross-section 0.4×0.4 and with magnetic permeability $\mu=10\mu_0$, distinct from the magnetic permeability μ_0 of the surrounding medium, was located. The following boundary conditions were set for the square area *abcd*: H_y=-1 on the upper side *dc*, $\varphi=0$ on the lower side *ab*, and $\partial\varphi/\partial x=0$ on both lateral boundaries. As a result of this calculation the distribution of magnetic field intensity $H_y(x)$ on the lower side *ab* was found.

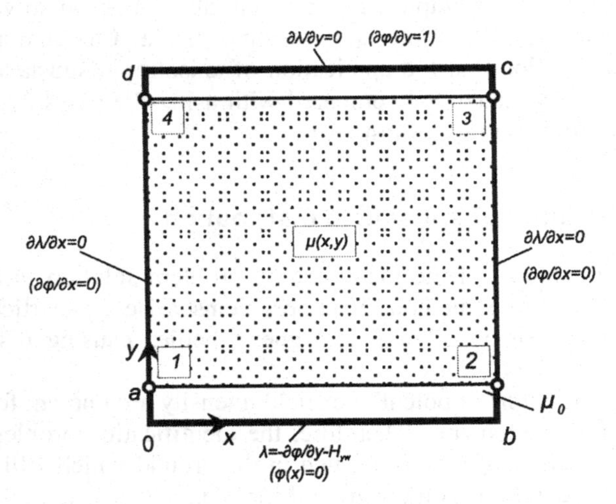

Fig. 4.18. The first stage of identification problem solution: field calculation at desired arrangement of body. Boundary conditions for the potential are shown in parenthesis

At the second stage, the problem of search of body optimum shape has been solved at the above-found distribution of field intensity H_y on the line *ab*. Comparison of the two body shapes, given at the first stage and calculated at the second stage, allows us to estimate the accuracy of the identification problem solution.

Thus, in this statement the identification problem involves the search for the shape of a body based upon the distribution of field intensity given on some surface enclosing the body.

The solution was searched in the class of homogeneous media when magnetic permeability could possess values $\mu_{\min} = \mu_0$ and $\mu_{\max} = 10\mu_0$. Boundary conditions for the solution of direct and adjoint problems are shown in Fig. 4.18. The admissible area, where a change of material is allowed, is confined inside the line *1234*. In areas *ab21* and *cd43* the medium properties are fixed and $\mu_{\min} = \mu_0$, and the permissible values of magnetic permeability inside the area *1234* are $\mu_{\min} = \mu_0$ or $\mu_{\max} = 10\mu_0$.

Distribution of the module H_y of field intensity along the lower boundary *ab* for the required (1) and found (2) body shapes is shown in Fig. 4.19.

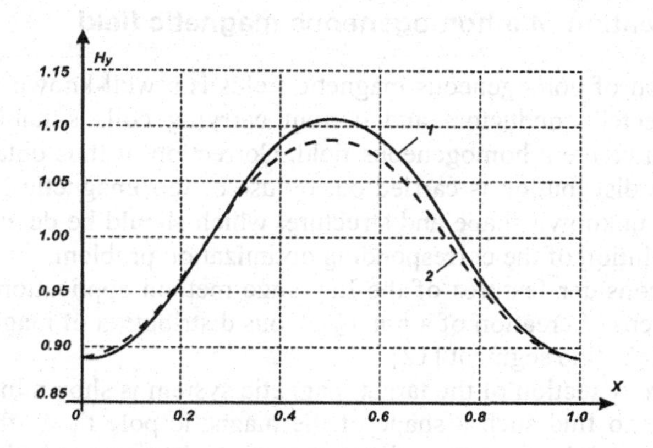

Fig. 4.19. Magnetic field strength on the line *ab* at desired (1) and obtained (2) distributions of medium

In Fig. 4.20, the square section 1 of the identified body is shown by dashed lines. The area 2 represents the cross section of the body obtained as a result of the identification problem solution.

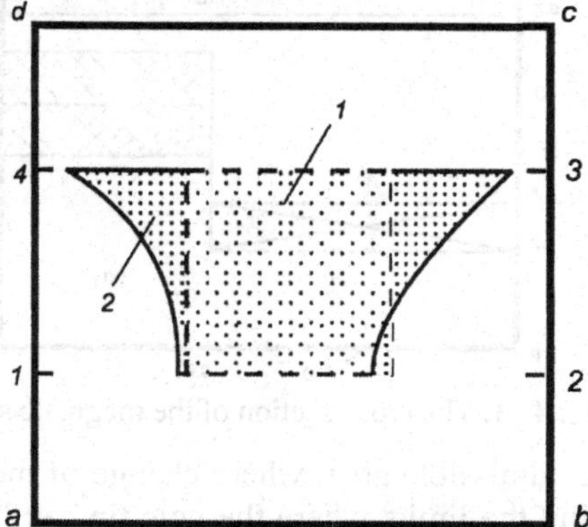

Fig. 4.20. Required (1) and obtained (2) distributions of medium when solving the identification problem

As can be seen, even for the requirement of closely obtained dependence $H_y(x)$ on the area boundary (Fig. 4.19) when the maximal error was less than 3%, there is significant distinction between the shapes of the body cross-sections both specified and obtained at the solution. Therefore, a solution for practical identification problems require several orders of higher accuracy.

4.4.5 Creation of a homogeneous magnetic field

Generation of homogeneous magnetic fields is a well-known practical problem. Selected conductors and current carrying coils suitably arranged in space can create a homogeneous field. Correction of thus obtained magnetic induction distribution is carried out by use of ferromagnetic bodies with beforehand unknown shape and structure, which should be determined as a result of solution of the corresponding optimization problem.

Let's consider features of the Lagrange method application when solving the problem of creation of a homogeneous distribution of magnetic induction on a straight line segment [12].

The cross-section of the target magnetic system is shown in Fig. 4.21. It is necessary to find such a shape of the magnetic pole tip S_μ that provides as small as possible deviation of the magnetic induction vertical component B_y, distributed along the segment *ab* from a constant value B_{yw}=0.1 T. The magnetic field is assumed to be plane-parallel.

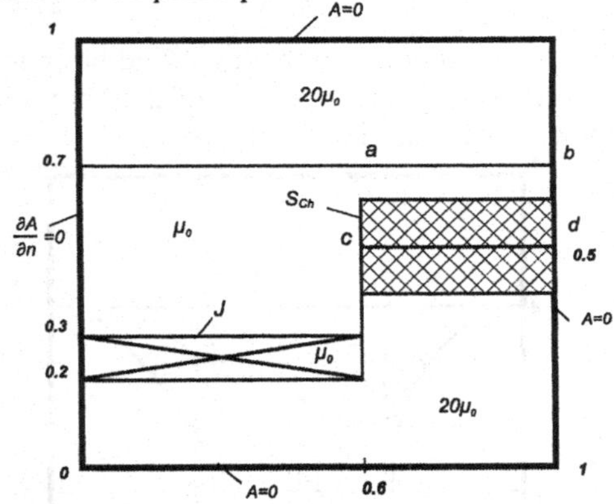

Fig. 4.21. The cross-section of the magnetic system

In Fig. 4.21 the admissible area, where change of medium is allowed, i.e. the area determining the limits where the pole tip can be situated, is marked by dashed lines and designated S_μ. As well, boundary conditions for the vector magnetic potential A used at solution of the direct problem are shown. On the left border of this area (at $x = 0$), the condition $\partial A/\partial n = 0$ is assumed. On the other borders of the area, the condition $A = 0$ is specified. The current density J in the winding is $5.6 \cdot 10^4\ \mathrm{A/m^2}$.

The objective function is assumed to be: $I = \frac{1}{2}\int_a^b (-\frac{\partial A}{\partial x} - B_{yw})^2 dx$.

Its variation with respect to potential A is:

$$\delta I = \int_a^b \left(-\frac{\partial A}{\partial x} - B_{yw} \right) \delta \left(-\frac{\partial A}{\partial x} \right) dx = \left(\frac{\partial A}{\partial x} + B_{yw} \right) \delta A \Big|_a^b - \int_a^b \left(\frac{\partial^2 A}{\partial x^2} + \frac{\partial B_{yw}}{\partial x} \right) \delta A dx.$$

Taking into account that $B_{yw} = \text{const}$ and $\partial B_{yw} / \partial x = 0$, we have:

$$\delta I = \left(\frac{\partial A}{\partial x} + B_{yw} \right) \delta A \Big|_{x=b} - \left(\frac{\partial A}{\partial x} + B_{yw} \right) \delta A \Big|_{x=a} - \int_a^b \frac{\partial^2 A}{\partial x^2} \delta A dx .$$

Since $A = 0$ in the point $x = b$ is specified as a condition, then $\delta A = 0$ in this point, and then we have:

$$\delta I = \left(\frac{\partial A}{\partial x} + B_{yw} \right) \delta A \Big|_{x=a} - \int_a^b \frac{\partial^2 A}{\partial x^2} \delta A dx .$$

Writing down δI as $\delta I = \int_V f(\delta A) dV$ we find the adjoint variable field sources. They form a current layer on the line *ab*. In the point *a*, electric current $i = -\left(\partial A / \partial x + B_{yw} \right)$ is located. The density of current distributed along the line *ab* is $\partial^2 A / \partial x^2$. Apparently, it is a function of potential A of the direct problem and the solution was sought in the class of homogeneous media.

The numerical solution for this problem has been obtained by the finite-element method. Triangular elements were applied both at linear and at quadratic interpolation of the potential.

As the solution was sought in the class of homogeneous media it was sufficient to calculate functions grad$A$ and gradλ only on the pole tip surface at each step of the process. This allowed reducing the amount of calculations in comparison with the case of search of its structure.

Several moveable vertexes (from 11 up to 41) have been chosen along the line *cd* determining the initial shape of the pole tip surface. Each vertex could move only downwards or upwards on a specified distance during any step of the process. The amount of change of the moved vertex coordinate on each step depended on the angle between vectors grad$A$ and gradλ . It was selected from the condition of solution convergence proceeding from experience of calculations.

At linear interpolation of the potential the whole area has been divided into a grid with 30314 elements. Along the pole border *cd*, 22 movable vertexes have been chosen.

As a result of the solution, the root-mean-square deviation of magnetic induction B_y from the specified 0.1T was equal 0.0066. However, the pole shape has turned out to be rough with sharp protrusions. Therefore, calculations were also carried out for the potential quadratic interpolation when the grid had 11204 elements, and the border *cd*, 41 movable vertexes.

After 24 iterations the root-mean-square deviation of the magnetic induction from the specified one $\sigma = \frac{1}{B_{want}}\sqrt{\int_a^b (B_y - B_w)^2 dl}$ was 0.0091. Figure 4.22 shows the substance distribution determining the pole tip shape, after completion of the 24th iteration.

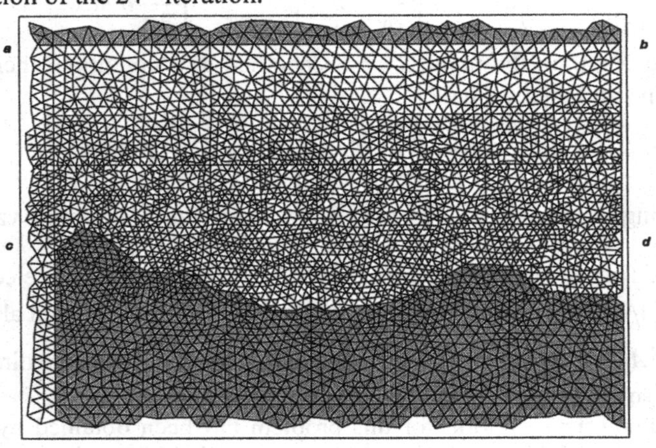

Fig. 4.22. The optimal shape of the polar tip at interpolation of the potential by a polynomial of first degree

The best shape for the pole tip has been found after 216 iterations (Fig. 4.23). The tip shape appeared to be close to results of other researchers, who have also solved this problem.

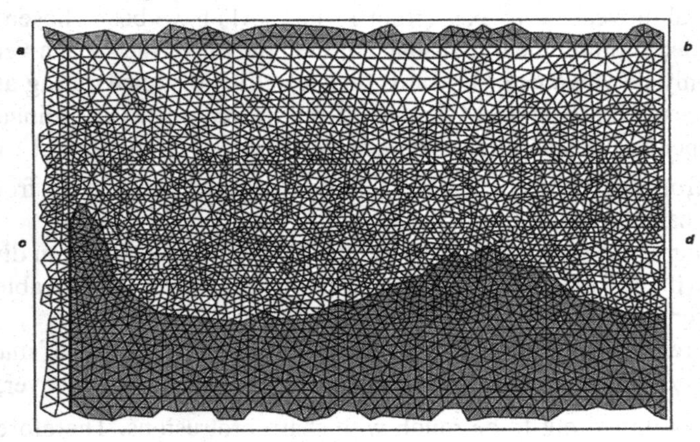

Fig. 4.23. The optimal shape of the polar tip at interpolation of the potential by a polynomial of second degree

The distribution of magnetic induction along the line *ab* for the obtained shape of the pole tip is shown in Fig. 4.24. The distribution of induction along the line *ab* in another scale is shown in an additional window in this figure. In this case the root-mean-square deviation of B_y from the required one is slightly increased (up to 0.0071), which may be attributed to the more smoothed surface of the pole tip.

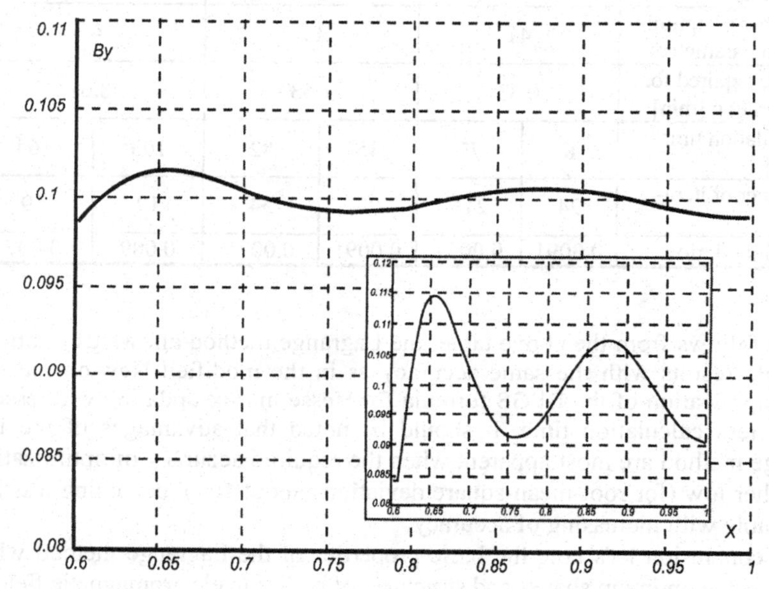

Fig. 4.24. Distribution of magnetic induction on the line *ab* at optimal shape of the polar tip. The small box shows the same dependence in another scale

For juxtaposition of the Lagrange method with other methods, and also with the purpose of comparison of obtained results, a solution has been derived also with the help of the gradient method. The modified Newton method was chosen with application of BFGS formula for Hesse matrix updating. It possesses one of the best characteristics when solving optimization problems in electromagnetic fields.

At variables number (movable vertexes on the pole border) equal to 41 the problem solution time by the Lagrange method appeared to be less on an order than the modified Newton method. At that, the root-mean-square error was the same.

Optimization results by the Lagrange method and the modified Newton method are summarized in the following comparison table.

Table 4.4. Initial data and characteristics of the problem solution by the Lagrange method and by the modified Newton method

Lagrange method			Modified Newton method with BFGS formula			
Number of grade elements	11204		8135		11204	
Number of optimization parameters	41		11		41	
Time, required for an iteration [min]	0.33		1.53		8.61	
Calculation time [min]	8	72	35	82	103	164
Number of iterations	24	216	23	54	12	19
RMS deviation	0.0091	0.0071	0.0091	0.0071	0.009	0.007

As follows from the above table, the Lagrange method allows us to obtain reliable results with the same accuracy as in the modified Newton method with application of the BFGS formula for Hesse matrix updating with essentially less calculation time. It should be noted that advantages of the Lagrange method are most apparent when the required accuracy of optimization is rather low (for root-mean-square deviations about 10^{-3}), becoming less appreciable with increasing of accuracy.

In conclusion let's note the basic properties of the Lagrange method when seeking the optimum shapes and structures of bodies in electromagnetic fields.

The solution can be sought in various classes of media, both isotropic and anisotropic, and each step of search demands solution of two boundary value problems. The direct problem is formulated for the main variable determining the field (for example, the scalar potential φ). The adjoint problem is solved with regard to an auxiliary variable λ. Functions gradφ and gradλ obtained as a result of solution of these problems determine the character of medium properties change during calculation.

4.5 Features of numerical optimization by the Lagrange method

Choice of numerical methods for solution of direct and adjoint boundary problems is determined by their specific nature, namely the type of boundary conditions, the accepted class of the optimized medium, presence of natural boundaries of calculations domain, etc. Methods of finite-elements and integral equations are most frequently applied to find numerical solutions of boundary problems in electromagnetic field theory.

Regardless of the method selected for a numerical solution, the search algorithm for substance optimum distribution stays the same. We find the value μ at a point and the directions of anisotropic axes (when seeking solution in the class of anisotropic media) on the found functions gradφ, gradλ. Features of search algorithm realization lie in the method used for calculating the vectors gradφ and gradλ, and in the approach used for deformation of the shape or change of the structure of the optimized body.

When solving optimization problems in the class of homogeneous media for calculation of potentials of direct and adjoint problems, the following integral equation for density of auxiliary sources σ on the surface S of the optimized body can be used:

$$\sigma + \frac{\lambda}{2\pi}\int_S \sigma \frac{\cos(\mathbf{r},\mathbf{n})}{r^2} ds = f . \tag{4.9}$$

Here, λ is a parameter dependent from magnetic permeabilities μ_0 and μ of the media, $\mathbf{r}$ is the radius vector determining the distance between points on the surface S, $\mathbf{n}$ is the normal vector for points on the surface of S, and f is a function determined by a given distribution of external sources in the domain V_s (Fig. 4.25).

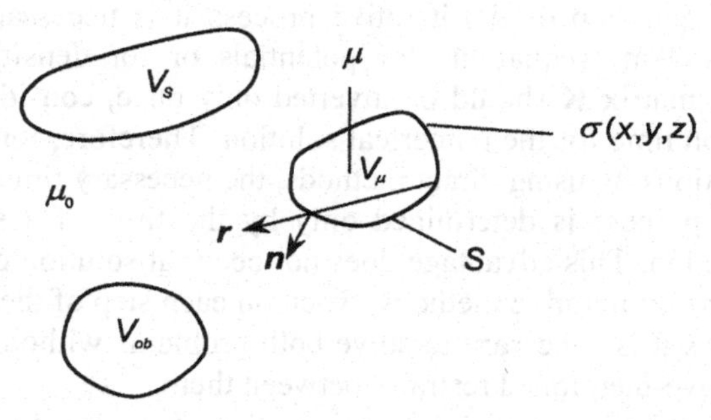

Fig. 4.25. Domains V_{ob}, where the objective function is specified, V_s - of field, and V_μ - of medium distributions

External sources in the domain V_s can be electric currents or equivalent magnetic charges of density ρ. In the latter case for calculation of the function gradφ, the following expression can be used:

$$\text{grad}\varphi = \text{grad}\left(\frac{1}{4\pi\mu_0}\int_S \frac{\sigma}{r}dS + \frac{1}{4\pi\mu_0}\int_{V_s}\frac{\rho}{r}dV\right) = \frac{1}{4\pi\mu_0}\left(\int_S \frac{\sigma\mathbf{r}}{r^3}dS + \int_{V_s}\frac{\rho\mathbf{r}}{r^3}dV\right).$$

If the objective function is given outside of the area V_μ, then the left member of the equation for density of scalar sources (we shall designate it as p) in the adjoint problem coincides with the left member of Eq. (4.9). However, the right members of these equations differ.

Indeed, external sources of fields φ and λ can generally have different spatial distributions and they can be located in various domains. But at the same time, secondary sources of density ρ of the adjoint problem are located on the same surface S as sources of density σ of the direct problem. Therefore, the left members of the equations for σ and ρ coincide, whereas the right members differ.

After solving the integral equations for densities σ and p of direct and adjoint problems, we further calculate gradφ, gradλ and solve the optimization problem in accordance with the algorithm discussed in Section 4.3.

When using the same expressions for approximation of integrals included in equations for $\boldsymbol{\sigma}$ and ρ, the algebraic equations of direct and adjoint problems $\mathbf{K\sigma=F}$, $\mathbf{K\rho=Y}$ have identical coefficient matrices $\mathbf{K}$.

We arrive at a similar conclusion when considering algebraic equations of the direct and adjoint problems in the case of seeking their solutions by using the finite-elements method. Indeed, at realization of any of these methods (finite-elements or integral equations) the direct and adjoint problems on each step should be solved using the same geometry of domains, and for the same physical properties of materials. In both cases, correction of the body surface S is carried out only after finding gradφ and gradλ.

Thus, on each step of the iterative process it is necessary to solve two boundary problems (equations for potentials or for density of secondary sources), but matrix **K** should be inverted only once, considerably reducing the calculation time for the numerical solution. Therefore, for solutions of algebraic equations by using direct methods, the necessary time on each step of the iterative process is determined only by the time of a single boundary problem solution. This advantage does not occur at solution of direct and adjoint problems by iterative methods, when on each step of the optimum shape search process it is necessary to solve both problems without taking into account the above-mentioned restraint between them.

However one cannot draw a conclusion that the direct method of equations, the solution on each step of the problem solution, is unconditionally preferable on the basis of this reasoning.

The process of finding the optimum shape of a body is an iterative one as the objective function depends nonlinearly on the body shape. At each step of the process, the body shape can change insignificantly, particularly at the final steps of the search. During iterative solution of algebraic equations effective use of the solution found by the previous step is possible if it can be considered an initial approximation. At the same time at solution of algebraic equations by direct methods, the initial approximate value of the coefficients matrix **K** is not used. In this connection the choice of the method of solution for the algebraic equations of direct and adjoint problems on each step of the iterative solution should be connected with convergence of the search process for the optimum body shape.

The following reasons of general character should be noted. At iterative solution of algebraic equations of direct and adjoint problems at a given step of the process, their values obtained during the previous step may expediently be accepted as initial an approximation for potentials (or sources densities). It demands storage of arrays of unknown quantities both for the direct and adjoint problems. When solving algebraic equations by direct methods it is advisable to accept the inverse matrix obtained during the previous step as a parent matrix, and then to specify values of its elements by iterative calculations.

To increase the accuracy of solution of optimization problems the grid FEM is accepted to be deformable near the surface, whose shape is subject to find.

Assume it is found during solution that the body surface should be moved in the direction of external normal at some point. On the following step of the iterative process we deform the surface not for the whole cell size, but only for a portion of it. Moving the corresponding node of the element lying on the surface carries this out. Thus we are permit the following: firstly, to enter small deformations of surface during search of its optimal shape, and secondly, always to place nodes on the surface. After deformation of elements, control of their degeneration with corresponding corrections should be carried out. The procedure of moving the surface nodes of an optimized body may expediently be carried out only at the final stage of optimization. In that case, the objective functional will be close to its minimum and the surface shape will be close to optimum.

The possibility of calculation of functions $\mathrm{grad}\varphi$, $\mathrm{grad}\lambda$, defining the medium properties at a point, not only on the body surface, but inside and outside of it as well, is a distinctive feature of optimization by the Lagrange method. According to the algorithm of solution it is possible to correct the function $\mu(x,y,z)$ in the whole domain at once on each step of the process. For this reason the structure, and therefore the surface shape as well, can change considerably during each step of the process. It essentially allows finding the solution more speedily, in particular during the initial steps when the body shape and structure are far from optimum.

4.6 Optimizing the medium and sources distribution in non-stationary electromagnetic fields

The Lagrange method allows solving optimization problems not only in stationary fields, but in non-stationary electromagnetic fields as well. As a result, an auxiliary adjoint problem for the variable λ is formulated. The equation and boundary conditions for it are derived from the necessary condition of the augmented functional extremum $\delta L = 0$. Solution of the adjoint problem allows finding the augmented functional gradient, and by virtue of it, constructing an effective search algorithm for optimization parameters.

Problems of search of the shape of bodies changing in time in a non-stationary field are not posed because of practical reasons. In a number of problems, external sources are considered as optimized variables with requirement to find their distribution in space and time. Finding the spatial distribution of material characteristics fulfilling the accepted criterion of optimality is also of interest. In some cases it is necessary to find such field distribution at the domain boundary, where the objective function reaches its minimal value.

Let's consider features of the Lagrange method application when solving optimization problems in a non-stationary electromagnetic field.

Let the substance penetrated by a flat electromagnetic wave be characterized by having a magnetic permeability μ and a specific electric conductivity σ. Inside the body, magnetic induction $B_y = B(x,t)$ is described by the equation $\frac{\partial}{\partial x}\left(\frac{1}{\mu\sigma}\frac{\partial B}{\partial x}\right) = \frac{\partial B}{\partial t}$. On the surface of the body (at $x = 0$) it changes under the law $B_y = B(0,t)$ (Fig. 4.26).

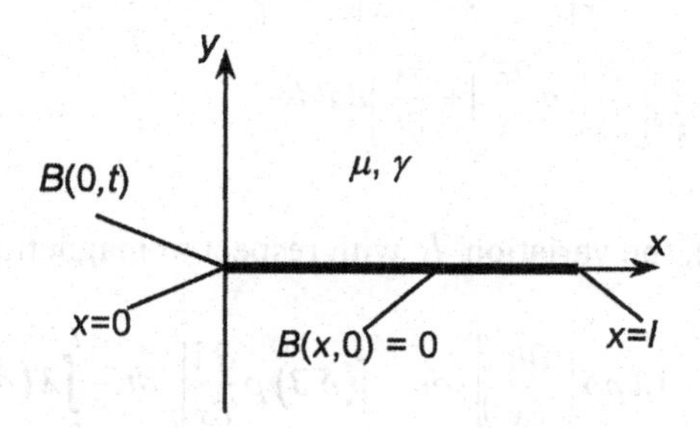

Fig. 4.26. One-dimensional domain $0<x<l$ of substance arrangement; initial and boundary values of magnetic induction

We accept the following initial conditions: $B(x)=0$ at $t=0$, and $\frac{\partial B}{\partial x}=0$ at $x=l$.

Let's search for such a function $\frac{1}{\mu(x)\sigma(x)}=p(x)$ determining the medium properties, at which the magnetic induction distribution $B(x,T)$ by the moment of time T becomes as close as possible to a given distribution $B_w(x)$ [3,13].

The objective functional can be written down as $I=0.5\int\limits_0^l (B(x,T)-B_w(x))^2 dx$.

We construct the augmented functional as

$$L=I+\int\limits_0^l\int\limits_0^T \lambda(x,t)\left[\frac{\partial}{\partial x}\left(p(x)\frac{\partial B}{\partial x}\right)-\frac{\partial B}{\partial t}\right]dtdx,$$

and, after double integration by parts, we have:

$$L=I+\int\limits_0^T \lambda p\frac{\partial B}{\partial x}\bigg|_0^l dt-\int\limits_0^l\int\limits_0^T\left[p\left(\frac{\partial B}{\partial x}\cdot\frac{\partial \lambda}{\partial x}\right)+\lambda\frac{\partial B}{\partial t}\right]dtdx \tag{4.10}$$

$$L = I + \int_0^T \lambda p \frac{\partial B}{\partial x}\bigg|_0^l dt - \int_0^T Bp \frac{\partial \lambda}{\partial x}\bigg|_0^l dt - \int_0^l \lambda B\big|_0^T dx + \\ + \int_0^l \int_0^T \left[\frac{\partial}{\partial x}\left(p \frac{\partial \lambda}{\partial x} \right) + \frac{\partial \lambda}{\partial t} \right] B dt dx. \tag{4.11}$$

Using Eq. (4.11), the variation L with respect to magnetic induction B can be written down as

$$\delta L = \delta I + \int_0^T \lambda p \delta \left(\frac{\partial B}{\partial x} \right)\bigg|_0^l dt - \int_0^T (\delta B) p \frac{\partial \lambda}{\partial x}\bigg|_0^l dt - \int_0^l \lambda (\delta B)\big|_0^T dx + \\ + \int_0^l \int_0^T \left[\frac{\partial}{\partial x}\left(p \frac{\partial \lambda}{\partial x} \right) + \frac{\partial \lambda}{\partial t} \right] (\delta B) dt dx.$$

From the condition $\delta L = 0$ follows that the adjoint variable λ satisfies the equation $\frac{\partial}{\partial x}\left(p \frac{\partial \lambda}{\partial x} \right) + \frac{\partial \lambda}{\partial t} = 0$. As the variation of the objective functional with respect to magnetic induction B is equal:

$$\delta I = \int_0^l (B(x,T) - B_w(x)) \ (\delta B(x,T)) dx,$$

then the following condition should be imposed on the adjoint variable: $\lambda(x,T) = B(x,T)\text{-}B_w(x)$.

As $B(0,t)$ is a given value at x=0, then variation of only $\frac{\partial B}{\partial x}$ at $x = 0$ is possible. Therefore, we have $\lambda(0,t) = 0$. Furthermore, $\frac{\partial B}{\partial x}(l,t) = 0$ is a given value at $x = l$, and variation of only B at $x = l$ is possible. Then, we have $\frac{\partial \lambda}{\partial x}(l,t) = 0$.

Using Eq. (4.10) we find the augmented functional variation with respect to the variable $p(x)$. The condition $\delta_p L = 0$ results in the following expression:

$$\int_0^l \int_0^T \left(\frac{\partial B}{\partial x} \cdot \frac{\partial \lambda}{\partial x} \right) (\delta p) dt dx = 0.$$

By virtue of arbitrariness of variation δp, the necessary condition of optimality $\frac{\partial B}{\partial x} \cdot \frac{\partial \lambda}{\partial x} = 0$ can be obtained, which should be valid in each point x inside the segment $0 < x < l$. The value of $\frac{\partial B}{\partial x} \cdot \frac{\partial \lambda}{\partial x}$ determines the augmented functional "material" derivative. Its calculation allows applying known gradient methods for seeking of a medium optimal distribution.

Thus, the process of solution includes the following steps. The dependence $B(x,t)$ is calculated by solving the equation $\frac{\partial}{\partial x}\left(p(x)\frac{\partial B}{\partial x}\right) = \frac{\partial B}{\partial t}$ under the conditions $B|_{x=0} = B(t)$, $\left.\frac{\partial B}{\partial x}\right|_{x=l} = 0$ and $B(x,0) = 0$. Solution of the equation for the adjoint variable $\frac{\partial}{\partial x}\left(p(x)\frac{\partial \lambda}{\partial x}\right) + \frac{\partial \lambda}{\partial t} = 0$ under conditions $\lambda(0,t) = 0$, $\frac{\partial \lambda}{\partial x}(l,t) = 0$, and $\lambda(x,T)=B(x,T) - B_w(x)$, allows finding the function $\lambda(x,t)$ in order to find the augmented functional gradient.

Problems of searching of such distribution of medium specific electric conductance $\sigma = \sigma(x)$ on the line segment $0 \le x \le x_0$, at which the maximum local thermal emission Q in a point of the conductor with coordinate x for all duration t_0 of the electromagnetic process, i.e. $Q(x) = \int_0^{t_0} \sigma(x)E^2(x,t)dt$ is minimum, are of practical interest. For such problems, the functional can be written down as $I = \max\left(\int_0^{t_0} \sigma(x)E^2(x,t)dt\right)$. A problem of this kind is discussed in Chapter 5.

Another possible statement of this problem involves finding a dependence $B(0,t)$ on the surface of the body, at which the magnetic induction $B(x,T)$ at a certain moment of time T is as close as possible to the required $B_w(x)$. Seeking this dependence is carried out by use of a similar expression for the augmented functional, and the same equation for the adjoint variable.

To find the function $B(0,t)$, we calculate the variation of both parts of the expression given in Eq. (4.11) with respect to $B(0,t)$. This results in the relationship

$$\delta_{B(0,t)}L = -\int_0^T p(0)\frac{\partial \lambda}{\partial x}(0,t)\delta B(0,t)dt = 0,$$

which gives the necessary condition of optimum $\frac{\partial \lambda}{\partial x}(0,t)=0$. The expression for $\delta_{B(0,t)}L$ determines the functional derivative with respect to the desired quantity $B(0,t)$ that, as above, allows using gradient methods for search of optimal law of magnetic induction change on the surface of the body.

A characteristic problem is the search of such spatial distribution of an electric current with density J_{ext} and a law of its change in time, at which the power of losses $p(x,y,z)$ in a conducting body located in the field of this current is distributed as close as possible to a given $q_w(x,y,z)$ at the moment of time *T*. Such problems arise for inductive heating of conducting solids [14,15,16], particularly when uniform heating of the body is important. In this special case we have $q(x,y,z)=\text{const}$.

In Fig. 4.27, the conducting body is designated as S_σ, the allowable domain of current distribution - as S_i, and the domain of existence of the electromagnetic field - as S (l is its border line).

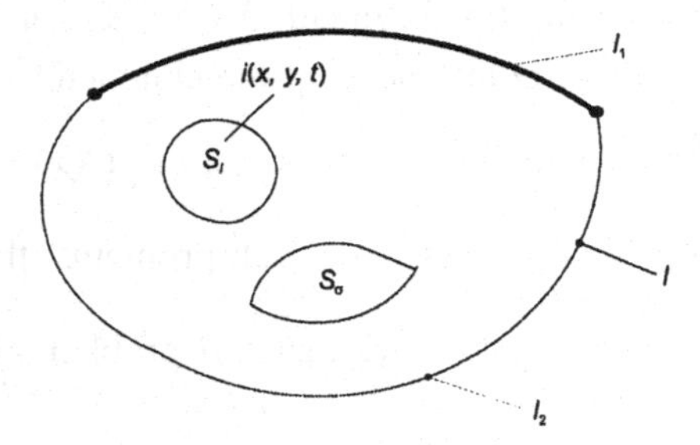

Fig. 4.27. Domains S_i and S_σ of desired current distribution and specification of the objective function, accordingly

Assume that the electromagnetic field is plane-parallel. For its description the vector magnetic potential *A*, being in this case a scalar, is used. The potential in the area S_i satisfies the equation

$$\text{div grad}\, A = -\mu J_{ext},$$

and in the area S_σ - the equation

$$\text{div grad}\, A - \mu\sigma \frac{\partial A}{\partial t} = 0 .$$

Assume the condition $A=0$ on the part l_1 of the border, and the condition $\frac{\partial A}{\partial n}=0$ on the part l_2 of the border. The objective functional can be written down as:

$$I = 0.5\int_S\left[\int_0^T \sigma\left(\frac{\partial A}{\partial t}\right)^2 dt - q_w(x,y)\right]^2 ds.$$

Here, the quantity $\sigma\left(\partial A/\partial t\right)^2 = \sigma E^2(x,y,t)$ determines the power emitted in the body element ds caused by induced electric current flowing in the conducting body.

Assuming that there is no field at $t \le 0$ and $J_{ext}=0$ at $t=0$, we can write down the initial condition for the potential as $A(x,y,0)=0$ throughout the whole domain. The desired distribution $J_{ext}(x,y,t)$ can be discontinuous and will therefore have discontinuous time derivatives, hence, the problem solution is searched in a generalized sense.

Let's find the equation, boundary and initial conditions for the adjoint variable $\lambda(x,y,t)$ included in the augmented functional

$$L(A,J_{ext},\lambda) = I +$$
$$+\int_S\int_0^T \lambda(x,y,t)\left[\text{divgrad}A - \mu\sigma\frac{\partial A}{\partial t} + \mu J_{ext}(x,y,t)\right]dtds. \quad (4.12)$$

We shall seek the equation for function $\lambda(x,y,t)$ and the algorithm of current density $J_{ext}(x,y,t)$ change in the domain S_i from conditions $\delta_A L=0$, $\delta_{J_{ext}} L=0$.

Let's transform the integral in Eq. (4.12):

$$\int_S\int_0^T \lambda(x,y,t)\text{divgrad}A dtds =$$
$$= \int_l\int_0^T (\lambda\frac{\partial A}{\partial n} - A\frac{\partial \lambda}{\partial n})dtdl + \int_S\int_0^T A\,\text{divgrad}\,\lambda dtds.$$

As $\int_0^T \lambda\mu\sigma\frac{\partial A}{\partial t}dt = \mu\sigma(\lambda A|_{t=T} - \lambda A|_{t=0}) - \int_0^T A\mu\sigma\frac{\partial \lambda}{\partial t}dt$, then the expression (4.12) can be written down as:

$$L(A,J_{ext},\lambda)=I+\int_{l}\int_{0}^{T}(\lambda\frac{\partial A}{\partial n}-A\frac{\partial \lambda}{\partial n})dtdl+\int_{S}\int_{0}^{T}A\mathrm{divgrad}\lambda dtds-$$

$$\int_{S}\mu\sigma\left(\lambda A\big|_{t=T}-\lambda A\big|_{t=0}\right)ds+\int_{S}\int_{0}^{T}A\mu\sigma\frac{\partial \lambda}{\partial t}dt+\int_{S}\int_{0}^{T}\lambda\mu J_{ext}dtds.$$

Taking into account that

$\delta_A\int_0^T\lambda\mu\sigma\frac{\partial A}{\partial t}dt=\mu\sigma\lambda(\delta A)\big|_{t=T}-\int_0^T(\delta A)\mu\sigma\frac{\partial \lambda}{\partial t}dt$, we have

$$\delta_A L=\delta_A I+\int_{l}\left\{\int_0^T[\lambda\delta\left(\frac{\partial A}{\partial n}\right)-\left(\frac{\partial \lambda}{\partial n}\right)\delta A]dt\right\}dl+$$

$$+\int_{S}\left[\int_0^T\mathrm{div}\,v\mathrm{grad}\lambda(\delta A)dt\right]ds-$$

$$-\int_{S}\mu\sigma\lambda(\delta A)\big|_{t=T}ds+\int_{S}\int_{0}^{T}\mu\sigma\frac{\partial \lambda}{\partial t}(\delta A)dtds.$$

The equation for the variable $\lambda(x,y,t)$ in the domain S_σ can be given as:

$$\mathrm{divgrad}\lambda(x,y,t)+\mu\sigma\frac{\partial \lambda}{\partial t}=f_\sigma.$$

Here, f_σ is a function determined by variation $\delta_A I$ of the objective functional.

It is necessary to supplement this equation with boundary conditions $\lambda=0$ on the part l_1 and $\partial\lambda/\partial n=0$ on the part l_2 of the domain contour. They should take into account boundary conditions for functions A and $\partial A/\partial n$. The condition for variable λ at $t=T$ is bound to the type of objective functional.

Outside of the domain S_σ, the adjoint variable satisfies the Laplace's equation $\mathrm{divgrad}\lambda=0$.

If the adjoint function λ satisfies this equation's initial and boundary conditions, then we have $\delta_A L=0$, and fulfillment of equation $\delta_{J_{ext}}L=0$ is necessary for observance of optimality conditions. Variation of Eq. (4.12) with respect to current density J_{ext} results in the relationship

$$\int_{S}\int_{0}^{T}\mu\lambda(x,y,t)\delta J_{ext}(x,y,t)dtds=0,$$

i.e. in the condition $\lambda(x,y,t)=0$. As the increment of the functional can be presented as $\Delta L(J_{ext})\cong-\int_{S}\int_{0}^{T}\mu\lambda(x,y,t)\Delta J_{ext}(x,y,t)dtds$, then the objective function gradient appears to be equal $\mu\lambda(x,y,t)$.

Thus, the optimization problem in a two-dimensional non-stationary electromagnetic field is reduced to boundary problems for potentials A and λ, which enables calculating the objective function gradient and constructing an iterative process for search of the optimum function $J_{ext}(x,y,t)$.

On the first step we solve the equation for potential A at a given initial function $J_{ext}(x,y,t)$ to find the function $A(x,y,t)$, and further the adjoint function $\lambda(x,y,t)$, which satisfies the above equations as well as the boundary and initial conditions. Using the objective function gradient found at solution of the adjoint problem, we can apply known gradient methods for search of current density $J_{ext}(x,y,t)$ optimum distribution.

It should be noted that a constraint $J_{ext}(x,y,t) \le J_{max}$ can be imposed on the desired current density, which means that for a given diameter of the wire in the winding located in the domain S_i, the winding current $i \le i_{max}$. Another possible constraint is the condition $J^2(x,y,t) \le K$ in points of the conducting body, which means limitation of instant power of losses in connection with infeasibility of local overheating of the material.

References

1. Reklaitis, G.V., A. Ravindran, and K.M. Ragsdell (1983). *Engineering Optimization*. New York: John Wiley and sons.
2. Demirchian, K.S., and V.L. Chechurin (1986). *Electromagnetic fields computing* (in Russian). Moscow: High School.
3. Vasiliev, F.P. (1981). *Methods for Extreme Problems Solution* (in Russian). Moscow: Nauka.
4. Korn, G.A., and T.M. Korn (2000). *Mathematical Handbook for Scientists and Engineers: Definitions, Theorems, and Formulas for Reference and Review*. New York: Dover Publication.
5. Bendsoe, M., and O. Sigmund (2003). *Topology Optimization. Theory, Methods, and Applications*. New York: Springer-Verlag.
6. Park, I.H., et al. (2002). Comparison of Shape and Topology Optimization Methods for HTS Solenoid Design. *Korea-Japan Joint Workshop Appl Supercond and Cryog*, Seoul, Nov. 14-15:32-33.
7. Allaire, G. (2002). *Shape Optimization by the Homogenization Method*. New York: Springer.
8. Yoo, J., N.Kikuchi, and J.L.Volakis (2000). Structural optimization in magnetic devices by the homogenization design method. *IEEE Trans Mag*, vol 36, no3:574-580.
9. Chechurin, V.L., and A.E.Plaks (1996). The procedure of optimization of magnetic structure. *IEEE Trans Mag*, vol 32, no 3:1278-1281.

10. Cherkaev, A. (2000). *Variational methods for structural optimization*. New York: Springer-Verlag.

11. Gibianski, L.V., K.A. Lure, and A.V. Cherkaev (1988). Optimization of thermal flux by the inhomogenious medium (in Russian). *J of Tech Phys*, vol 58:67-74.

12. Chechurin, V.L., M.V. Eidemiller, and A.G. Kalimov (2001). Comparison of Lagrange Multipliers and Constrained Quasi-Newton Methods in Magnetic Shape Optimization. *Rec 13th Compumag Conf Comp Electromag Fields*, vol IV:1332-1333.

13. Haug, E.J., K.K. Choi, and V. Komkov (1985). *Design Sensitivity Analysis of Structural Systems*. New York: Academic press.

14. Byun, J., and S. Hahn (1997). Optimal Design of Induction Heating Devices Using Physical and Geometrical Sensitivity. *Proc Fourth Japan-Korea Symp Elect Eng*, Seoul: 55-58.

15. Park, I.H., et al. (1996). Design sensitivity analysis for transient eddy current problems using finite element discretization and adjoint variable method. *IEEE Trans Mag*, vol 32:1242.

16. Kwak, I., et al. (1998). Design Sensitivity of Electro-Thermal Systems for Exciting-Coil Positioning. *Intern Journ of Appl Electromag and Mech*, vol 9, no 3:249-261.

Chapter 5. Solving Practical Inverse Problems

5.1 Search for lumped parameters of equivalent circuits in transmission lines

In this section we shall consider problems of synthesis of equivalent circuits of transmission lines. As it was already noted repeatedly, synthesis of an equivalent circuit is a typical inverse problem of circuit theory.

The problem of synthesis consists of two subproblems: synthesis of the equivalent circuit structure and synthesis of its parameters. Synthesis of structures of lines' equivalent circuits is fairly well-known. Therefore, classical results will be considered rather briefly, however the main focus will be given to newly obtained equivalent circuits. In this section we shall consider synthesis in the time domain. The purpose of synthesis of equivalent circuits (their optimization) consists in the definition of such parameters of equivalent circuits, at which their transient responses approximate transient responses of a line in the best way.

Processes in transmission lines are described by the system of telegrapher's equations. For operator representation of vectors of currents $\boldsymbol{I}(p,x)$ and voltages $\boldsymbol{U}(p,x)$ at zero input conditions, they are given by [1,2]:

$$-\frac{dU(p,x)}{dx}=Z(p)I(p,x), \qquad -\frac{dI(p,x)}{dx}=Y(p)U(p,x) \tag{5.1}$$

where $\mathbf{Z}(p)$ and $\mathbf{Y}(p)$ are matrixes of parameters. Problems of propagation of electromagnetic waves in systems of horizontal wires, coaxial cables, transformer windings or in cylindrical screens can be reduced to telegraph-type equations [1].

Classical open-chain ladder-type equivalent circuits of lines are derived by conversion from partial derivatives to finite differences in Eqs. (5.1):

$$\left.\frac{\partial f(x)}{\partial x}\right|_{x=\frac{x_n+x_{n+1}}{2}} \approx \frac{f(x_{n+1})-f(x_n)}{\Delta x}.$$

The system of difference equations together with boundary conditions forms the difference scheme. Difference schemes corresponding to T- or U-shaped n-link equivalent circuits are graphically represented in Fig. 5.1.

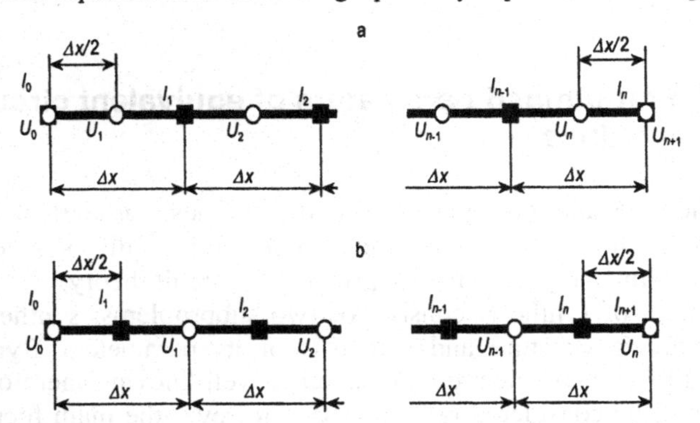

Fig. 5.1. Stencils for deriving of n-link T-shaped **a)** and U-shaped **b)** equivalent circuits

Let's consider in more detail the process of construction of an equivalent circuit with use of the stencil represented in Fig. 5.1a. We divide the line of length l into n equal sections each of length Δx, and then substitute partial derivatives in Eq. (5.1) by finite differences:

$$\begin{cases} I_k(p)-I_{k-1}(p)=-\Delta x Y(p)U_k(p), & k=\overline{1,n}, \\ U_{\nu+1}(p)-U_\nu(p)=-\Delta x Z(p)I_\nu(p), & \nu=\overline{1,n-1}, \\ U_1(p)-U_0(p)=-\dfrac{\Delta x}{2}Z(p)I_0(p), \\ U_{n+1}(p)-U_n(p)=-\dfrac{\Delta x}{2}Z(p)I_n(p). \end{cases} \tag{5.2}$$

We shall supplement Eq. (5.2) with boundary conditions $U_0=U(0,p)$ and $U_{n+1}=U(l,p)$. Then Eq. (5.2) will correspond to the n-link equivalent circuit made of T-shaped links. Each link of the equivalent circuit has the form shown in Fig. 5.2a. This equivalent circuit of the line is named as T-shaped. Similarly, the stencil shown in Fig. 5.1b can be applied to finite-difference approximation of derivatives in Eq. (5.1). The set of equations similar to Eq. (5.2) will have the following form:

$$\begin{cases} U_k(p)-U_{k-1}(p)=-\Delta x Z(p)I_k(p), & k=\overline{1,n}, \\ I_{\nu+1}(p)-I_\nu(p)=-\Delta x Y(p)U_\nu(p), & \nu=\overline{1,n-1}, \\ I_1(p)-I_0(p)=-\dfrac{\Delta x}{2}Y(p)U_0(p), & \\ I_{n+1}(p)-I_n(p)=-\dfrac{\Delta x}{2}Y(p)U_n(p). & \end{cases} \tag{5.3}$$

Equation (5.3) will be supplemented with boundary conditions $I_0 = I(0,p)$ and $I_{n+1} = I(l,p)$. Then the set of Eq. (5.3) will correspond to the n-link equivalent circuit made of U-shaped links. Each link of the equivalent circuit has the form shown in a Fig. 5.2b. Further, this equivalent circuit of the line is named as U-shaped.

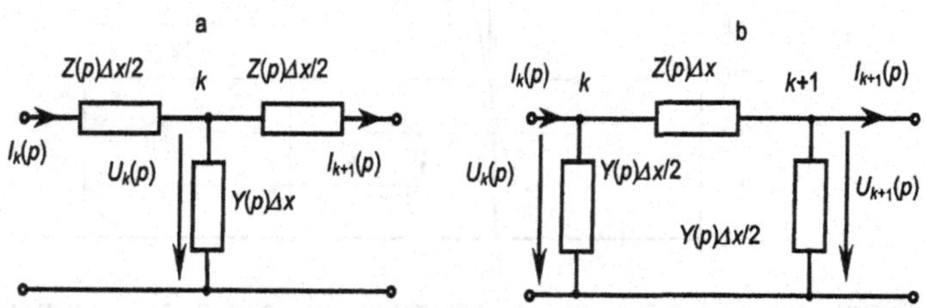

Fig. 5.2. Links of T-shaped **a)** and U-shaped **b)** equivalent circuits of a line

Let's compare the processes in a line and in its open-chain equivalent circuit. For this purpose we shall consider transient at energizing the T-shaped m-link equivalent circuit of the line with a constant voltage U_0. The equivalent circuit models reproduce processes in a lossless line of length l with linear parameters L' and C' loaded on wave impedance $z_w = \sqrt{L'/C'}$. In this case, we have $Z(p) = pL'$, $Y(p) = pC'$. This problem has an analytical solution. The load current of an m-link T-shaped equivalent circuit is given by:

$$i_m(t) = \frac{U_0\omega_0}{z_w}\int_0^{\omega_0 t} J_{2m}(\omega_0 t)\,dt\ ,$$

where $\omega_0 = 2/\sqrt{L_s C_s}$ is the link cutoff frequency, $L_s = L'\Delta l/2$ and $C_s = C'\Delta\ell$ are the link inductance and capacitance, accordingly, and J_{2m} is a Bessel function of the first kind of the second order, $2m$. This expression for the current allows us to write down the voltage transfer characteristic $k_U(t)$ for the line equivalent circuit as:

$$k_U(t) = \frac{z_w i_m(t)}{U_0} = \omega_0 \int_0^{\omega_0 t} J_{2m}(\omega_0 t)\ dt.$$

Dependences of $k_U(t)$ for equivalent circuits of lines containing three (curve 2) and eight (curve 3) T-shaped links are shown in Fig. 5.3. The step curve 1 in Fig. 5.3 represents the dependence $k_U(t)$ for a lossless line. The transmission factor on voltage is equal to zero up to the moment $t_0 = \frac{l}{v} = l\sqrt{L'C'}$, equal to the wave travel time along the line (v is the velocity of wave propagation along the line).

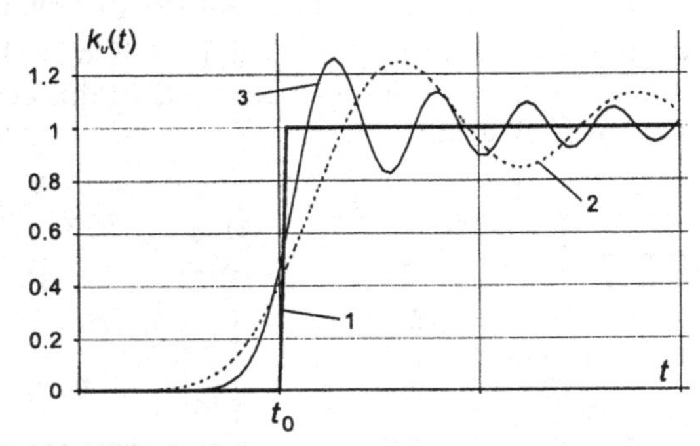

Fig. 5.3. Transfer characteristics of a line (1), and its 3-link (2) and 8-link (3) equivalent circuits

Comparison of curves in Fig. 5.3 shows that equivalent circuits reproduce processes in a real line with a large error. Modeling of processes in the line by means of equivalent circuits shows weakly damped oscillations that are not present in the real transient. It can be shown that increasing the number of equivalent circuit links does not result in diminution of the amplitude of these oscillations, but only increases their frequency. The first maximum of oscillations is about 27% of the established value. Such errors of modeling are inadmissible in the majority of electrotechnical calculations.

Examination of non-distorting lines closer to practice, as well as lines with frequency-dependent linear parameters, shows that oscillations introduced by equivalent circuits are damped. Therefore their equivalent circuit will show reproduction of real transients with not so large error. Nevertheless, the problem of synthesis of an equivalent circuit reproducing the properties of a lossless line in the best way represents fundamental interest. Equivalent circuits obtained when solving this problem will correspond to its solution for the worst case. Therefore, they will allow modeling of processes in lines with losses with a greater accuracy.

Let's consider an approach to optimization of lines' equivalent circuits based on improvement of finite-difference approximation of Eq. (5.1). It can be shown [4,5,6], that the approximation error of derivatives in Eq. (5.1) differs for the internal, and extreme points of the stencil. For internal points of both stencils shown in Fig. 5.1, the principal member of the error is given by $\frac{\Delta x^2}{24}\frac{\partial^3 U}{\partial x^3}$. As for extreme points of stencils, the principal member of error on variable I for the stencil shown in Fig. 5.1a is $\frac{\Delta x}{4}\frac{\partial^2 U}{\partial x^2}$, and for the stencil shown in Fig. 5.1b it is $\frac{\Delta x}{4}\frac{\partial^2 I}{\partial x^2}$ on variable U. This imperfection of classical approximation of derivatives on x in Eq. (5.1) can be overcome by use of three-point approximation for extreme points of stencils.

Let's show the way of deriving the difference equation for the extreme left point of the stencil shown in Fig. 5.1a. We search the expression for the derivative in the following form:

$$\left.\frac{dU(x,p)}{dx}\right|_{x=0}=\frac{\alpha_0 U_0(p)+\alpha_1 U_1(p)+\alpha_2 U_2(p)}{\Delta x}. \tag{5.4}$$

We employ the Maclaurin formula to represent functions $U_1(p)$ and $U_2(p)$ in the point $x=0$

$$U_1(p)=U_0(p)+\frac{\Delta x}{2}\left.\frac{dU(x,p)}{dx}\right|_{x=0}+\frac{1}{2}\left(\frac{\Delta x}{2}\right)^2\left.\frac{d^2U(x,p)}{dx^2}\right|_{x=0}+$$
$$+\frac{1}{6}\left(\frac{\Delta x}{2}\right)^3\left.\frac{d^3U(x,p)}{dx^3}\right|_{x=0}+O(\Delta x^4),$$

$$U_2(p)=U_0(p)+\frac{3\Delta x}{2}\left.\frac{dU(x,p)}{dx}\right|_{x=0}+\frac{1}{2}\left(\frac{3\Delta x}{2}\right)^2\left.\frac{d^2U(x,p)}{dx^2}\right|_{x=0}+$$
$$+\frac{1}{6}\left(\frac{3\Delta x}{2}\right)^3\left.\frac{d^3U(x,p)}{dx^3}\right|_{x=0}+O(\Delta x^4).$$

After substitution of the expressions for $U_1(p)$ and $U_2(p)$ in Eq. (5.4), we come to the following set of equations for definition α_0, α_1 and α_2:

$$\begin{cases}\alpha_0+\alpha_1+\alpha_2=0,\\ \alpha_1+3\alpha_2=2,\\ \alpha_1+9\alpha_2=0,\end{cases} \quad \text{from which} \quad \begin{cases}\alpha_0=-8/3,\\ \alpha_1=3,\\ \alpha_2=-1/3.\end{cases}$$

Substitution of expressions for α_0, α_1, α_2 in Eq. (5.4), after some transformations gives:

$$\left.\frac{dU(x,p)}{dx}\right|_{x=0} = \frac{8\left(U_1(p)-U_0(p)\right)-\left(U_2(p)-U_1(p)\right)}{3\Delta x}+O\left(\Delta x^2\right). \qquad (5.5)$$

The principal member of error $\frac{\Delta x^2}{8}\frac{\partial^3 U}{\partial x^3}$ has the infinitesimal order $O\left(\Delta x^2\right)$. It means that the relationship (5.5) ensures the same accuracy of approximation in the extreme point of the stencil, as in its internal points.

Similar approximation of the derivative can be fulfilled for the extreme right point of the stencil. Using thus found expressions for derivatives in extreme points of the stencil, we have the following, instead of Eq. (5.2):

$$\begin{cases} I_k(p)-I_{k-1}(p)=-\Delta x Y(p)U_k(p), \quad k=\overline{1,n}, \\ U_{v+1}(p)-U_v(p)=-\Delta x Z(p)I_v(p), \quad v=\overline{1,n-1}, \\ \frac{8}{3}(U_1(p)-U_0(p))-\frac{1}{3}(U_2(p)-U_1(p))=-\Delta x Z(p)I_0(p), \\ \frac{8}{3}(U_{n+1}(p)-U_n(p))+\frac{1}{3}(U_n(p)-U_{n-1}(p))=-\Delta x Z(p)I_n(p). \end{cases} \qquad (5.6)$$

Sets of Eqs. (5.2) and (5.6) differ only by the form of the last two equations. Therefore, modifications in equivalent circuits will occur only in extreme links. So the extreme left link takes the form shown in Fig. 5.4 when using Eq. (5.6).

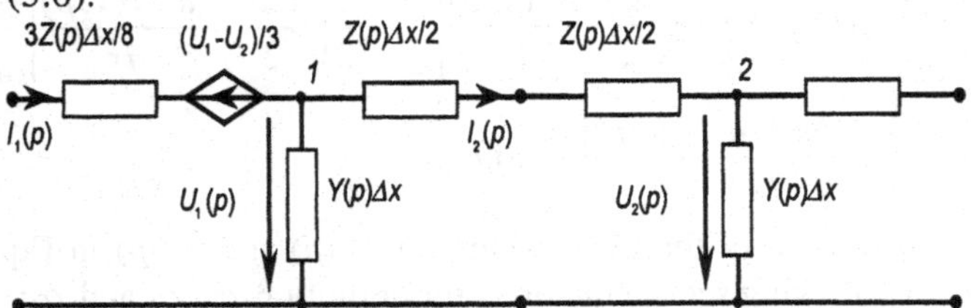

Fig. 5.4. The equivalent circuit extreme left link according to equations (5.6)

A set of equations similar to Eqs. (5.2) and (5.6) can also be obtained for the stencil shown in Fig. 5.1b. The corresponding equivalent circuit consists of U-shaped links, and extreme links also contain controlled sources (in this case – current sources).

The transient characteristic $k_U(t)$ of a lossless line loaded on its wave impedance is shown in Fig. 5.5. We compare this transient characteristic with characteristics of two five-link, T-shaped equivalent circuits. One of the equivalent circuits is the traditional one corresponding to Eq. (5.2). The other for the improved circuit with controlled sources in extreme links corresponds to Eqs. (5.6).

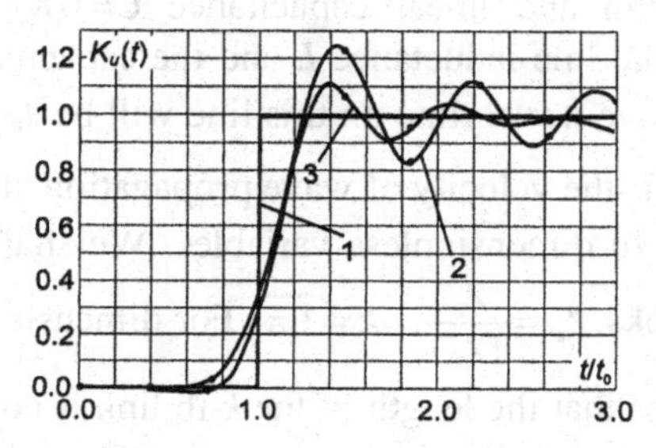

Fig. 5.5. Dependences $k_U(t)$ for:1- line, 2- equivalent circuit corresponding to Eqs. (5.2), 3- equivalent circuit corresponding to Eqs. (5.6)

As may be seen from Fig. 5.5, line transient characteristic's representation by the transient characteristic of the equivalent circuit with controlled sources is better in comparison with the traditional circuit. It shows diminution of magnitudes of extremums of oscillations with increased damping of oscillations and a steeper pulse front. Modifications have concerned only extreme links of the equivalent circuit. Therefore, equivalent circuits with controlled sources can be used for any number of links. It must be noted as well that factors 3/8 at extreme resistances $Z(p)\Delta x$ 1/3 next to controlled sources in extreme links can be subject to further optimization.

Another concept of line equivalent circuit synthesis will be discussed further. Let the modeled line be divided into unequal sections, each modeled by a single T-shaped link. We consider lengths of the line sections λ_k, $k = \overline{1,m}$ modeled by a link as parameters of optimization. At that, we assume that the following conditions are satisfied:

- the sum of lengths of all sections is equal to the line length - $\sum_{k=1}^{k=m} \lambda_k = \lambda$;

- sections arranged symmetrically with respect to the middle of the line are have identical lengths that necessary for the equivalent circuit symmetry $\lambda_k = \lambda_{m-k+1}$, $k = \overline{1,n}$ $m = 2n+1$, or $m = 2n$; and

- T-shaped links of the equivalent circuit are symmetrical.

Results derived below can be used for lines modeled by equivalent circuits consisting of U-shaped links as well.

The inductance $L_k = K_1 L' \lambda_k / 2\ell$ and capacitance $C_k = K_2 C' \lambda_k / \ell$ of each link of the equivalent circuit are proportional to the length λ_k of the modeled line section. Here, K_1 and K_2 are constants depending on the line geometry. We can assume constants K_1L and K_2C to be equal to 1 without belittling the generality of the further results, which will correspond to a line with a linear inductance $L=1/K_1$ H/m and linear capacitance $C=1/K_2$ F/m. Let the line length $\ell=1$ m, then the line inductance L and the capacitance C will also be equal to 1. Thus, a wave travel time on this line will be $t_0 = 1/v = 1$ s, where $v = 1/\sqrt{L'C'} = 1$ m/s is the velocity of wave propagation along the line.

Let's change over to dimensionless variables. We shall introduce dimensionless lengths of links $\ell_k = \dfrac{\lambda_k}{1/m}$, $k = \overline{1,n}$. For dimensionless quantities the equality $\ell_k = 1$ means that the length of the k-th link is equal to the length of links in the case of dividing the line into equal sections. After normalization, the vector of optimization parameters becomes $\mathbf{l} = (\ell_1, \ell_2, \ldots \ell_n)^T$, where $n = entier\left((m-1)/2\right)$ owing to the equivalent circuit symmetry.

We shall consider the module of difference of the lossless line's characteristic $k_U(t)=1(t-1/v)$ and the transient characteristic $k_{UC}(t)$ of its equivalent circuit as the minimized functional F:

$$F(\mathbf{l}) = \frac{1}{T}\int_0^T \left| k_{UC}(t,\mathbf{l}) - 1(t - 1/v) \right| dt \xrightarrow[\mathbf{l}]{} \min. \tag{5.7}$$

Integration of upper limit T needs to be chosen equal to $3t_0/v=3$, which corresponds to the moment of arrival of the first reflected wave to the line end.

The reason for choice of such a statement of this problem is that comparison of a lossless line and its equivalent circuit most clearly shows the discrepancy between processes in a line and in its equivalent circuit. Dimensionless values of length and linear parameters of the line can be converted to real parameters by multiplication to corresponding factors. Using unit line parameters and $1(t)$ as input action ensures simplicity of calculation and obviousness of results. It should be noted that problems of connection of a line closed on its wave impedance z_w (or close to it) are typical for electric transmission and communication lines.

For calculation of each value of the functional F it is necessary to solve the following set of condition equations:

$$\begin{cases} \dfrac{di_k}{dt} = \dfrac{u_k - u_{k+1}}{L_{k+1} + L_k}, & k = \overline{0,m} \quad u_0 = U_0, \quad L_0 = 0, \quad L_{m+1} = 0, \\ \dfrac{du_p}{dt} = \dfrac{i_{p-1} - i_p}{C_p}, & p = \overline{1,m} \quad k_{UC}(t) = i_m(t) z_w / U_0, \end{cases} \tag{5.8}$$

for the equivalent circuit shown in Fig. 5.6 at zero entry conditions. Parameters of the equivalent circuit $L_1 \ldots L_m$ and $C_1 \ldots C_m$ are determined by elements of the vector **l**. Solution of Eq. (5.8) can be carried out by any numerical-analytical or numerical method, e.g. by the Runge-Kutta method. The integral in Eq. (5.7) can be calculated, for example, by the method of trapezoids.

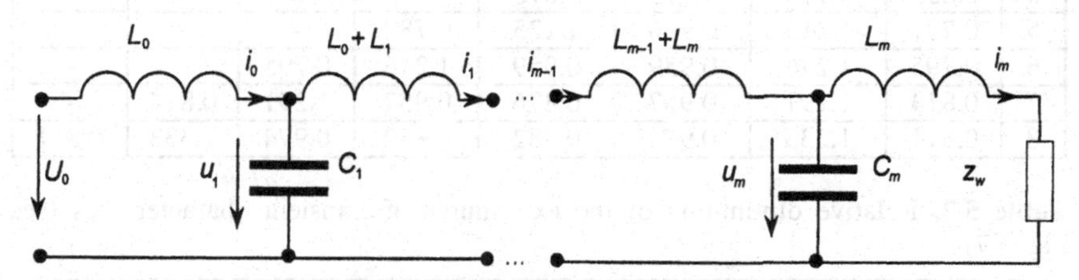

Fig. 5.6. The equivalent circuit of forming line

To minimize the functional F at rather small number of variables (n<5) Powell's method, using information on gradient F for advancement to the minimum, has been applied. For narrowing the search area of a minimum with a view of prevention of process convergence to local minima of the functional, which correspond to optimal equivalent circuits with smaller numbers of links, restrictions given by $0.5 \le \ell_k \le 1.5,\ k = \overline{1,n}$ have been superimposed on optimization variables. This restriction was ensured by adding to the functional of penalty functions in the form of:

$$\varphi_k(\ell_k) = \begin{cases} 0, & \text{under } 0.5 \le \ell_k \le 1.5, \\ A_k \cdot (\ell_k - 1.5)^2, & \text{under } \ell_k > 1.5, \qquad k = \overline{1,n}, \\ A_k \cdot (\ell_k - 0.5)^2, & \text{under } \ell_k < 0.5. \end{cases}$$

Weight factors A_i are positive and have been chosen empirically from previous experience of calculations. In view of this, at use of the gradient method the minimized functional becomes

$$F(\mathbf{l}) = \frac{1}{T}\int_0^T \left| k_{UC}(t,\mathbf{l}) - 1(t - 1/v) \right| dt + \sum_{k=1}^{n} \varphi_k(\ell_k) \xrightarrow[\mathbf{l}]{} \min. \tag{5.9}$$

Results of solution of the problem (5.9) are listed in Table 5.1, where optimum partitions of the line to T-shaped links for m=3-8 are shown. Effectiveness of thus obtained solutions is illustrated in Table 5.2. Data in this table show that equivalent circuits obtained by solution of the problem (5.9) reproduce processes in a line better than traditional ones. Indeed, relative diminution of the first maximum is 18-33 % and there are practically no subsequent extremums.

Table 5.1. Relative diminution of extremums obtained as the result of solution of the problem (5.9) for m=3-8

m	1st	2nd	3rd	4th	5th	6th	7th	8th
3	0.852	1.296	0.852	-	-	-	-	-
4	0.830	1.170	1.170	0.870	-	-	-	-
5	0.787	1.215	0.996	1.125	0.787	-	-	-
6	0.795	1.246	0.959	0.959	1.246	0.795	-	-
7	0.814	1.291	0.957	0.876	0.957	1.291	0.814	-
8	0.811	1.333	0.974	0.882	0.882	0.974	1.333	0.811

Table 5.2. Relative diminution of the extremums of transient characteristics (see Fig. 5.7)

m	1st $(b_1-a_1)/b_1$	2nd $(b_2-a_2)/b_2$	3rd $(b_3-a_3)/b_3$
		%	
3	33.7	95.4	63.8
4	23.1	59.6	72.1
5	22.4	61.4	96.0
6	21.5	60.4	97.6
7	17.9	60.6	98.5
8	18.5	59.3	99.2

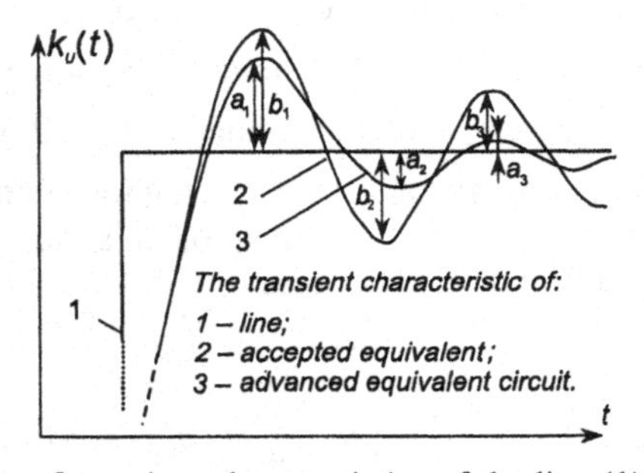

Fig. 5.7. Comparison of transient characteristics of the line (1), accepted equivalent circuit (2) and advanced equivalent circuit (3)

When increasing the number of variables n, the use of gradient methods in synthesis problems rapidly becomes ineffective. Therefore, at $n>5$ it was expedient to use the genetic algorithm. Restrictions on optimization parameters ($0.5 \leq \ell_k \leq 1.5,\ k = \overline{1,n}$) have been maintained. The size of the population has been assumed to be equal to $150n$, and the number of parents – to $20n$.

Figure 5.8 shows comparison of transient characteristics of a line (curve 1) and T-shaped equivalent circuits for n=10 (Fig. 5.8a) and for n=15 (Fig. 5.8b). Curve 2 defines the traditional equivalent circuit transient characteristic (equal apportionment of the line on sections), and curve 3 – transient characteristic of the equivalent circuit obtained at minimization of Eq. (5.9) by means of genetic algorithm.

Figure 5.8a,b also show relative lengths of links $\ell_k,\ k = \overline{1,n}$ that are the solution of the problem (5.9). One must pay attention to the fact that the functional minimum value is reached at oscillating distribution of lengths of links relative to their average value. It is practically impossible to choose entry conditions for the gradient method leading to such solution without prior information on the character of this distribution. Therefore, for $n>5$ even the best solutions obtained by the gradient method were always essentially worse than solutions obtained by means of the genetic algorithm. Moreover, solution of the problem by the gradient method requires much more computational time in comparison with the evolutionary method.

It should be also noted that though calculations were carried out for a lossless line and at U_0=const, obtained optimum partitions of the line into sections will remain optimal for undistorting lines with losses. They will be close to optimum in case of lines with monotone dependence of parameters on frequency, as well as for arbitrary voltages of the source.

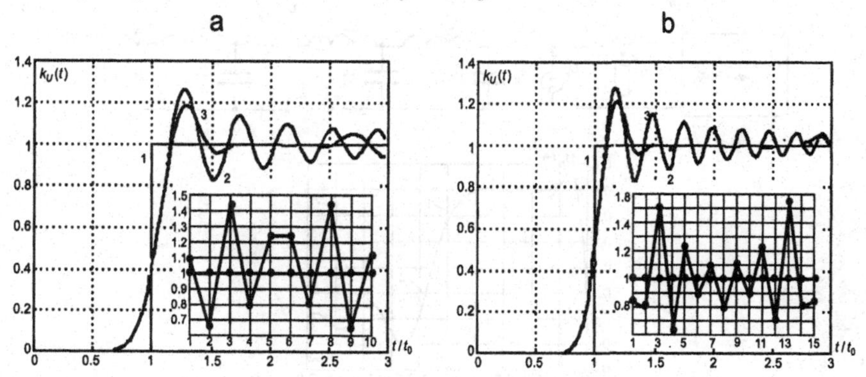

Fig. 5.8. Transient characteristics of the line (1), accepted equivalent circuit (2) and advanced equivalent circuit (3) and relative lengths of links $\ell_k,\ k = \overline{1,n}$, obtained from solution of the problem (5.9) for n=10 **a**) and n=15 **b**)

5.2 Optimization of forming lines

In this section we shall consider parametrical synthesis of optimum artificial forming lines. In practice, forming circuits (Fig. 5.9a) are often applied for shaping of voltage or current impulses of prescribed forms on electrophysical loads. They store electromagnetic energy in condensers and inductance coils at the stage of charging of the accumulating device, then return it into the load at discharge. Energy is slowly transferred into the accumulator during the charging period (the switch is in position 1), and then the reserved energy is transmitted into the load R_L at discharge (the switch is in position 2). At that, the shape of current and voltage impulses in the load do not depend on the strength or stability of the source U_0. It is defined only by parameters of the forming circuit. This allows forming power pulses of highly stable shape and duration in the load.

A line of length ℓ with characteristic impedance z_w, equal to the load resistance, can be used as a forming circuit. In this case a rectangular impulse (Fig. 5.9c, curve 1) with amplitude $U_L = U_0/2$ and duration $\tau_{imp} = 2\frac{\ell}{v} = 2\ell\sqrt{L'C'}$ equal to wave double travel time along the line is formed in the load at discharge. A long line is an ideal forming device from the point of view of the impulse shape. However, use of lines when designing sources of high-power impulses in practice results in expensive structures of large dimension. Therefore, accumulating devices' modeling properties of lines, namely artificial forming lines, are created using reactive elements (Fig. 5.9b) to generate high-power impulses. In practice, these lines consist of high-voltage impulse condensers and connecting buses that act as inductances.

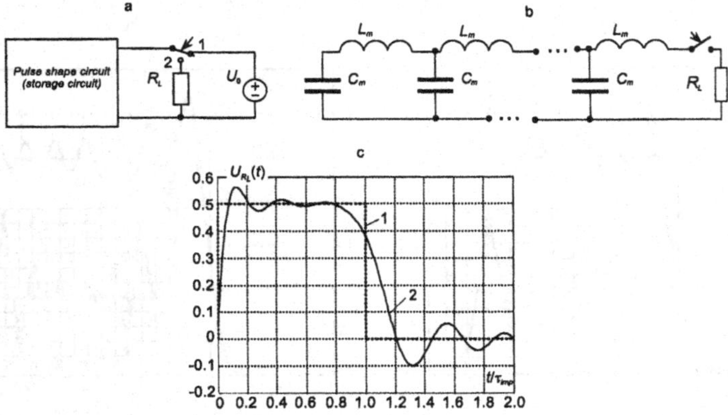

Fig. 5.9. Forming circuit **a**), simplified diagram of a forming circuit **b**), and voltage impulse formed on the load by a five-link artificial forming line consisting of identical links **c**)

Artificial forming lines are extensively used as impulse power supplies for lasers and in other electrophysical installations. Thus, the optimum form of voltage or current impulse on the load is a priory known to the researcher. Further, we assume that a voltage impulse of rectangular form is optimum, if not specified otherwise. Obtained results will be valid for other shapes of impulses as well. The problem of parametrical synthesis of these particular accumulators will be discussed below.

Assume the artificial forming line (Fig. 5.9b) consists of m identical links. At a given load resistance R_L, and impulse duration τ_{imp}, the capacitance and inductance of a link can be found from the following relationships:

$$C_0 = \frac{\tau_{imp}}{2mR_{н}}, \qquad L_0 = \frac{\tau_{imp} R_L}{2m}.$$

Impulse $U_{R_L}(t)$ formed by a five-link circuit (or five-link artificial forming line) in the load $R_L = z_w$ at $U_0 = 1$, is shown in Fig. 5.9c. One can easily see that a five-link line generates impulses considerably differently from the rectangular shape. Therefore, artificial lines consisting of identical links are rarely used for creating high-power forming lines. To improve the shape of impulse, parameters of links should be optimized. Before proceeding to the optimization problem, we shall consider the equivalent circuit of a line in more detail. It will allow us to find results of optimization considerably close to practical objectives.

A rough diagram of an artificial forming line is shown in Fig. 5.10a. The circuit diagram shown in Fig. 5.9b does not reflect in sufficient precision all properties of high-voltage impulse condensers modeled by capacitances $C_1, \ldots, C_m$. As experience demonstrates, when modeling impulse condensers their inductance and losses should be taken into account. In view of the aforesaid, the artificial line circuit diagram will become the equivalent circuit, as shown in Fig. 5.10b.

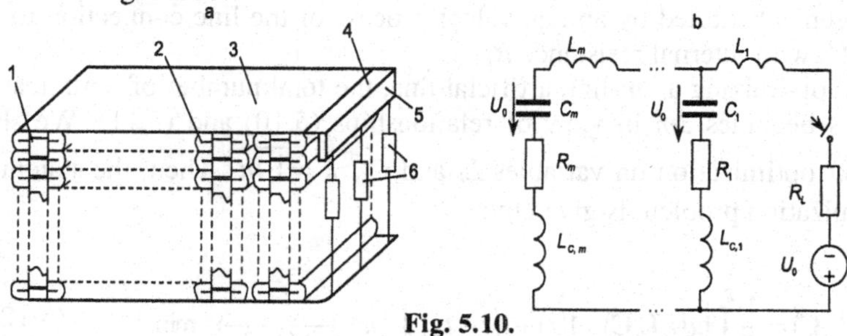

Fig. 5.10.

Diagram of artificial forming line.
1 - connection of condenser sections in a cell
2 - a cell of condensers
3 - connecting buses; 4 - line terminal lead
5 - semiconductor switch
6 – load, n_c - number of sections in a cell

Circuit diagram of artificial forming line.
R_L - load resistance
$C_1, \ldots, C_m$ - condenser capacitances
$L_1, \ldots, L_m$ - inductances of connecting buses
$L_{C,1}, \ldots, L_{C,m}$ - transverse inductances of condensers
$R_1, \ldots, R_m$ - resistors corresponding to losses in condensers

The transverse inductance of series connected cells (vertical packages of sections) $L_{C,1}$ ... $L_{C,m}$ depends on sizes, design and shape of sections and section terminals and methods used for connecting sections in a package. It can be shown [7,8] that the inductance of connecting buses and the transverse inductance of condensers are connected by the following relation:

$$k_p = \frac{L_n}{L_{C,n}} = \text{const}(n), \tag{5.10}$$

where the coefficient $k_p \in [0.25,1]$. The active resistance R_n of the n-th section can be calculated by the cell capacitance as follows:

$$R_n = \frac{\rho C_n d}{3\varepsilon\, \mathrm{d}_f}\left(\frac{n_c}{b}\right)^2. \tag{5.11}$$

Here ρ is the specific resistance of the foil material, ε is permittivity of the dielectric located between the foil sheets, d_f and d are the thicknesses of the foil and the dielectric. Proceeding from dimensions and materials used in real forming lines design, the coefficient of calculation of the active resistance has been assumed R_n Ohm$=PC_n$ [μF] (P=55 Ohm/μF).

At the moment of switching on the load (see Fig. 5.10a), condensers of the artificial forming line are charged up to the working voltage U_0. However it is more convenient to begin the transient calculation at zero initial conditions. Therefore, to calculate the process of line discharge on the load resistance R_L, it has been substituted by an equivalent process of the line connection to the source U_0 with internal resistance R_L.

When optimizing an m-link artificial line, the total number of optimization variables becomes $2m$ in view of relationships (5.10) and (5.11). We shall carry out optimization on variables L_k and C_k, $k=\overline{1,m}$. Then, the statement of optimization problem is given by:

$$F(\mathbf{L},\mathbf{C}) = \frac{1}{T}\int_0^T \left|u(t,\mathbf{L},\mathbf{C}) - 1(t) + 1(t-\tau_{imp})\right| dt \xrightarrow[\{\mathbf{L},\mathbf{C}\}_m \in G]{} \min, \tag{5.12}$$

where $\mathbf{L}=(L_1, \ldots, L_m)^T$, $\mathbf{C}=(C_1, \ldots, C_m)^T$ are vectors of optimization variables and τ_{imp} is the given duration of impulse. The upper limit of integration in Eq. (5.12) is assumed to be $T=2t_{imp}$. In the beginning we shall seek the solution of Eq. (5.12) using the gradient method, and then by the genetic algorithm.

Use of the gradient method requires establishing constraints on the search domain for a minimum value of G and applying a special algorithm for choice of initial conditions. Domain G is given by the following inequalities:

- $L_k>0$ and $C_k>0$ $k=\overline{1,m}$;
- $L_k \geq L_{k-1}$, $C_k \geq C_{k-1}$, $k=\overline{2,m}$, that excludes large numbers of local minima corresponding to circuits with the number of links smaller than m;
- $\frac{1}{q}<\frac{C_k^{(j)}}{C_k^{(0)}}<q$, $\frac{1}{q}<\frac{L_k^{(j)}}{L_k^{(0)}}<q$, $k=\overline{1,m}$, that sets constraints on changing of link parameters in relation to their initial approximations. Here, $L_k^{(0)}$, $C_k^{(0)}$ are initial approximations of the k-th link parameters, $L_k^{(j)}$, $C_k^{(j)}$ are values of the k-th link parameters at the j-th iteration of solution, and p and q are a priori given magnitudes of maximum deviations. Experience in solution of Eq. (5.12) has shown that $p,q \in [5,\ 20]$.

Despite rather stiff constraints on the search domain for a minimum, solution of Eq. (5.12) by the gradient method requires a special algorithm defining the initial approximation. Without this algorithm described in [8], only local minima that are far away from the global minimum of functional F may be calculated from Eq. (5.12). As our calculations have shown, application of the gradient method appears to be effective only for $m<9$.

Values of parameters of links for lines with the number of links from 3 up to 8, resulting from solution of problem (5.12) by the gradient method, are listed in Tables 5.3 and 5.4. In both tables, C_0 and L_0 are respectively the capacitance and inductance of a forming line with identical parameters of links, and the coefficient $k_p=0$. Figure 5.11 shows the voltage impulse in the load, generated by a 5-link homogeneous (curve 1) and optimized (curve 2) forming lines. Apparently, optimization of line parameters has resulted in significant improvement of the impulse shape.

Table 5.3. Values C_k/C_0 for optimal forming lines

k	C_k/C_0	C_k/C_0	C_k/C_0	C_k/C_0	C_k/C_0	C_k/C_0
1	0.242	0.177	0.139	0.114	0.097	0.085
2	0.253	0.177	0.139	0.114	0.097	0.085
3	0.462	0.214	0.151	0.118	0.098	0.085
4		0.402	0.191	0.135	0.107	0.090
5			0.359	0.172	0.124	0.099
6				0.328	0.158	0.114
7					0.304	0.147
8						0.283
C_Σ/C_0	0.957	0.969	0.978	0.982	0.984	0.987

Table 5.4. Values L_k/L_0 for optimal forming lines

k	L_k/L_0	L_k/L_0	L_k/L_0	L_k/L_0	L_k/L_0	L_k/L_0
1	0.232	0.172	0.137	0.113	0.097	0.084
2	0.234	0.173	0.137	0.114	0.097	0.084
3	0.306	0.188	0.141	0.114	0.097	0.084
4		0.264	0.166	0.125	0.102	0.087
5			0.235	0.149	0.114	0.094
6				0.214	0.137	0.105
7					0.197	0.127
8						0.184
L_Σ/L_0	0.772	0.796	0.816	0.828	0.841	0.848

Analysis of data in Tables 2 and 3 shows that values of cell capacitances and inductances increase with rise of the cell number, with a sharp increase of C_k and L_k values for the last cells of the line. The gradient method does not allow finding of satisfactory solutions for Eq. (5.12), for which inequalities $L_k \geq L_{k-1}$, $C_k \geq C_{k-1}$, $k = \overline{2,m}$ are not valid. They can, however, be found by means of the genetic algorithm applied for problems discussed below.

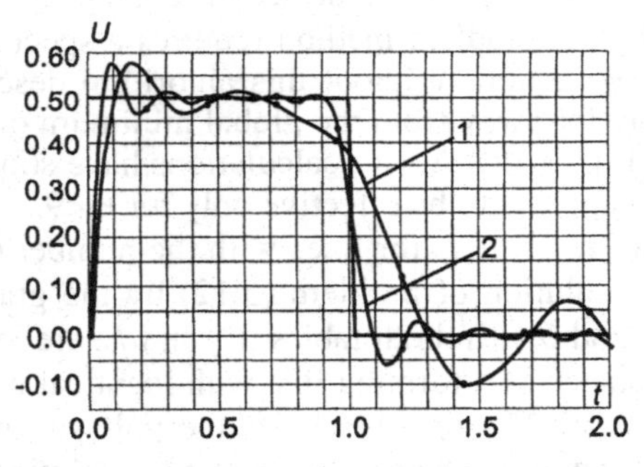

Fig. 5.11. Voltage impulse on the load formed by homogeneous (1) and optimized (2) lines at m=5

Also, one can easily see that for circuits with a number of cells m>5, parameters of several first cells are practically equal. Hence, the impulse shape is improved (in comparison with the impulse shape in a homogeneous line) only by the last 3-4 cells. Therefore, the inhomogeneous line part located near the load can consist of identical cells. For the most part, calculations will be conducted for five-link lines that are more frequently used in practice.

As noted above, the coefficient k_p changes within the interval [0.25,1] depending on the line design. Tables 5.5 and 5.6 show how parameters of cells in a five-link forming line vary depending on k_p value. Obviously, capacitances and inductances of sections change noticeably for various values of factor

k_p. This circumstance is rather important when designing and manufacturing forming lines.

Table 5.5. Values C_k for optimum forming lines at various k_p.

k_p	C_1	C_2	C_3	C_4	C_5
0	0.693	0.694	0.754	0.954	1.794
1/4	0.506	0.507	0.624	0.895	2.381
1/3	0.476	0.477	0.598	0.878	2.495
1/2	0.421	0.435	0.555	0.849	2.665
1	0.323	0.367	0.480	0.788	2.980

Table 5.6. Values L_k for optimum forming lines at various k_p.

k_p	L_1	L_2	L_3	L_4	L_5
0	0.683	0.686	0.704	0.830	1.177
1/4	0.488	0.489	0.547	0.705	1.132
1/3	0.450	0.451	0.519	0.678	1.131
1/2	0.397	0.397	0.477	0.627	1.027
1	0.310	0.310	0.394	0.523	0.848

The load of the lines has been assumed to be linear, active and constant in time in the problem under consideration. However, characteristics of real loads of high-power forming lines (e.g. in the case of volume discharge for gas laser pumping) can be much more complex. Real loads can be nonlinear or variable in time, with inductive or capacitive properties. Besides the load properties, characteristics of the switch may influence the shape of impulse as well. All these features can be approximated in functional (5.12) during optimization. With the above-described approach, the line optimization will not change.

In a number of problems impulse shapes distinct from rectangular are required. The above-described approach for optimization of parameters of inhomogeneous forming lines thus remains unchanged. Let the desired shape of impulse be given by the following relation:

$$U_{imp}(t)=\begin{cases} U_0(\alpha+\beta t), & \text{при} \quad t\le\tau_{imp} \\ 0, & \text{при} \quad t>\tau_{imp} \end{cases}. \tag{5.13}$$

As the number m of optimized line links is equal to 10, a functional similar to Eq. (5.12) was minimized by the genetic method. Impulses obtained as a result of optimization of forming line for ramp-up (α=0.4, β=0.2/τ_{imp}) and ramp-down (α=0.6, β=-0.2/τ_{imp}) ideal impulses are shown in Fig. 5.12.

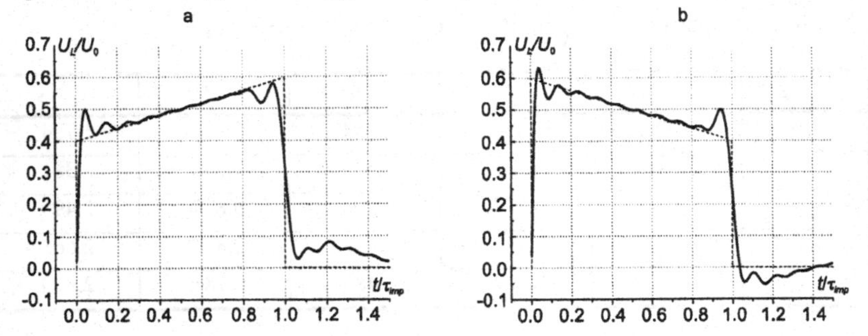

Fig. 5.12. Voltage impulses of 10-link forming line at the optimum impulse given by the expression (5.13) at β>0 – **a**), and β<0 – **b**)

In some areas of high-power impulse engineering, forming of impulses with maximum steep fronts is required. As a result, the following problem of optimization of parameters of inhomogeneous forming lines can be formulated: to obtain impulses with as steep as possible fronts on the load of an inhomogeneous line, at the expense of some deterioration of the impulse shape at the top and droop of the impulse. This problem can be solved by means of a time-dependent weight factor introduced under the sign of integral in expression (5.12), and given by:

$$K(t) = \frac{1}{1 + A\cdot\left(t/\tau_{imp}\right)^2}.$$

Here, the factor A defines the degree of influence of the instantaneous shaping the impulse has on the load. Factor A has been assumed equal to 30. Constraints on positivity of parameters were taken into account by the introduction of penal functions in the functional. Then the statement of optimization problem becomes:

$$F(\mathbf{L},\mathbf{C}) = \frac{1}{T}\int_0^T K(t)\cdot\left|U_L(t,\mathbf{L},\mathbf{C}) - U_{imp}(t)\right| dt + \sum_{i=1}^{n}\varphi_i \xrightarrow[\mathbf{L},\mathbf{C}]{} \min .$$

Our calculations have shown that an increase of front steepness can be achieved only for circuits with a small number of cells. Thus, one can conclude that increasing the steepness of the impulse front is possible only by increasing the number of cells in the line.

So-called double forming lines frequently find application in practice. The diagram of a double forming line, consisting of two identical lines, is shown in Fig. 5.13a. Both lines are charged up to voltage U_0 (the switch is in position 1). When putting the switch in position 2, a rectangular impulse is formed in the active load equal to double wave impedance of lines. Its duration is 2τ, where τ is the wave travel time along one of the lines. Diagrams of voltage distribution along the lines in various instants of time are shown in Fig. 5.13b. Double forming lines allow creating impulses of amplitude U_0 on the load, that is with amplitude equal to charge voltage of condensers. This is an important advantage.

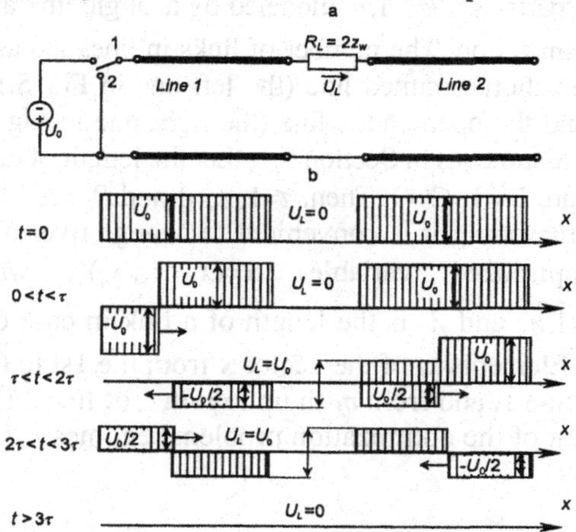

Fig. 5.13. Diagram of a double forming line **a**) and voltage distributions along the lines in various instants of time **b**)

Double lines can also be created on the basis of artificial lines. In this case, each of the lines in the double line design is replaced by a multilink equivalent circuit. However, such replacement results in significant deterioration of the formed impulse shape that is a basic disadvantage of artificial double forming lines. The voltage impulse formed on the load by a double line consisting of two five-link circuits with identical links is shown in Fig. 5.14. One can easily see that the shape of the impulse is rather far from the ideal. Nevertheless, the possibility to form impulses of twice higher amplitude is the governing factor in practice. Therefore, optimization of artificial double forming lines' parameters is an actual problem.

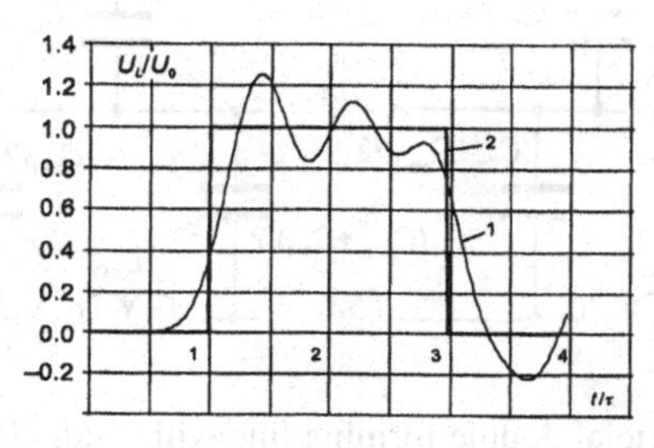

Fig. 5.14. Voltage impulse formed by a double artificial line consisting of 2×5 links (1), ideal impulse (2)

The purpose of optimization of double line parameters consists of creation of impulses on the load which minimally differ from the rectangular impulse. We shall consider this problem in the statement described in Section 5.1, when lines are divided into unequal parts and each of these parts is modeled by a single link. Lengths of line parts λ_k, $k=\overline{1,n}$ modeled by a single link are considered as parameters of optimization. The number of links in lines are assumed to be identical. Let also the short-circuited line (the left one in Fig. 5.13a) be made of U-shaped links, and the open-ended line (the right one in Fig. 5.13a) of T-shaped links. We shall assume, as in Section 5.1, that the length of each line l=1, and lineal parameters are L'=1, C'=1. Then, τ=1, z_w=1 and R_L=2z_w=2.

For further reasoning it is convenient to change over to the dimensionless vector of optimized variables $\mathbf{x}=(x_1, \ldots, x_n)^T$, where $x_k=\lambda_k/\lambda_0$, $\lambda_0=1/m$, $k=\overline{1,n}$, and λ_0 is the length of a link in case of partitioning lines to equal parts. Elements x_k of the vector $\mathbf{x}$ from the 1st to (m-1)-th correspond to the parts of line 1, and from m-th up to n-th – of line 2 (Fig. 5.13,a).

The statement of the optimization problem becomes:

$$F(\mathbf{x}) = \int_0^{5\tau} \left| u(t,\mathbf{x}) - 1(t-\tau) + 1(t-3\tau) \right| dt \xrightarrow[\mathbf{x}]{} \min, \tag{5.14}$$

where $u(t,\mathbf{x})$ is the load voltage created by the artificial line, τ is the wave travel time along the line, and $1(t-\tau)+1(t-3\tau)$ gives the impulse of ideal shape (see Fig. 5.14). Parameters of the equivalent circuit (Fig. 5.15) are calculated by x_k, i.e. $L_k=x_k/m$ and $C_k=x_k/m$.

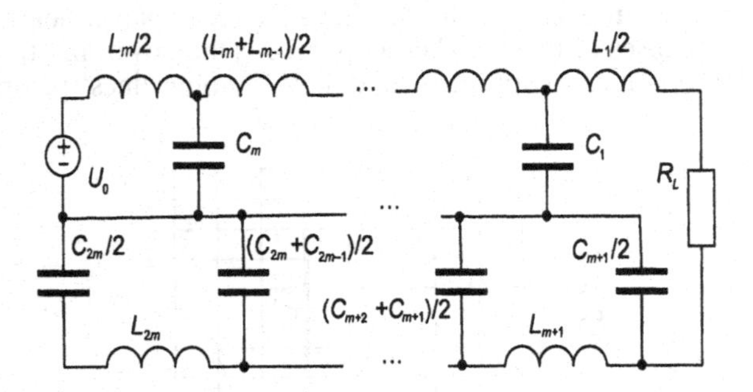

Fig. 5.15. Artificial double forming line with T-and U-shaped links

Minimization of Eq. (5.14) has been carried out by means of the genetic algorithm. Just as for problems considered in Section 5.1, optimization of a double line by use of the genetic algorithm has allowed finding previously unknown solutions. These solutions are characterized by nonmonotonic change of x_k at changing of k. Optimization of double lines by the gradient method does not give an oscillating solution. At that, solutions with monotonically changing parameters correspond to large values of the functional.

The problem of parametrical synthesis under consideration has a large number of local extremums. However, its solution by the genetic algorithm does not require assignment of special initial conditions and/or constraints on the search domain for minimum. The important advantage of the genetic algorithm to overcome local extrema allows obtaining the solution at the first run of the optimization program.

As a result, optimum partitions of lines for numbers m of parts from 3 up to 10 have been obtained. Values of optimum lengths of parts for double forming lines at various m are listed in Table 5.7. For both lines, the counting of parts begins from the load resistance to the short-circuited end (for the line 1), or to the open end (for the line 2).

Table 5.7. Optimum partition of double forming lines

Line 1

m	x_{10}	x_9	x_8	x_7	x_6	x_5	x_4	x_3	x_2	x_1
3	-	-	-	-	-	-	-	1.119	1.153	0.728
4	-	-	-	-	-	-	1.130	1.043	1.348	0.479
5	-	-	-	-	-	0.981	0.855	1.030	1.596	0.538
6	-	-	-	-	1.200	0.961	0.776	1.271	1.328	0.464
7	-	-	-	1.143	0.816	0.903	0.709	1.406	1.538	0.485
8	-	-	1.332	0.942	0.673	1.201	1.528	0.499	1.262	0.563
9	-	1.244	0.862	0.855	0.861	0.943	0.720	1.461	1.582	0.472
10	1.330	0.689	0.836	0.696	0.788	1.264	0.776	1.355	1.711	0.556

Line 2

m	x_{10}	x_9	x_8	x_7	x_6	x_5	x_4	x_3	x_2	x_1
3	1.379	0.657	0.964	-	-	-	-	-	-	-
4	1.072	1.613	0.744	0.571	-	-	-	-	-	-
5	1.250	1.853	0.438	0.516	0.963	-	-	-	-	-
6	1.185	0.489	0.755	1.925	1.156	0.490	-	-	-	-
7	1.000	1.172	0.449	2.014	0.661	1.260	0.444	-	-	-
8	1.464	0.668	1.220	2.332	0.618	0.458	0.710	0.530	-	-
9	1.407	0.918	0.560	1.525	2.248	0.582	0.627	0.688	0.445	-
10	1.421	0.756	1.032	1.685	0.498	1.993	0.629	0.615	0.767	0.601

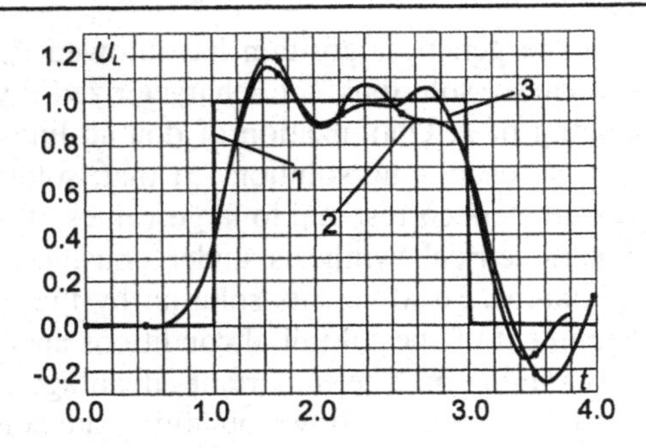

Fig. 5.16. Voltage impulse on the load formed by double lines (1) - ideal impulse, (2) - impulse formed by an artificial line with identical length links (m=5), (3) - impulse formed by an artificial line partitioned to links of optimum length (m=5)

The shape of impulse formed on the load by optimum double 2×5-link lines is shown in Fig. 5.16. The impulse formed by an artificial line with links of identical length is shown in the same figure. One can easily see that the impulse produced by the optimized line is much closer to the rectangular shape. The front of the impulse, as well as the amplitude of oscillations at its top, has decreased. However, the shape of the impulse has appeared neverthe-less to be far from ideal.

Let's consider how a changeover from T-and U-shaped links to L-shaped links influences the shape of impulse (Fig. 5.17). Thus, both inductances L_k and capacitances C_k of each link in each line should be optimized. In this case, the total number of optimization variables will increase up to $4m$. Optimization has been carried out by means of the genetic method. As before, the minimized functional was given by Eq. (5.14).

Results of the solution are shown in Tables 5.8 and 5.9 for 2×3 ... 2×10-link double forming lines. Numeration of links begins from the load R_L. It should also be noted that inductances of the first cells of lines (L_1 and L_{m+1}) are series connected and form a single variable inductance.

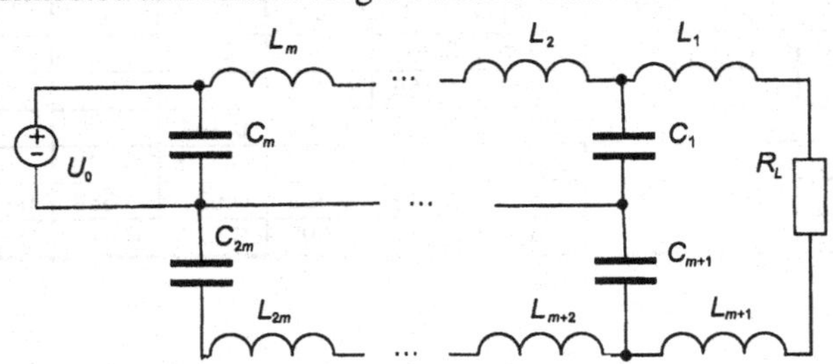

Fig. 5.17. Artificial double forming line with L-shaped links

Table 5.8. Values C_k for optimum double forming lines with L-shaped links

Line 1

m	C_{10}	C_9	C_8	C_7	C_6	C_5	C_4	C_3	C_2	C_1
3	-	-	-	-	-	-		2.879	0.963	1.428
4	-	-	-	-	-	-	2.530	0.810	0.897	1.439
5	-	-	-	-	-	1.685	0.864	1.129	1.065	1.283
6	-	-	-	-	1.839	0.992	0.836	1.161	1.059	1.240
7	-	-	-	1.908	0.884	1.120	1.305	0.902	0.931	1.278
8	-	-	1.277	0.909	0.954	1.024	0.923	1.015	0.955	1.388
9	-	1.579	0.704	0.921	0.927	1.018	0.978	1.045	0.708	1.723
10	1.923	0.698	0.941	0.916	0.975	1.024	0.898	1.071	0.996	1.587

Line 2

M	C_1	C_2	C_3	C_4	C_5	C_6	C_7	C_8	C_9	C_{10}
3	0.850	1.250	1.257	-	-	-	-	-	-	-
4	0.616	1.240	1.471	1.285	-	-	-	-	-	-
5	0.881	1.115	0.840	1.200	0.929	-	-	-	-	-
6	0.819	1.103	1.113	0.800	1.330	1.048	-	-	-	-
7	0.976	0.977	0.990	1.048	0.586	1.382	0.961	-	-	-
8	0.800	1.154	1.283	0.900	1.096	0.529	1.292	1.146	-	-
9	0.973	1.097	1.178	1.064	0.907	0.990	0.832	1.149	1.008	-
10	0.923	1.030	1.035	0.981	1.061	0.865	1.195	0.495	1.276	1.315

Table 5.9. Values L_k for optimum double forming lines with L-shaped links

Line 1

m	L_{10}	L_9	L_8	L_7	L_6	L_5	L_4	L_3	L_2	L_1
3	-	-	-	-	-	-	-	0.709	1.255	0.698
4	-	-	-	-	-	-	0.800	0.953	1.151	0.412
5	-	-	-	-	-	0.962	1.082	1.313	0.750	0.685
6	-	-	-	-	1.046	1.043	0.897	1.285	0.558	0.552
7	-	-	-	1.000	1.117	1.204	1.925	1.063	0.482	0.718
8	-	-	1.200	0.936	1.070	1.006	0.881	1.116	0.623	0.714
9	-	1.246	0.916	0.873	0.976	0.988	0.990	0.943	0.980	0.512
10	1.225	0.721	1.021	0.981	1.028	0.940	0.958	1.119	0.682	0.821

Line 2

m	L_1	L_2	L_3	L_4	L_5	L_6	L_7	L_8	L_9	L_{10}
3	0.774	1.488	0.384	-	-	-	-	-	-	-
4	0.802	1.190	1.471	0.351	-	-	-	-	-	-
5	0.858	0.916	0.853	1.182	0.346	-	-	-	-	-
6	0.656	0.974	1.059	0.799	1.535	0.410	-	-	-	-
7	0.856	0.860	0.983	0.939	1.075	1.107	0.485	-	-	-
8	0.526	0.963	1.093	1.070	0.787	1.223	1.000	0.410	-	-
9	0.559	0.678	1.300	1.113	1.031	0.869	1.040	0.873	0.280	-
10	0.659	0.914	0.910	0.906	0.955	1.006	0.876	1.112	1.283	0.475

Voltage impulses formed on the load by optimum 2×5-link double lines and a line with links of identical length are shown in Fig. 5.18.

Analyzing curves in Fig. 5.18, one can draw the following conclusions:

- the impulse formed by the optimum double line is essentially closer to the rectangular shape in comparison with the impulse formed by a homogeneous line. Durations of the impulse front and droop, as well as the amplitude of oscillations at its top, has decreased. It should also be noted that at $m>5$ the maximum amplitude of oscillations at the impulse top is no more than 3-5% of the impulse amplitude;
- comparing curve 3 in Fig. 5.18 with the similar one obtained by optimization of partitions' lengths (Fig. 5.16), one can easily see that optimization of L_k, and C_k (at twice the number of variable parameters) yields noticeably better results. Distributions of capacitances and inductances versus k obtained by optimization of L_k, and C_k cannot be obtained by optimization of lines' partition lengths. Thus, optimization of L_k, and C_k, in despite of much greater volume of calculations, yields in fundamentally better results.

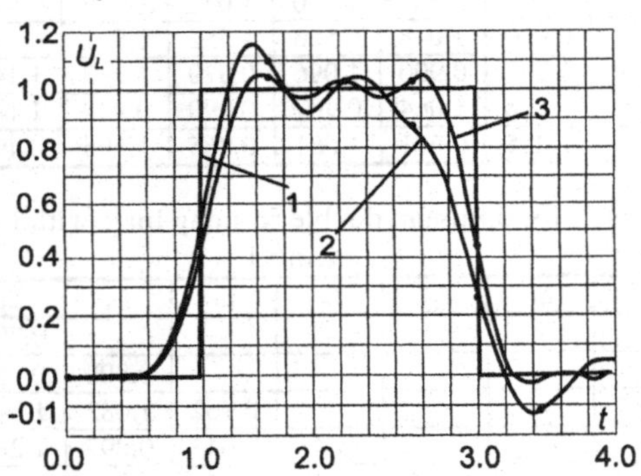

Fig. 5.18. Voltage impulse of the load formed by double lines of double forming lines with m=5. (1) - ideal impulse, (2) - impulse formed by an artificial line with links of identical length (m=5), (3) - impulse formed by an artificial line partitioned to optimum length links (m=5)

Evaluation of dependences of both total capacitance and total inductance of optimum lines on the number of links is of interest. Results are presented in Table 5.10. The table shows that for circuits with any number of cells m, total capacitances C_Σ of optimum lines are greater than the total capacitance C_0=1 of the homogeneous line by 10% on average. The total inductance L_Σ is less than the total inductance L_0=1 of the homogeneous line, also by ~10 %. Accordingly, wave impedances z_w of both line 1 and line 2 are ~0.9 of the homogeneous line wave impedance.

Table 5.10. Total capacitances C_Σ, inductances L_Σ, and wave impedance z_w of optimum double forming lines

m	C_Σ		L_Σ		$z_w = \sqrt{L/C}$	
	Line 1	Line 2	Line 1	Line 2	Line 1	Line 2
3	1.40	1.12	0.87	0.87	0.79	0.88
4	1.42	1.15	0.83	0.95	0.77	0.91
5	1.21	1.00	0.96	0.83	0.89	0.91
6	1.19	1.04	0.90	0.91	0.87	0.93
7	1.19	1.00	0.93	0.88	0.88	0.94
8	1.06	1.03	0.94	0.88	0.94	0.93
9	1.07	1.02	0.94	0.86	0.94	0.92
10	1.10	1.02	0.95	0.91	0.93	0.94

In Sections 5.1 and 5.2, we have discussed problems of parametrical synthesis and characteristic problems arising at their solution in time domain. As noted in Chapter 1, problems of parametrical synthesis can be stated and solved in frequency domain as well. We shall consider these problems in such statement in the following section.

5.3 The problems of synthesis of equivalent electric parameters in the frequency domain

In this section we shall consider problems of calculation of equivalent circuits' parameters for devices according to their frequency characteristics, or problems of parametrical synthesis in frequency domain. At parametrical synthesis in frequency domain the minimized functional is created on the basis of the circuit's frequency characteristics. Deriving of circuits' frequency characteristics in whole is easier than deriving their transient responses.

At first we shall consider a rather simple problem of definition of the equivalent circuit modeling the internal resistance of a round cylindrical wire. In Sections 5.1 and 5.2, solution of the synthesis problem of substitution of long lines in time domain was discussed. In this section, solution of a similar problem in the frequency domain is described. Apart from the fact that these problems are important by themselves, comparison of synthesis methods in different domains is of interest as well.

Let's define parameters of an equivalent circuit for calculation of internal resistance of the round cylindrical wire with frequency characteristics as follows:

$$Z(\omega) = r(\omega) + jx(\omega) = \frac{\ell}{2\pi\gamma a^2} \cdot \frac{a\sqrt{j\omega\mu\gamma} \cdot I_0(a\sqrt{j\omega\mu\gamma})}{I_1(a\sqrt{j\omega\mu\gamma})},$$

where a, l are the wire radius and length, accordingly, and $I_0(x)$ and $I_1(x)$ are modified Bessel functions of zero and first order.

For approximation of the frequency characteristic $Z(\omega)$, various equivalent circuits can be used. Let's consider the ladder circuit shown in Fig. 5.19, having parameters $r_1, \ldots, r_m$, $L_1, \ldots, L_m$, which are parameters of optimization. The frequency characteristic of internal resistance for an aluminum wire with length 1 km and radius 5 mm, for which we shall define parameters of equivalent circuit, is also shown in Fig. 5.19.

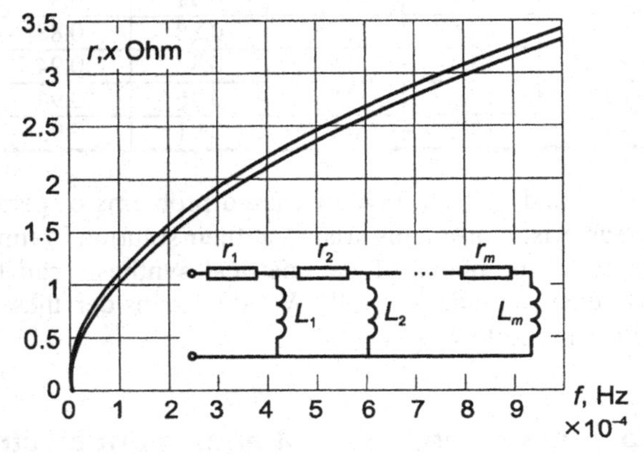

Fig. 5.19. Equivalent circuit and frequency characteristics $x(\omega)$ and $r(\omega)$ of a round cylindrical wire

If we assume that the number of links m in the ladder circuit can be infinite, then values of its parameters are as follows:

$$r_n = (2n-1)\cdot r_0, \quad L_n = \frac{L_0}{n}, \quad n = 1,2,3,\ldots,$$

where $r_0 = \dfrac{\ell}{\gamma\pi a^2}$, and $L_0 = \dfrac{\mu\ell}{8\pi}$ are the wire resistance and internal inductance in the case of uniformly distributed current across the wire's cross-section (direct current). The circuit shown in Fig. 5.19 defines the form (structure) of the equivalent circuit. To use equivalent circuits in practice, it is desirable to have a small number of links in the equivalent circuit at the given approximation error for the frequency characteristic. Therefore, we shall consider this problem for number of links m=4 and m=6.

Assume parameters of the first link are equal to their values at direct current (r_1=r_0, L_1=L_0). Therefore, if the number of links is m, the number of modified variables should be n=2(m-1). Then the problem can be formulated as follows:

$$F(\mathbf{p})=\left[\int_{\omega_1}^{\omega_2}(r(\omega)-r_c(\omega,\mathbf{p}))^2+(x(\omega)-x_c(\omega,\mathbf{p}))^2\frac{d\omega}{\omega_2-\omega_1}\right]^{1/2}\xrightarrow[\mathbf{p}]{}\min, \quad (5.15)$$

where $\mathbf{p}=(r_2, \ldots, r_m, L_2, \ldots, L_m)^T$ is the vector of optimization parameters, and $r_c(\omega,\mathbf{p})$, $x_c(\omega,\mathbf{p})$ are equivalent active and reactive resistances of the ladder circuit. Limits of integral correspond to the frequency interval (0.1-10^2) kHz. To find a solution, the genetic method is applied.

Parameters of equivalent circuits with four and six links obtained for result of minimization of Eq. (5.15) are shown in Table 5.11. Frequency characteristics in both cases practically coincide with the characteristics of the wire.

Table 5.11. Parameters of ladder-type equivalent circuits at m=4 and m=6

Number of link	1	2	3	4	5	6
r, Ohm	0.354	1.225	3.737	0.048	-	-
L, μH	50	25.44	14.40	7.19	-	-
r, Ohm	0.354	0.250	2.409	0.960	10.70	0.0015
L, μH	50	98.32	14.43	18.89	15.32	7.66

The form of frequency characteristic $Z(\omega)$ defines the choice of limits of integral (or the frequency range). In the case of monotonic dependence of $Z(\omega)$, the upper limit of the integral can be defined by means of repeated optimization for its several values. However, at nonmonotone dependences of frequency characteristics such an approach needs additional substantiations.

The problem considered above shows that for frequency characteristics of a simple form and without extremums, calculation of equivalent circuit parameters does not represent any difficulties. In practice, however, devices quite often have nonmonotone characteristics with numerous extremums. Such problems are much more complex than the above-considered one. We shall solve one of such problems noting the basic properties of synthesis problems in the frequency domain.

Synthesis of a long line's equivalent circuit shall be carried out, as before, on the basis of a lossless line problem, closed on its characteristic wave impedance switching on to a constant voltage source U_0=1(t). The diagram of the equivalent circuit is shown in Fig. 5.9b. Voltage gain of such a line in the frequency domain is given by:

$$\underline{K}_U=\frac{\underline{U}_2}{\underline{U}_1}=e^{-j\omega\ell\sqrt{L'C'}},$$

where $\underline{U}_1$ and $\underline{U}_2$ are voltages accordingly in the beginning and in the end of the line, ℓ is its length, and L' and C' are lineal inductance and capacitance, accordingly.

The module of voltage gain of a line is equal to 1 and does not depend on frequency. On the contrary, the module of voltage gain $K_{U,Cir}(\omega)=|\underline{K}_{U,Cir}(\omega)|$ of the multilink equivalent circuit is a strongly varying function of frequency. This function for equivalent circuits of a line consisting of m identical T-shaped links at m=3-6 is shown in Fig. 5.20. One can easily see significant distinction of frequency characteristics of a line and its equivalent circuit. Transient characteristics of a line and its equivalent circuit similarly differ in time domain, as considered above.

Reasons of these distinctions can be explained as follows. Processes in a long line are described by a system of differential equations with partial derivatives that has an infinite spectrum of eigenvalues. Line frequency characteristics (as well as its transient characteristics for time domain) are obtained from these equations. Similar characteristics of the equivalent circuit are obtained from the system of Kirchhoff equations. Hence, the equivalent circuit spectrum of eigenvalues is finite and only approximates the beginning of eigenvalues spectrum of the long line. Therefore, both frequency and transient characteristics of a line and its equivalent circuit cannot coincide in the whole range of frequencies (or in the whole range of time) at any number of links in the equivalent circuit.

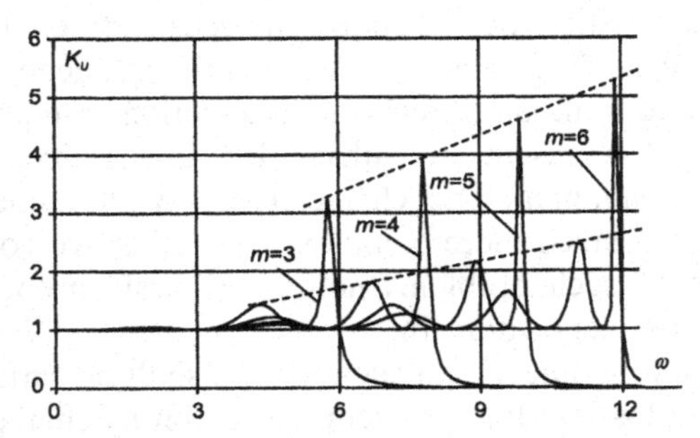

Fig. 5.20. Frequency characteristics of a line and its equivalent circuit at various m

Let each link of the equivalent circuit model be a part $\Delta\ell$ of the line. The line equivalent circuit can be considered a low-pass filter with cutoff frequency $\omega_c = 2/\sqrt{L_{ch}C_{ch}}$, where $L_{ch}=L'\cdot\Delta\ell$ and $C_{ch}=C'\cdot\Delta\ell$ are the inductance and capacitance of the equivalent circuit link, accordingly. As $K_{U,Cir}(\omega)\xrightarrow[\omega>\omega_c]{}0$, then harmonics with frequencies exceeding cutoff frequency are practically zero on the output of the equivalent circuit that does not agree with the line frequency characteristic. Besides, the characteristic $K_{U,Cir}(\omega)$ has maximums at $\omega < \omega_c$, magnitudes of which are proportional to ω (see dashed lines in Fig. 5.20). Those harmonics of input action that correspond to maximums of $K_{U,Cir}(\omega)$ will be amplified on the output. Here, the input action amplitude spectrum is U_0=1(t) $\sim 1/\omega$. Therefore, frequencies corresponding to these maximums at any m will be present on the output of the equivalent circuit that also distinguishes it from a long line. Thus, there is a significant distinction between frequency characteristics of a line and its equivalent circuit.

Further, we consider optimization of line equivalent circuits on frequency characteristics that will allow us to improve the approximation of the frequency characteristics of the line.

Assume parameters of optimization are lengths of line parts λ_k $k=\overline{1,m}$ that are modeled by one link of equivalent circuit each. Then, the optimization problem becomes

$$F(\mathbf{l})=\frac{1}{\Omega}\int_0^{\Omega}\left|K_{U,Cir}(\omega,\mathbf{l})-e^{-j\omega\ell\sqrt{L'C'}}\right|d\omega \xrightarrow[\mathbf{l}]{} \min, \tag{5.16}$$

where $\mathbf{l}=(\lambda_1,..,\lambda_m)^T$ is the vector of varied variables, and Ω defines the frequency range within which minimizations of functional F are carried out.

Let's find an expression for $K_{U,Cir}(\omega)$. For this purpose we shall represent the equivalent circuit by a four-terminal network and write down its equations in A-parameters:

$$\begin{bmatrix}\underline{U}_1\\ \underline{I}_1\end{bmatrix}=\begin{bmatrix}A & B\\ C & D\end{bmatrix}\cdot\begin{bmatrix}\underline{U}_2\\ \underline{I}_2\end{bmatrix}=\prod_{i=1}^{i=m}\mathbf{A}_i(\omega)\cdot\begin{bmatrix}\underline{U}_2\\ \underline{I}_2\end{bmatrix}=\aleph\begin{bmatrix}\underline{U}_2\\ \underline{I}_2\end{bmatrix},\ \text{where } \aleph=\prod_{i=1}^{i=m}\mathbf{A}_i(\omega).$$

Here, $\mathbf{A}_i(\omega)$ is the matrix of A-parameters of the line equivalent circuit i-th link, which in case of T-shaped links is given by:

$$\mathbf{A}_i(\omega)=\begin{bmatrix}1-\frac{1}{2}\omega^2 L_{i,ch}C_{i,ch}, & j\omega L_{i,ch}-j\frac{1}{4}\omega^3 L_{i,ch}^2 C_{i,ch}\\ j\omega C_{i,ch}, & 1-\frac{1}{2}\omega^2 L_{i,ch}C_{i,ch}\end{bmatrix}.$$

A resistance equal to the wave impedance z_w of the initial long line is connected to the output of the equivalent circuit. We assume L=1, C=1, ℓ=1, then z_w=1 and

$$\underline{K}_{U,Cir}(\omega)=\frac{1}{A+B}.$$

Let's change over to dimensionless parameters of optimization. Parameters $L_{i,ch}$ and $C_{i,ch}$ of the equivalent circuit's i-th link are numerically equal to the relative length $x_i = \Delta\ell_i / \ell$ of the line $\Delta\ell_i$ part corresponding to this link, i.e. $L_{i,ch} = C_{i,ch} = x_i$. We shall assume that the line equivalent circuit is symmetric, that is $x_i = x_{m-i+1}$, $(i = \overline{1,m})$. It can also easily be shown that $\sum_{i=1}^{i=m} x_i = 1$. To decrease calculation errors we shall change-over from relative lengths of parts x_k to relative deviations $\delta_k = mx_k$ of lengths. So, when $x_k = 1/3$ and $m=3$ (uniform partition of the line into three parts), we have $\delta_k = mx_k = 1$.

The main contribution into the value of functional F according to the form of integrand (Fig. 5.20) is due to signals of frequencies near the cutoff frequency ω_c. Therefore, it is expedient in this problem to changeover from the frequency ω to dimensionless frequency ϖ through the following relation:

$$\varpi = \frac{\omega}{\omega_c} = \frac{\omega\ell\sqrt{LC}}{2m} = \frac{\omega}{2m}.$$

In dimensionless variables, the matrix of A-parameters of the T-shaped link becomes

$$\mathbf{A}_i(\varpi, d_k) = \begin{bmatrix} 1-2d_k^{\,2} & j2d - j2d_k^{\,3} \\ j2d_k & 1-2d_k^{\,2} \end{bmatrix}, \quad \text{where } d_k = \varpi\delta_k.$$

Matrix $\aleph$ can be derived by multiplication of A-parameters' matrixes for all links of the circuit. To reduce error, we carry out these multiplications analytically. Elements $\pi_{i,j}$ $(i, j = 1,2)$ of the matrix $\aleph$ are given by polynomials with integer-valued coefficients $P_{i,j}$:

$$\pi_{i,j} = P_{i,j} \cdot \varpi^q \cdot \prod_{k=1}^{n} d_k^{\,s_k}. \tag{5.17}$$

Expressions for elements of matrix $\aleph$ for the three-link T-shaped circuit (in this case there is only one optimized variable d_1) are shown in Table 5.12. Recall that $\pi_{1,1}\pi_{2,2} - \pi_{2,1}\pi_{1,2} = 1$. The number of members in polynomials $\pi_{i,j}(\varpi, d_k)$ $i, j = 1,2$ grows rapidly when m increases. So at $m=10$, the number of members in $\pi_{1,2}$ (ϖ, d_k) is ~104. Therefore, a computer program working with analytical expressions has been used for their calculation.

In dimensionless variables the optimization problem (5.16) becomes:

$$F(\mathbf{d}) = \frac{1}{\Theta}\int_0^{\Theta} f(\varpi,\mathbf{d})\,d\varpi \xrightarrow[\mathbf{d}]{} \min$$

$$\text{where } f(\varpi,\mathbf{d}) = \sqrt{\left(\frac{\pi_{1,1}}{\pi_{1,1}^2+\pi_{1,2}^2} - \cos\varpi\right)^2 + \left(\frac{\pi_{1,2}}{\pi_{1,1}^2+\pi_{1,2}^2} - \sin\varpi\right)^2}, \qquad (5.18)$$

where $\mathbf{d} = (d_1, \ldots, d_n)^T$ is the vector of dimensionless variables of optimization and $\Theta=\Omega/\omega_c$.

Research shows that the integrand in Eq. (5.18) varies considerably with small changes of δ_k (see Fig. 5.21), and has high and narrow peaks which move along the frequencies' axis when the vector of parameters changes. It can be shown that the parametrical synthesis problem under consideration is a stiff problem.

Table 5.12. Expressions for coefficients of matrix $\aleph$ at m=3 in the form (5.17)

$\pi_{1,1}$			$\pi_{1,2}$			$\pi_{2,1}$		
$P_{1,1}$	q	s_1	$P_{1,2}$	q	s_1	$P_{2,1}$	q	s_1
1	0	0	6j	1	0	6j	1	0
-18	2	0	-54j	3	0	-72j	3	1
108	4	1	36j	3	1	48j	3	2
-72	4	2	-24j	3	2	-8j	3	3
12	4	3	4j	3	3	216j	5	2
-216	6	3	216j	5	2	-288j	5	3
288	6	4	-216j	5	3	120j	5	4
-120	6	5	72j	5	4	-16j	5	5
16	6	6	-8j	5	5			
			-216j	7	4			
			288j	7	5			
			-120j	7	6			
			16j	7	7			

At numerical integration of Eq. (5.18) there are several problems that are characteristic for synthesis in the frequency domain in general. Let's consider these in more detail.

Numerical calculation of the integral in Eq. (5.18) by using the method of trapezoids requires extremely small steps of integration and, accordingly, significant time in performing the calculation. This is because of acute peaks in the integrand close to ϖ_c that give essential contribution to the value of the integral. Positions of these peaks are unknown beforehand. Moreover, positions of these peaks are different for various **d** found during minimization of Eq. (5.18). This property of the integrand is typical for synthesis upon frequency characteristics and is strongly exhibited at synthesis of high-Q circuits. Therefore calculation of the functional should often be carried out by means of special methods that take into consideration properties of each specific problem.

Another feature of synthesis in the frequency domain is the substantial dependence of optimization results on the choice of Θ - the upper limit of integration. If we assume $\Theta < \varpi_c$, then after several steps of minimization process, the highest peaks of the integrand will be transferred into the area $\Theta > 1$. That is to say they will not be taken into account at minimization of F. If, on the other hand, we choose Θ with a "margin" (for example $\Theta = 2$), then the main contribution into the functional will come from frequencies higher than ϖ_c. So this "uninteresting" part of the curve $K_{U,Cir}(\omega,\mathbf{d})$ will be optimized, whereas its "interesting" part, namely from zero up to the cutoff frequency, will have a weak influence upon the value of the functional. In both cases, not the best voltage gain frequency characteristic will correspond to the minimum of the functional. Choice of Θ varying during minimization is an alternative to the two possibilities considered above. However, problems cannot be avoided in that case too. It requires comparison of functionals calculated at various Θ during minimization.

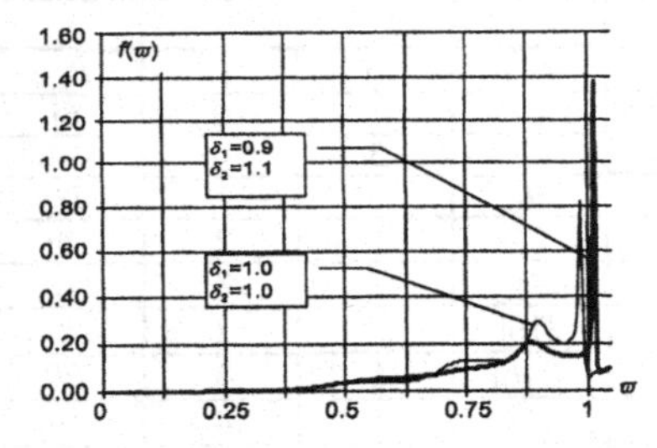

Fig. 5.21. Typical dependences of integrand $f(\varpi,\mathbf{d})$ from the relative frequency ϖ in (5.18) at various δ_1 and δ_2

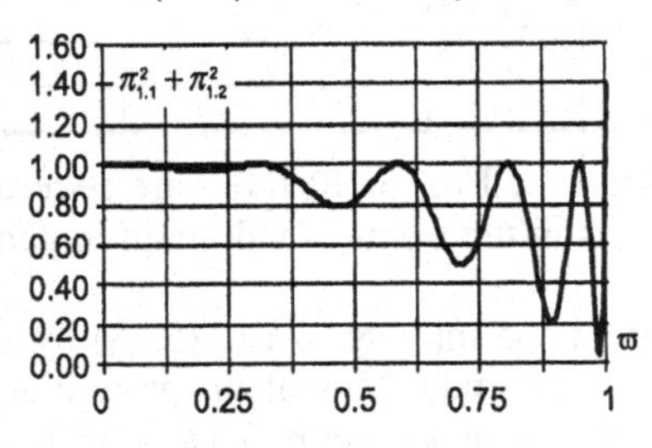

Fig. 5.22. Typical dependence of function $\pi_{1,1}^2 + \pi_{1,2}^2$ from the relative frequency ϖ

We will show below how the problem properties can be used to overcome the above-mentioned difficulties. Before each calculation of the integral in

Eq. (5.18), we shall analyze beforehand properties of the integrand $f(\varpi,\mathbf{d})$. At that we shall focus our attention on properties of its denominator $\pi_{1,1}^2+\pi_{1,2}^2$, that will allow localizing positions of maximums of the integrand. A typical form of $\pi_{1,1}^2+\pi_{1,2}^2$ dependence versus frequency is shown in Fig. 5.22. Maximums of $f(\varpi,\mathbf{d})$ are located in the vicinity of points of minimum of this curve.

To define integration intervals by frequency, upon which the integrand has high peaks, let's find roots of the equation $\pi_{1,1}^2+\pi_{1,2}^2=\varepsilon<<1$. They allow selecting intervals on the frequency axis, on which the integrand has peaks $f(\varpi)>1/\varepsilon$. The algorithm of numerical calculation of the integral in Eq. (5.18) uses this property of functions $f(\varpi,\mathbf{d})$ and $\pi_{1,1}^2+\pi_{1,2}^2$ for adaptive modification of the integration step. It allows reducing the functional F calculation time some tens of times.

The value of Θ - the upper limit of integration - can be found as the maximum real root of the equation $\pi_{1,1}^2+\pi_{1,2}^2=1$. As the analysis demonstrates, the accepted upper limit of integration ensures smooth modification of the functional at modification of its parameters and essentially reduces the stiffness of the problem.

As the method of optimization, the gradient method (method of secants) has been used. To assure better convergence to the global minimum some constraints have been superimposed on variables of optimization, realization of which were ensured by means of the following penal functions φ_i:

$$\varphi_i(\delta_i)=\begin{cases}0, \text{ when } 0.5\le\delta_i\le1.5\\ A_i(\delta_i-1.5)^2, \text{ when } \delta_i>1.5\\ A_i(\delta_i-0.5)^2, \text{ when } \delta_i<0.5\end{cases},\quad i=\overline{1,n},$$

where weight factors A_i were selected during calculations.

In view of the aforesaid, the problem (5.18) becomes

$$F(\mathbf{d})=\int_{\varpi_d}^{\Theta}\sqrt{\left(\frac{\pi_{1,1}}{\pi_{1,1}^2+\pi_{1,2}^2}-\cos\varpi\right)^2+\left(\frac{\pi_{1,2}}{\pi_{1,1}^2+\pi_{1,2}^2}-\sin\varpi\right)^2}\,\frac{d\varpi}{\varpi}+ \qquad (5.19)$$

$$+\sum_{i=1}^{n}\varphi_i(\mathbf{d}) \xrightarrow[\mathbf{d}]{} \min.$$

The factor $1/\varpi$ is introduced into the integrand in Eq. (5.19) to strengthen the influence of low frequencies on the value of functional. The limit inferior of integration ϖ_d is assumed to be equal 0.1 to exclude division by zero and to reduce the calculation time of F.

Results of Eq. (5.19) solution are shown in Table 5.13.

Table 5.13. Relative parameters of the long line equivalent circuits, obtained from Eq. (5.19) at various m.

m	δ_1	δ_3	δ_3	δ_4	δ_5	δ_6	δ_7	δ_8	δ_9	δ_{10}
3	0.918	1.164	0.918	-	-	-	-	-	-	-
4	0.894	1.106	1.106	0.894	-	-	-	-	-	-
5	0.875	1.118	1.014	1.118	0.875	-	-	-	-	-
6	0.866	1.116	1.018	1.018	1.116	0.866	-	-	-	-
7	0.868	1.114	1.013	1.010	1.013	1.114	0.868	-	-	-
8	0.875	1.110	1.016	0.999	0.999	1.016	1.110	0.875	-	-
9	0.862	1.145	1.034	0.986	0.946	0.986	1.034	1.145	0.862	-
10	0.879	1.101	1.027	1.038	0.955	0.955	1.038	1.027	1.101	0.879

Transient characteristics of a long line (curve 1), 8-link T-shaped traditional (curve 2) and optimized equivalent circuits (curve 3) are shown in Fig. 5.23. Apparently, the transient characteristic of the optimized equivalent circuit reflects the real properties of a long line somewhat better than the transient characteristic of the circuit with uniform partition. The improvement however, is insignificant. The calculation process of the minimized functional in the frequency domain is also more complicated than in the time domain. In Sections 5.1 and 5.2, we have considered solution of similar problems in the time domain, and have encountered a series of difficulties. However, we had achieved more significant improvement of the line equivalent circuits' properties. As a whole, optimization in the time domain with reference to this problem is much more effective.

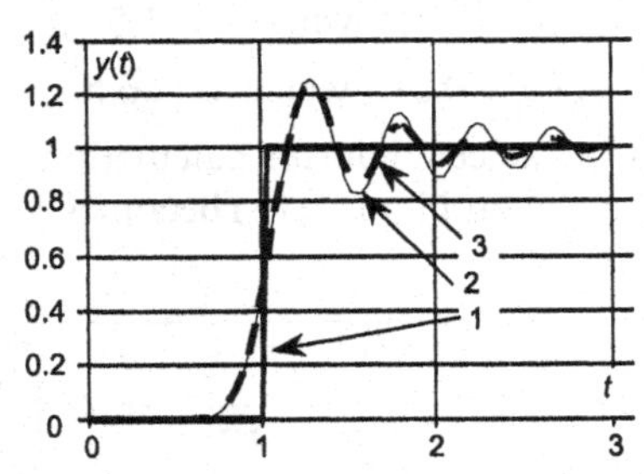

Fig. 5.23. Transient characteristics: - of a line (1), - of equivalent circuit with links of identical length (2) at m=8, - of optimized equivalent circuit (3) at m=8

Synthesis of parameters of equivalent circuits in the frequency domain for rather simple problems do not represent any significant difficulties. Here "simple" refers to problems of synthesis of dipoles (two-terminal networks) with monotone frequency characteristics and a small number of varied parameters in the equivalent circuit. Thus, as shown in the first of the above-considered problems, the synthesized equivalent circuit reproduces the device frequency characteristic with a small error and in a wide frequency range. At that, the number of varied parameters should be 3-6.

For more complex problems, such as synthesis of four-terminal networks with complicated frequency characteristics, parametrical synthesis in frequency domain represents a rather complicated problem for solution. To reduce complexity of its solution it is necessary to consider specific properties of the problem. Thus, each problem (or a class of problems – for example above-considered class of problems of long line's equivalent circuit synthesis) generally requires a specific approach to its solution.

5.4 Optimization of current distribution over conductors of 3-phase cables

This section is concerned with the solution of discrete inverse problems, i.e. problems in which optimization variables can take on only discrete values. Such problems are often referred to as combinatorial problems. Here we shall consider optimization of a multicore three-phase cable on the basis of various criteria of optimality. One of the distinctive features of these problems is the impossibility of using gradient methods for their solution. Therefore, all problems considered in this section were solved by means of the genetic algorithm.

Let's consider a multicore three-phase cable with cross-section shown in Fig. 5.24. Cable strands connected to various phases are marked with white, grey and black colors. Conductors of each phase are of circular section and have individual insulating coatings. They are arranged within the cable in concentric layers. If n_c is the number of layers, then at such disposition the total number of conductors in each phase is equal to $n_{ph}=n_c(n_c-1)$. The conductor radius thus will be equal to $r=R/(2n_c+1)$, where R is the radius of the cable's external sheath.

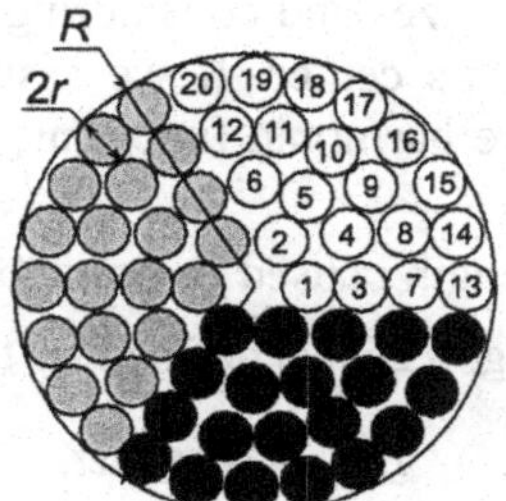

Fig. 5.24. The structure of a three-phase cable

Layout of conductors across the cable is invariable, and the subject of optimization is to what phases (a, b or c) each of the conductors is connected to.

Condition of the cable symmetry from the electromagnetic point of view will be the constraint zone of possible solutions. It corresponds to the symmetric arrangement of conductors in each third of the cable, i.e. the pattern of phase distribution over conductors of one third of the cable completely defines the internal structure of the whole cable. Moreover, one of the conductors can be connected arbitrarily to a certain phase (for example, to *a*) that will reduce the number of optimization variables by unit. Thus, the total number of optimization variables becomes n_{ph}-1.

Let's consider the minimality of external magnetic field generated by the cable current as optimization criterion. Such a cable will create minimum disturbance at transmission of three-phase current through it. Due to the reciprocity principle such a cable will be best protected from disturbances in it, generated by a three-phase current field.

The minimized functional is defined by a loop integral on the circle of radius R_0 from the module of magnetic field strength

$$F(\mathbf{p}) = \frac{3}{2\pi}\frac{1}{T}\int_0^{\frac{2\pi}{3}}\int_0^T \left|\boldsymbol{H}\left(R_0, \varphi, \mathbf{p}, t\right)\right| dt d\varphi, \tag{5.20}$$

where the magnetic field strength $\boldsymbol{H}$ is calculated as the vector sum of fields created by currents of all cable conductors, and elements p_k of optimized variables' vector $\mathbf{p}$ can take on integer values 1, 2, 3. Assume p_k=1 if the conductor is connected to phase *a*, p_k =2 if the conductor is connected to phase *b*, and p_k=3 if the conductor is connected to phase *c*.

Owing to symmetry of the cable structure, distribution of the magnetic field module is symmetric on each third of the circle, therefore we carry out integration by φ within the sector $[0, 2\pi/3]$. Radius R_0, on which the field is calculated, can have an arbitrary value; so it has been assumed R_0=1.5*R*. It allows neglecting conductor sizes and considering that the current of each conductor is concentrated in its center. If we assume that phase currents and magnetic field strength are harmonic functions, they can be written down in complex form.

The magnetic field strength caused in point (*x*,*y*) by current $\underline{I}_k$, $k = \overline{1,3n}$, located in point (x_k,y_k) (Fig. 5.25), is defined by the following expressions:

$$\underline{H}_k = \frac{\underline{I}_k}{2\pi r_k}, \qquad \underline{H}_{x,k} = \underline{H}_k \cos\alpha, \qquad \underline{H}_{y,k} = \underline{H}_k \sin\alpha.$$

Total components of the strength vector defined as sums $\underline{H}_{x,k}$ and $\underline{H}_{y,k}$, accordingly, for all currents in the cable are:

$$\underline{H}_x = \sum_{k=1}^{3n} \underline{H}_{x,k}, \quad \underline{H}_y = \sum_{k=1}^{3n} \underline{H}_{y,k}.$$

After changeover to pre-images and fulfilling integration for H, we have

$$H_x(t) = H_{mx}\sin(\omega t + \varphi_x),\ H_y(t) = H_{my}\sin(\omega t + \varphi_y),\ H(t) = \sqrt{H_x^2(t) + H_y^2(t)},$$

$$H = \sqrt{\frac{1}{T}\int_0^T H^2(t)dt} = \sqrt{H_x^2 + H_y^2}.$$

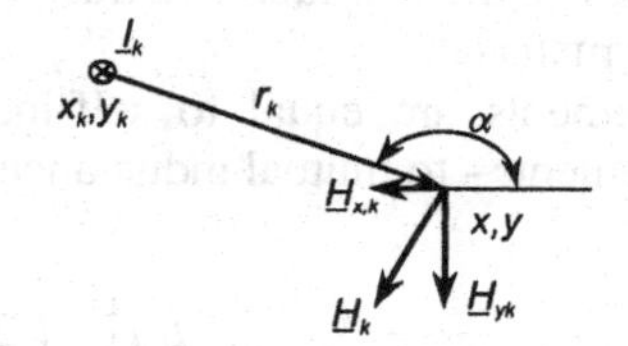

Fig. 5.25. Magnetic field strength $\underline{H}_k$ in the point (x,y), created by current located in the point (x_k,y_k)

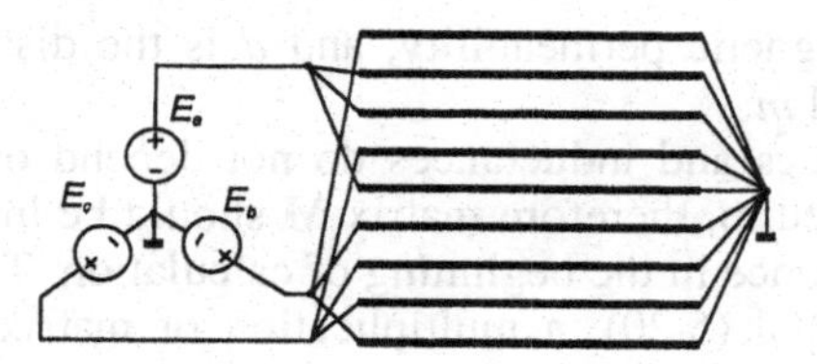

Fig. 5.26. Diagram for calculation of currents in cable conductors

Furthermore, it is necessary to find currents in the cable conductors. We assume that the cable is fed from a symmetric three-phase emf system: $E_a = 1,\ E_b = e^{-j2\pi/3},\ E_c = e^{j2\pi/3}$. In case of symmetry of the cable and its load, the current distribution in each phase does not depend on the load value. Therefore, the diagram shown in Fig. 5.26 with zero-resistance loads can be used for calculation of currents. For simplicity of the picture, only three conductors in each phase are shown. We shall find currents in the circuit by solving the following system of equations:

$$(\mathbf{R} + j\omega\mathbf{M})\cdot\mathbf{I} = \mathbf{Z}\cdot\mathbf{I} = \mathbf{E}, \tag{5.21}$$

where **R**=diag($R_{k,k}$) is the diagonal matrix of conductors' active resistance, **M** is the matrix of self- and mutual inductances of conductors, **I** is the vector of currents in conductors, **E** is the vector of emf sources connected to each conductor, and ω is the circular frequency.

All cable conductors are identical and their active resistance is

$$R_{k,k} = \frac{l}{\gamma\pi r^2}, \quad k = \overline{1,3n_{ph}}, \tag{5.22}$$

where l is the cable length, r is the conductor radius, and γ is the specific conductance of the conductor material.

Matrix **M** diagonal elements are equal to self-inductances of conductor wires, and off-diagonal elements - to mutual inductances of two parallel wires:

$$M_{k,m} = \frac{\mu_0 l}{2}\left[\ln\left(\frac{2l}{d}\right)-1\right], \quad k,m=\overline{1,n_{ph}}, \; M_{k,k} = \frac{\mu_0 l}{2}\left[\ln\left(\frac{2l}{r}\right)-1\right]+\frac{\mu l}{8\pi}. \tag{5.23}$$

Here, μ is the magnetic permeability, and d is the distance between the centers of wires k and m.

Values of resistances and inductances do not depend on which phase a conductor is connected to, therefore matrix **M** should be inverted of the conductor material only once in the beginning of calculation. Then at each calculation of the functional (5.20), a multiplication of matrix $\mathbf{M}^{-1}$ by a vector should be performed. Owing to the symmetry of the problem, finding the currents in only one phase of the cable will be enough. To normalize current values they should be divided by the phase total current.

Let's assume the following geometrical and physical parameters: length of the cable l=100 m, cable radius R=1 cm, conductor material is aluminum (γ=3.57·10^7 Sm/m, $\mu=\mu_0$), and ω=500 rad/s. The circular frequency ω=500 rad/s corresponds to depth of penetration

$$\delta = \sqrt{2/\omega\mu\gamma}, \tag{5.24}$$

that is approximately equal to the cable radius R.

Results of optimization by criterion (5.20) of a three-phase multicore cable with number of conductors n equal to 12, 20, 30 and 42 are shown in Table 5.15 and in Fig. 5.27.

Values of the minimized functional (5.20) for traditional F_{ref} (each phase completely occupies a third of the cable cross-section (see Fig. 5.24)) and for optimum configuration of phases are shown in Table 5.14. One can easily see that minimization results in a significant diminution of the functional value. Figure 5.27 shows optimum configurations of phases at various numbers of conductors n. Distributions of H along the arc $[0, 2\pi/3]$ on the radius R_0 for traditional (Fig. 5.24) and optimum distribution of conductors over the cable phases are shown in the same figure.

Table 5.14. Values of the functional (5.20) at minimization of the cable external field

Number of conductors	12	20	30	42
Reference value of the functional F_{ref}, corresponding to the arrangement of conductors shown in Fig. 5.24	5.545	4.618	4.184	3.958
Minimal value of the functional F_{opt}	0.910	0.562	0.237	0.282
$(1-F_{opt}/F_{ref})\cdot 100\%$	83.6	87.8	94.3	92.9

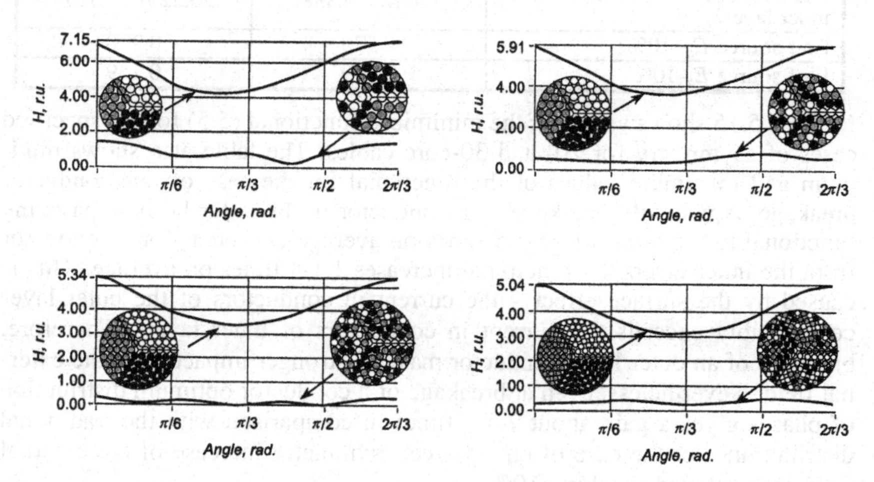

Fig. 5.27. Optimum arrangement of conductors in cable phases by the criterion of the external field minimum for n=12, 20, 30, 42, and distribution of H along the arc $[0, 2\pi/3]$ for optimal and traditional arrangement of conductors

Table 5.15 and graphs in Fig. 5.27 show that relative diminution of the minimized functional has averaged up to 90% that corresponds to diminution of the external field 10 times in average as a result of thus performed minimization. One might say that by connection of cable conductors to a three-phase system in accordance with this optimum, an increase of its noise immunity about 10 times may be achieved. If there are conducting structures or screens near the cable, the magnitude of induced currents in them will decrease 10 times as well.

Analysis of sensitivity of the functional to the three-phase system asymmetry represents practical interest. Let's consider two cases:

- asymmetry of currents caused by one-conductor breakage in the cable;
- asymmetry of currents caused by asymmetry of voltage in the three-phase emf source.

Table 5.15. Values of the functional (5.20) at analysis of its sensitivity to the system asymmetry and optimization of the cable by criterion of minimum of its external field

Number of conductors n in a phase of the cable	Minimal value of the functional (5.20)	
	20	30
Symmetric system	0.562	0.237
One-conductor breakage (in the outer layer)	0.871...1.750	0.683...1.351
One-conductor breakage (in the inner layers)	0.642...0.838	0.525...1.020
Emf source E_a+10%	0.570	0.290
Emf source E_a-10%	0.619	0.259

Table 5.15 shows values of the minimized functional (5.5) for the specified cases of asymmetry for 20- and 30-core cables. The table also shows minimum and maximum values of the functional for the case of one-conductor breakage. Apparently breakage of a conductor in the outer layer impairs the functional to the utmost (1.5-5.5 times on average), whereas for a conductor from the inner layers the functional increases 1.1-4 times on average. This is caused by the surface effect – the current in conductors of the outer layer considerably exceeds the current in conductors of inner layers. Therefore, breakage of an outer layer conductor makes a stronger impact upon the external field. Nevertheless, even at breakage of a conductor optimum distribution of phases gives a gain about 2.5-4 times in comparison with the traditional distribution. For the case of emf source, asymmetry increase of the external field has averaged roughly ~10%.

The structure of three-phase multicore cable can be optimized on the basis of other criteria of optimality as well. Let's consider cable optimization by the criterion of minimum active losses in cable conductors. At a given total phase current, the minimum of active losses can be reached at uniform distribution of current over the conductors. Therefore, optimization requires minimization not of the currents, but of their deviation from the average value.

The multicore cable structure was the same as in the problem considered above. Currents in conductors were also calculated by Eq. (5.21). After changeover to dimensionless currents of conductors we have:

$$\tilde{I}_k = \frac{I_k}{\max(I_k)}, \quad k = \overline{1,n} .$$

The problem of minimization of losses in the cable was solved in the following statement:

$$F(\mathbf{p}) = \sqrt{\frac{1}{n}\sum_{k=1}^{n}\left(\tilde{I}_k - \tilde{I}_{ave}\right)^2} \xrightarrow[\mathbf{p}]{} \min, \qquad (5.25)$$

where $\tilde{I}_{ave}$ is the average current in the cable conductor:

$$\tilde{I}_{ave} = \frac{1}{n}\sum_{k=1}^{n}\tilde{I}_k .$$

Minimum values of functional (5.25) in comparison with its value for the configuration of cable in Fig. 5.24 are shown in Table 5.16. One can easily see that the functional has essentially decreased for all values of *n*.

Table 5.16. Values of the minimized functional at minimization of losses in cable conductors

Number of conductors	12	20	30	42
Reference value of the functional F_{ref}, corresponding to the arrangement of conductors shown in Fig. 5.24	0.3247	0.3294	0.3270	0.3227
Optimized functional F_{opt}	0.0620	0.0627	0.0572	0.0593
$\left(1 - F_{opt}/F_{ref}\right)\cdot 100\%$	80.9	81.0	82.5	81.6

Table 5.17. Relative diminution of losses in cable conductors at minimization of the functional (5.25)

Number of conductors	12	20	30	42
$\Delta_p = \frac{\sum_{k=1}^{n}\tilde{I}^2_{k,opt}}{\sum_{k=1}^{n}\tilde{I}^2_k}\cdot 100\%$	85.0	83.0	83.4	81.2

Optimum cable configurations from the condition of minimum of losses and diagrams of currents distribution over the conductors are shown in Fig. 5.28. As can be seen from graphs of currents, distribution in phase conductors optimization has led to appreciable leveling of current values, and the functional has decreased approximately 5 times for all values of *n*. To estimate diminution of losses, let's assume the total phase current to be equal to 1 and then calculate the power ratio as the sum of quadrates of currents multiplied by the conductor resistance. Values of ratio of losses in the optimized cable to the losses in traditional cable for various numbers of phase conductors are shown in Table 5.17. It is apparent from this table that diminution of losses in conductors corresponds to the degree of minimized functional diminution.

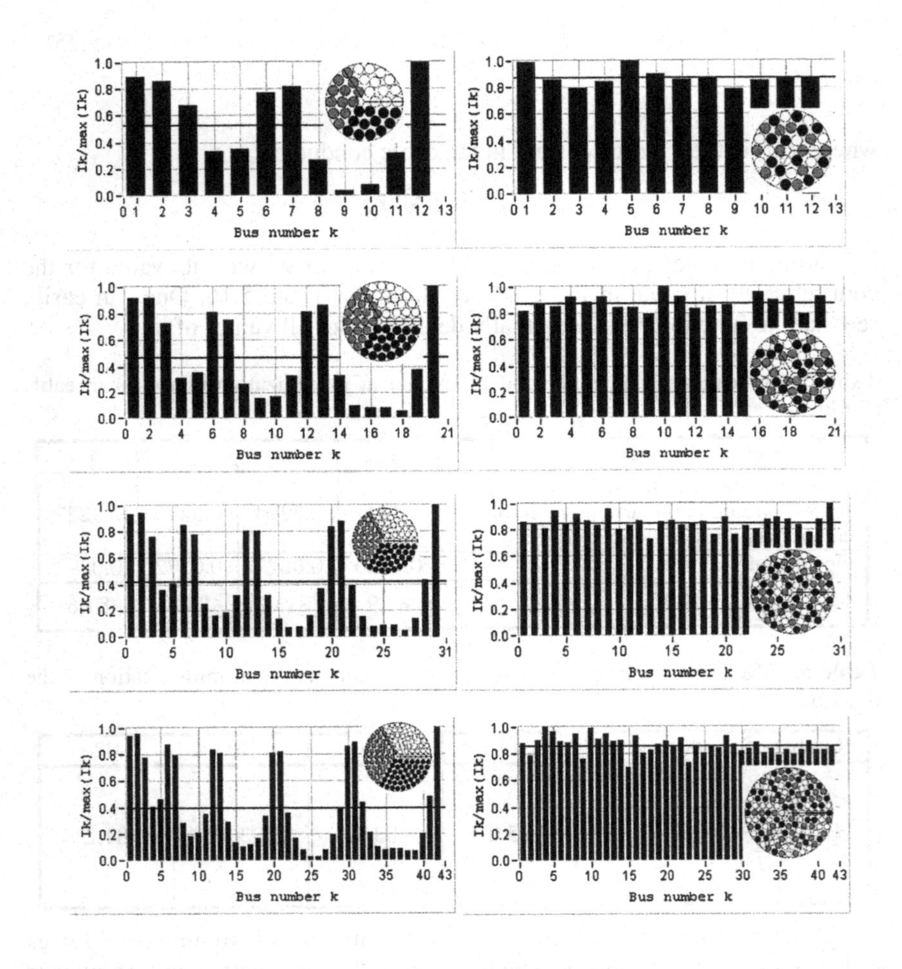

Fig. 5.28. Distributions of conductors over the cable cross-section (n=12, 20, 30, 42) and corresponding distributions of currents over conductors in one third of the cable resulting from optimization by (5.25)

By analogy with the first problem let's consider the influence of cable asymmetry (one-conductor breakage) and asymmetry of emf source on these results. Values of the minimized functional (5.25) at analysis of its sensitivity for 20- and 30-core cables are shown in Table 5.18.

Table 5.18. Values of the functional (5.25) at the analysis of its sensitivity to the system asymmetry

Number of conductors n in a phase of the cable	20	30
Symmetric system	0.0627	0.0572
One conductor breakage	0.185…0.202	0.166…0.172
Emf source E_a+10%	0.0643	0.0602
Emf source E_a-10%	0.0662	0.0616

Apparently, breakage of one conductor increases the functional 3 times on average. A similar result has been obtained for the cable optimized by the minimum of external field. Asymmetry of emf source barely affects the quality of optimization results (at 10% asymmetry of source the functional increases only by 2-7%).

The problem presented above of obtaining a three-phase multicore cable with minimum losses can be converted into the problem of determination of the optimum shape of conductors in a three-phase cable at a given filling factor. In this case, the cable cross-section is divided into segments by the same principle as in the case of separate conductors that were arranged in previous problems. Each segment is filled either by conductor material or by dielectric, at that ratio of the number of segments occupied by the conductor to the total number of segments, which should be equal to the set filling factor k_i. The number of segments into which the cable cross-section is divided is equal to the number of conductors $3n_{ph}$ in the cable of the previous problem and in this case acts as the degree of digitization, when defining the shape of a conductor. This approach leads to the problem of topological optimization, methods of solution of which are intensively developed nowadays.

The following condition was superimposed on each conductor of the cable: each conductor should be entirely located in its third of the cable. This condition has simplified the problem statement and has accelerated the calculation of functional. The obligatory condition of connectivity was not superimposed on segments occupied by conductor, i.e. each segment could be filled with conductor or dielectric irrespective of the condition of adjoining segments. Similar to previous problems, the cable should be symmetric from the electromagnetic point of view, therefore conductors in each phase should have identical shapes, and the total of optimization variables is equal to the number of segments into which one third of the cable is divided, i.e. equal to the number of conductors in each phase in the previous problem: n=n_{ph}.

Minimum loss in the cable is reached when the current is most uniformly distributed over the segments occupied by conductors. Calculation of currents in this problem is more laborious than in previous problems. Elements of matrixes **R** and **M** are calculated by formulas (5.22) and (5.23). Then during each calculation of the functional, new matrixes $\mathbf{R}_{new}$ and $\mathbf{M}_{new}$ should be derived that are composed of those rows and columns of matrixes **R** and **M** which correspond to the cable segments occupied by a current on the given step of solution. A system of equations of dimensionality $m=6k_i n_{ph}$ is set up according to Eq. (5.21). The right members' vector **E** is filled sequentially by values E_a, E_b, and E_c. The vector of optimization variables **p**, which defines the conductor shape, is determined not by the right members' vector, but by the matrix of coefficients which should be inverted at each calculation of currents.

The minimized functional was defined as mean square deviation from the average value of the current in the cable conductor according to Eq. (5.25).

The optimum shape of cable conductors depends on the value of circular frequency ω in connection with the degree of manifestation of the surface effect. Let's consider three cases:

- ω=300 rad/s: The surface effect is weakly expressed; equivalent depth of current penetration (5.9) is larger than the cable radius R;
- ω=5000 rad/s: The surface effect is sharply expressed; equivalent depth of current penetration is approximately equal to the size of the cable cross-section digitization element;
- ω=1500 rad/s: The average degree of surface effect; equivalent depth of current penetration is equal to $0.3R$.

Shapes of cable conductors were optimized for two values of the filling factor k_i - 0.5 and 0.7.

Figures 5.29, 5.30 and 5.31 show results of optimization of the shape of three-phase cable cross-section for three values of frequency. Results are presented for two various numbers of digitization of the cable cross-section: n=20 and n=42. Current distributions over the cross-section of conductors corresponding to these designs of conductors, and values of the functional (5.25) are shown in the same figures.

In case of weak manifestation of the surface effect the optimum arrangement and the shape of the cable conductors represent a "three-leaved" pattern. Sharp manifestation of the surface, which effect each conductor is horseshoe-shaped whereas the intermediate variant corresponds to some transitional form.

These results show that the genetic method effectively solves optimization problems of combinatorial type in electrical engineering. The above described problem can be formulated as a problem of searching of optimum shape or as a problem of topological optimization when the shape of the phase conductors is unknown beforehand. Its particular feature involves assigning specific constraints for desired distribution of cable currents. It should be noted that at its first stage this problem can be solved by search of continuous distribution of current over the cable cross-section as well.

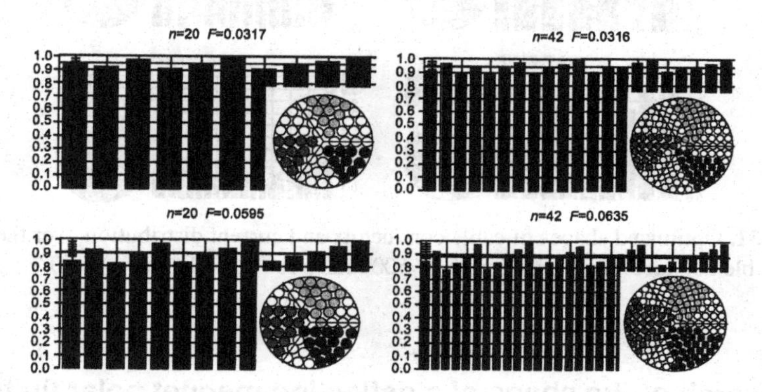

Fig. 5.29. Optimized shapes of cable conductors and current distribution over the cable conductor cross-section at ω=300 rad/s for a) k_i=0.5 and b) k_i=0.7

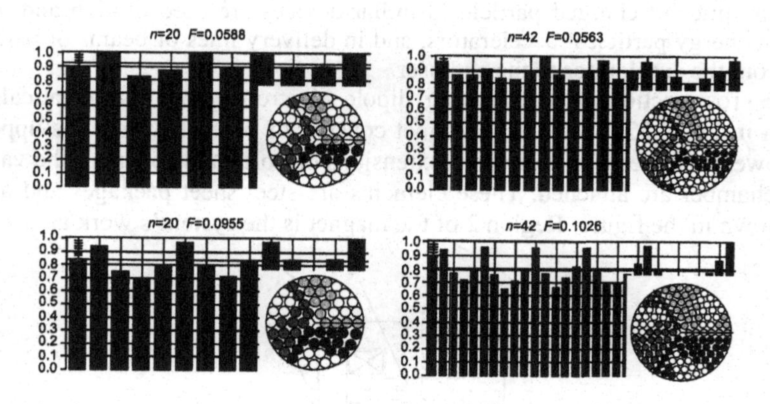

Fig. 5.30. Optimized shapes of cable conductors and current distribution over the cable conductor cross-section at ω=1500 rad/s for a) k_i=0.5 and b) k_i=0.7

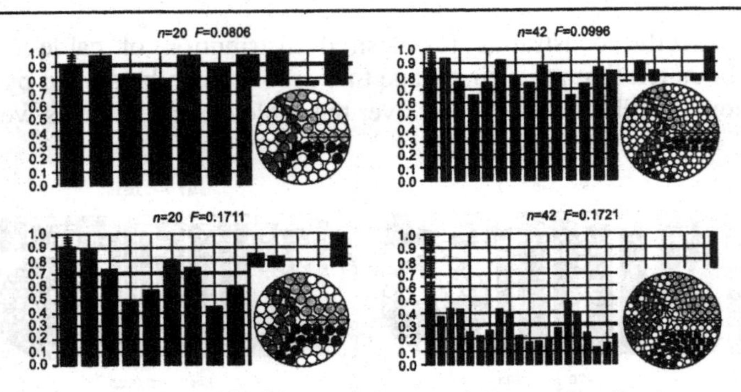

Fig. 5.31. Optimized shapes of cable conductors and current distribution over the cable conductor cross-section at ω=5000 rad/s for a) k_i=0.5 and b) k_i=0.7

5.5 Search of the shape of a deflecting magnet polar tip for producing homogeneous magnetic field

Let's consider the optimization problem of the polar tip for electromagnets used in optics of charged particles. Similar devices are used in high and ultrahigh energy particles' accelerators, and in delivery lines of beams of particles from the accelerator to the end user.

The cross-section of a deflecting dipole electromagnet is schematically shown in Fig. 5.32. This electromagnet consists of poles 1, where the upper and lower pole pieces that are both extensions of poles and covers of the vacuum chamber are attached. These elements are steel sheet packages and are not shown in the figure. Region 2 of the magnet is the system's work area.

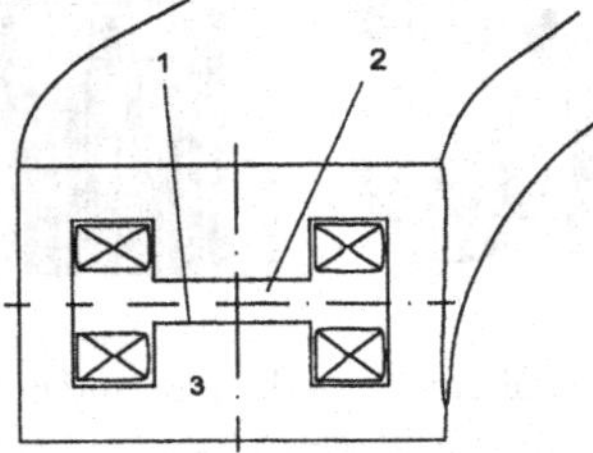

Fig. 5.32. The cross-section of a deflecting dipole electromagnet

In modern technical specifications the mean-root-square error for installation works is 1-2 mm at characteristic dimension of the whole structure up to 10 m, and mean square deviation $\sigma = \frac{1}{B_w}\sqrt{\int_S (B - B_w)^2 ds}$ of the magnetic induction B from its nominal value B_w is approximately $0.5 \cdot 10^{-3}$-$1.0 \cdot 10^{-3}$. At that, much larger local deviations are allowed. Correct arrangement of blocks of packages, of which the electromagnet is assembled, and correct allocation of iron sheets before assembly of packages allow achieving rather low mean square deviations of magnetic induction in the work area at relatively large local deviations.

Designing of an electromagnet of minimum weight, ensuring specified parameters of magnetic field in the working area, is a rather complex problem. It can be solved by calculations only as a first approximation. When designing an electromagnet up to ten models, often full-scale ones are made and investigated. Only after that, successful selection of the best configuration for the sheet 3 of laminated ferromagnetic core is possible (Fig. 5.32).

Static and dynamic distortions of the field in the work area are usually compensated by means of special correcting windings mounted on the poles' surfaces. At use of correcting windings the width of the working area can be increased from 35 cm up to 80 cm, which is considered to be sufficient.

The quarter of cross-section of an electromagnet is shown in Fig. 5.33. Here, 1 is the electromagnet core (made of steel sheets), 2 - the winding, 3 - air gap, 4 is the work area (air), and 5 is the magnet polar tip. The problem consists in determining a shape for the pole tip, at which magnetic induction in the work area 4 has the required degree of homogeneity.

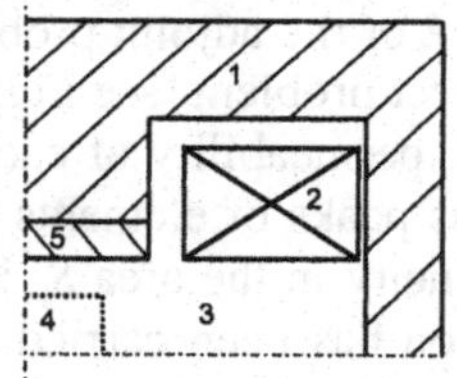

Fig. 5.33. Quarter of electromagnet cross-section

In various variants of calculation, various constant values of required induction B_w in work area S_0 have been accepted.

The magnetic field in the calculation area was assumed to be plane-parallel. For its analysis the vector magnetic potential was used. The calculation area of the investigated device and accepted boundary conditions for the vector magnetic potential are shown in Fig. 5.34. The winding current density has been set equal to J=$2.5 \cdot 10^6$ A/m^2. The shaded area in Fig. 5.34 shows the permissible region S_μ for change of the polar tip shape.

When searching for the optimal shape of the polar tip, the Lagrange method described in Chapter 4 was used. The numerical solution has been obtained by use of the finite-element method.

The objective function has been assumed to take the form of $I = 0.5 \int_{S_0} (B_y - B_w)^2 ds$. Here, B_y, is the vertical component of magnetic induction in the region S_0. The direct problem was solved relative to the vector magnetic potential A. The augmented functional was given by $L = I + \int_S \lambda(x, y) \text{divgrad} A ds$.

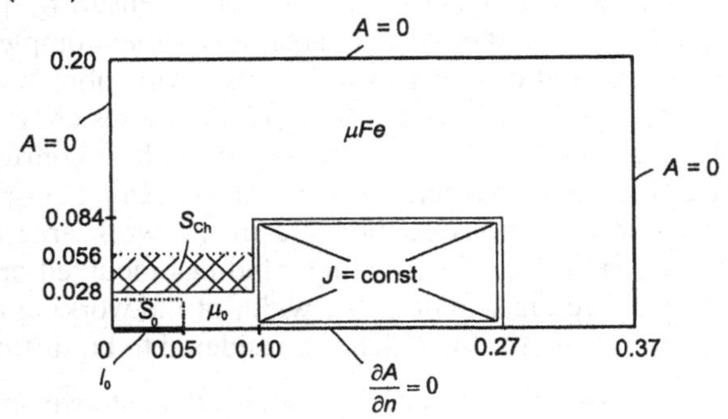

Fig. 5.34. Calculation area and boundary conditions for the vector magnetic potential

Electric currents, which are sources of the field in the adjoint problem, have been arranged within the region S_0 and on its boundaries. Boundary conditions for the potential λ of the adjoint problem coincide with conditions for the potential A of the direct problem (see Fig. 4.1).

Modification of magnetic permeability of each element in the area S_μ during calculation at motionless peaks of elements does not give substantial improvement of field homogeneity in the area S_0 in comparison with the initial field. Therefore, optimization has been carried out by means of moving the pole boundary. Twenty one moving peaks with coordinates that were parameters of optimization have been chosen on it. Movements of peaks were set in the direction of the normal to the initial rectangular pole shape shown in Fig. 5.34. Estimation of the obtained solution was carried out by evaluation of magnetic induction mean square deviation on the lower boundary l_0 of the area S_0.

Optimization was performed for cases when the core and poles were assumed to be made of a material with constant magnetic permeability μ_{Fe}=1000μ_0 or steel with a nonlinear characteristic μ=$\mu(H)$, i.e. taking into account the saturation phenomena.

At the beginning of the calculation of the first case, the mean square deviation of magnetic induction from the required value in the area S_0 was σ=0.00017. The required value B_w of the normal component of induction has been set equal to 1.475 T. The obtained optimum shape for the pole at use of quadratic interpolation of the potential at μ=1000μ_0 is shown in Fig. 5.35.

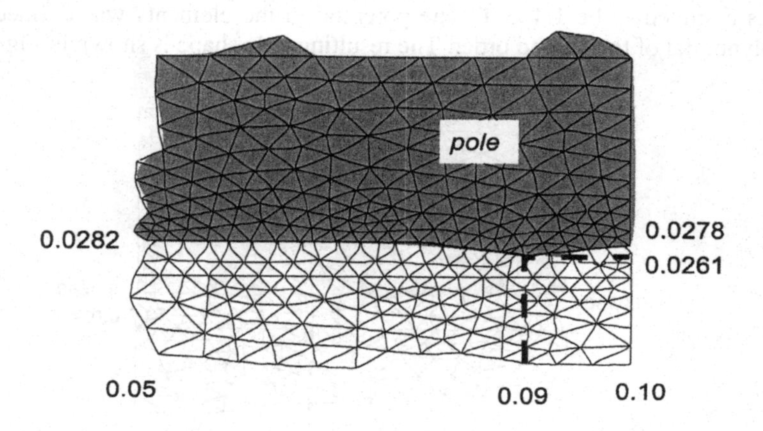

Fig. 5.35. The optimum pole shape at use of quadratic interpolation of the potential

Distribution of the magnetic induction for the obtained pole tip shape along the line l_0 and in the area S_0 are shown in Fig. 5.36a and Fig. 5.36b, accordingly.

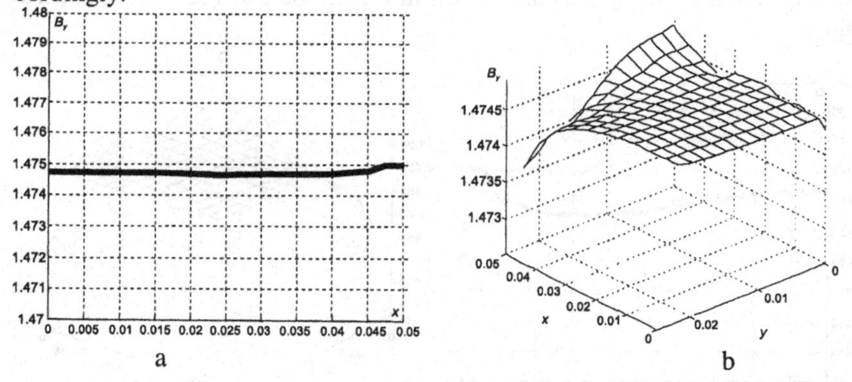

Fig. 5.36. Distribution of magnetic induction for the optimum pole tip shape along the line l_0 **a**), and in the area S_0 **b**)

The mean square deviation of induction for the pole of this shape was σ=0.0000248.

When optimizing the shape of the pole made of steel, nonlinear dependence of its magnetic permeability from magnetic induction has been taken into account, as noted above. The account of saturation was carried out by the relaxation method. The initial mean square deviation of induction in the area S_0 was σ=0.000907. The required value of the induction normal component was assumed to be 1.456 T. The potential in the elements was defined by a polynomial of the second order. The resulting pole shape is shown in Fig. 5.37.

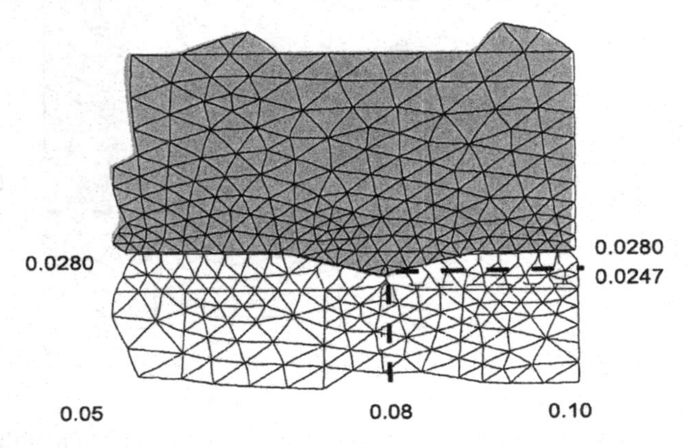

Fig. 5.37. The optimum pole shape at account of nonlinear properties of the material

Distribution of magnetic induction for the obtained pole tip shape along the line l_0 and in the area S_0 are shown in Fig. 5.38a and Fig. 5.38b, accordingly.

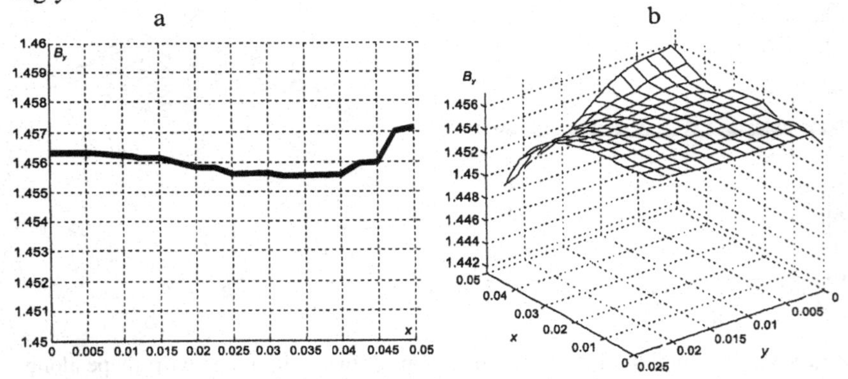

Fig. 5.38. Distribution of magnetic induction for the optimum pole tip shape along the line l_0 **a**), and in the area S_0 **b**) at account of nonlinear properties of the material

In this case, the mean square deviation of induction was σ=0.00021.

The obtained shape is in good agreement with known results given in publications, and with designs used in practice.

5.6 Search of the shape of magnetic quadrupole lens polar tip for accelerating a particle

Let's consider the optimization problem of a polar tip for quadrupole lenses used in accelerators of particles, applying for its solution the Lagrange method described in Chapter 4.

In practice of work with accelerators there is a necessity of focusing a charged particles' beam, since the cross-section of the beam can increase at rectilinear motion of particles by virtue of various reasons. Such focusing is carried out by means of magnetic lenses.

Cross-sections of two magnetic lenses are shown in Fig. 5.39. Four poles 3 with windings 4 are attached at the surface 1 of core 2. Magnetic fields created by the currents in the windings of both lenses are shown. If a beam of positively charged particles moves perpendicular to the plane of figure out the page, then the left lens focuses the beam in the horizontal direction whereas the right lens focuses it in the vertical direction. Two such lenses arranged (as shown in Fig. 5.40), focus a moving particle in both directions. In this figure the cross-section of two lenses by the plane rz is shown.

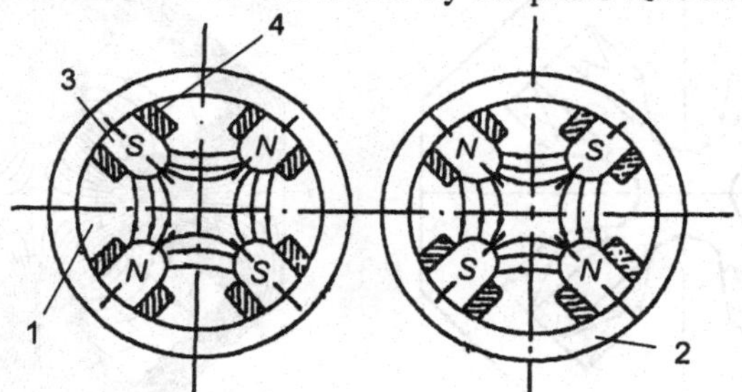

Fig. 5.39. Cross-sections of magnetic lenses

The thickness of each lens is l_1, and the distance between their adjacent sides is l_2. Curve 1 represents the trajectory of a beam of particles. It moves between poles of lenses through their central apertures.

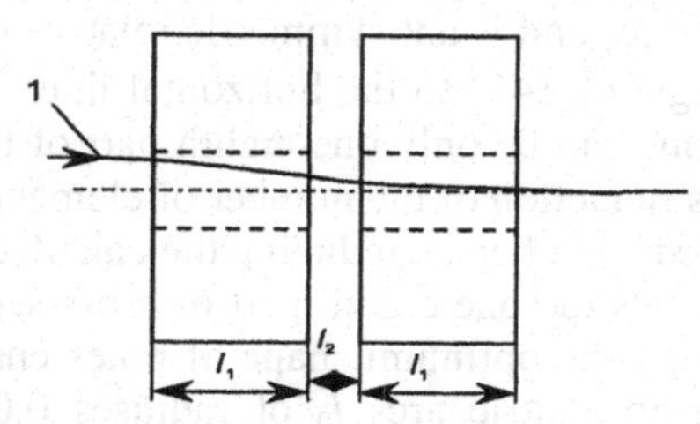

Fig. 5.40. The cross-section of two magnetic lenses by the plane rz. Line 1 is the trajectory of a beam of particles

With deviation of the beam in either the horizontal or vertical directions, the acting forces should return it to its initial position near the axis of lenses where the strength of magnetic field is zero. The required strength of magnetic field created by external sources should vary linearly both in the horizontal, and in vertical directions from the axis of lenses.

Application of electromagnetic lenses allows moving accelerated particles to distances of tens and hundreds of meters without substantial increase in the beam diameter. In some cases, elements of focusing magnets design six-field lenses are used, as well.

By optimum we imply such a shape of quadrupole lens pole that provides constant value of magnetic induction module on a circle enveloping the beam of particles. Let's consider the problem of search of the shape of such a pole.

The cross-section of quadrupole lens with optimized curvilinear profile of the poles is shown in Fig. 5.41. Here, 1 is the magnet steel core, and 2 is the winding with current.

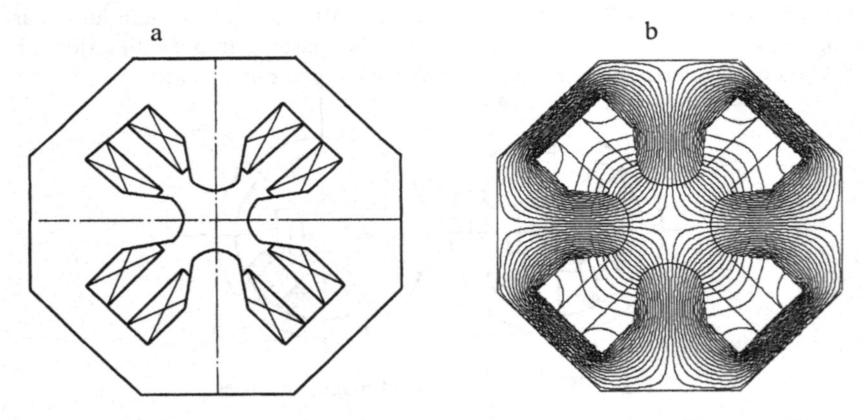

Fig. 5.41. Cross-section of a quadrupole lens with optimized curvilinear profile of the 3rd pole **a**) and lines of magnetic induction **b**) for some shape of the pole

As may be seen from this figure, the field is symmetric relative to the horizontal and vertical axes and is antisymmetric relative to axes passing through the center at an angle of ±45° to the horizontal line. Therefore the magnetic field calculation zone can be only one eighth part of the full cross-section of the lens that allows reduction of the number of elements into which the calculation area is divided, and, hence, reducing the calculation time.

Figure 5.42 presents the one eighth part of cross-section of the accelerator quadrupole for which the optimum shape of poles ensuring magnetic induction constancy on concentric arcs l_0 of radiuses 0.025 m or 0.03 m was searched. The quadrupole lens under consideration should provide a gradient of 27 T/m for the magnetic field at the radius of the pole's extreme point equal to 0.037 m.

The shaded area in Fig. 5.42 indicates a region S_μ in which modification of material properties is possible. $I = \int_{l_0} (|B| - B_w)^2 dl_0$ was used as the objective functional.

The augmented functional is given by:

$$L = \int_{l_0} (|B| - B_w)^2 dl_0 + \int_S \lambda \operatorname{div}\left(\frac{1}{\mu} \operatorname{grad} A\right) ds .$$

Boundary conditions for the vector magnetic potential at the solution of the direct problem are also shown in Fig. 5.42. Boundary conditions for the potential λ of the adjoint problem coincide with conditions for the potential A. Current density in the winding is assumed equal to J=5.07·10^6 A/m^2.

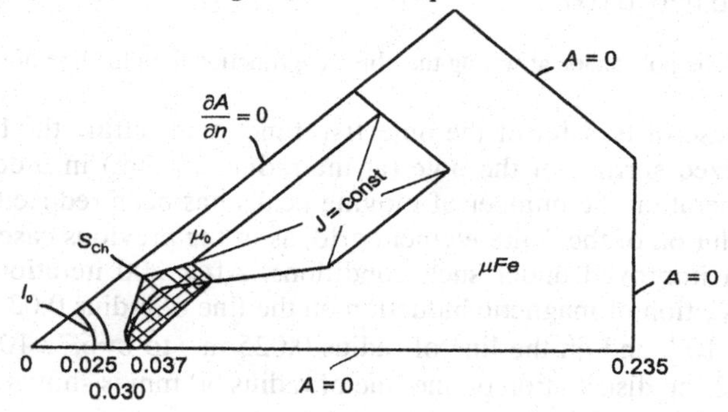

Fig. 5.42. Area of calculation and boundary conditions for the vector magnetic potential

The number of moving nodes arranged on the surface of the polar tip is 22. For each of the nodes the direction of movement is conterminous, as a rule, where the normal to the pole initial surface has been set. Calculations were carried out in view of nonlinear properties of the ferromagnetic material of the pole. It should be noted that when searching for solution, optimum parameters of additional correcting coils also to be used for obtaining the desired field, were not defined.

After 100 iterations, a shape of polar tip has been found that ensures mean square deviation of magnetic induction σ=1.26·10^{-2} from its average value on the target line l_0 of radius 0.03 m. The initial mean square deviation was σ=5.40·10^{-2}. If the target line is of radius 0.025 m, the mean square deviation of magnetic induction from its average value was σ=6.74·10^{-3} after 100 iterations (at initial mean square deviation σ=3.74·10^{-2}). The pole shape is shown in Fig. 5.43 (for the target line of radius 0.025 m).

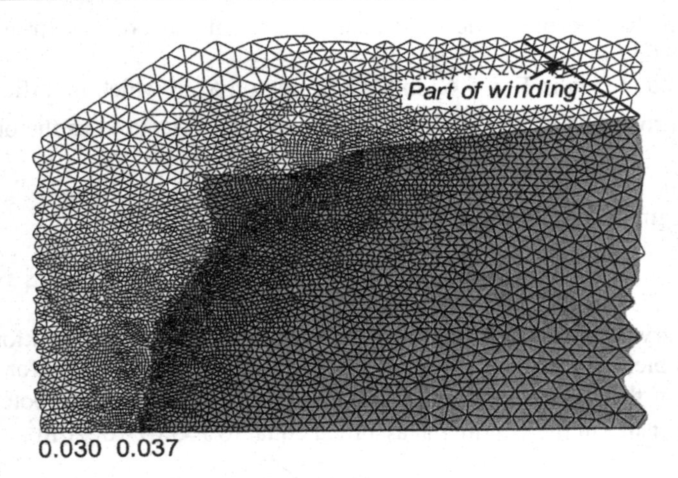

Fig. 5.43. The pole shape at setting the objective functional on the line of radius 25 mm

In the case of transfer of the objective functional setting the line closer to the optimized surface of the pole (at the radius 30 mm) in order to prevent grid degeneration the number of moving peaks has been reduced to 19 at the same resolution of the finite-element grid, as in the previous case. The results have been improved under such conditions. After 200 iterations, the mean square deviation of magnetic induction on the line of radius 0.03 m was equal to $\sigma=6.55\cdot10^{-3}$, and on the line of radius 0.025 m - to $\sigma=6.72\cdot10^{-3}$. The magnetic induction distribution on the line of radius 30 mm is shown in Fig. 5.44.

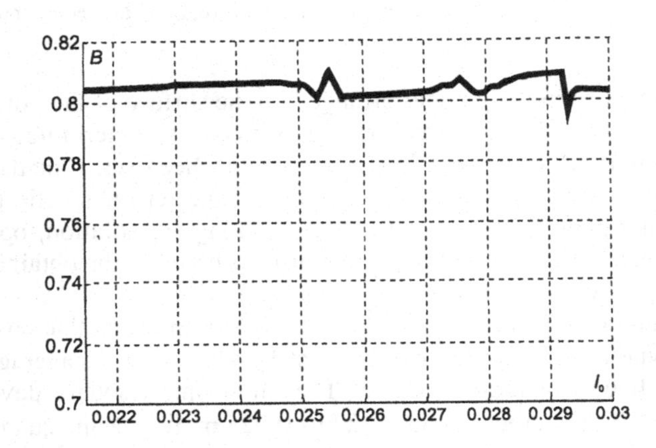

Fig. 5.44. Distribution of magnetic induction on the line of radius 30 mm

For this case the process of modification of the shape of quadrupole lens' magnetic pole tip can be seen in Fig. 5.45.

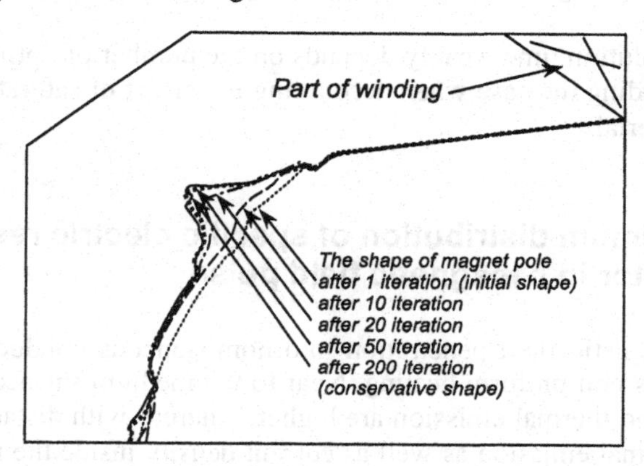

Fig. 5.45. The shape of quadrupole lens' magnetic pole tip at realization of various numbers of iterations

As a result of analysis of various approaches for the realization of the Lagrange method as well as the solution of the applied problems and comparison with results obtained by other methods of optimization, one can arrive at the following conclusions:

- an important property of the method is the possibility of determination of both the optimum shapes of bodies and their structure;
- as distinct from other methods of optimization, the Lagrange method allows calculating the objective function gradient by a single iteration in all points of the region at once on the basis of medium characteristics. This allows reducing the solution time of optimization problems;
- use of the Lagrange method allows finding close to optimum shape for bodies as a first approximation with short calculation times. As can be seen by the example of the problem of finding homogeneous distribution of magnetic induction along a line (see Section 4.4), the pole shape characterizing optimum distribution of magnetic material as a first approximation is obtained after 24 iterations when calculating by the Lagrange method. Finding similar results at solution of this problem by the modified Newton method will require approximately 10 times more calculation time;
- the solution convergence and form essentially depend on the parameters defining the process of optimization. Minimum and maximum values of boundaries' relocation Δn, the number of iteration i_{min} at which the value of relocation reaches its minimum, the number of iterations i_{max} with maximum value of relocations, and the chosen direction of shift for each movable node are among such parameters;

- it follows from the results described in this paragraphs and the previous that the Lagrange method allows solving practical problems with high accuracy;
- the solution time weakly depends on the number of optimization parameters, including the case when accounting the effect of saturation of ferromagnetic materials.

5.7 Optimum distribution of specific electric resistance of a conductor in a magnetic field pulse

Pulse magnetic field penetration into homogeneous conducting medium results in its non-uniform heating. Near to the medium surface, electric current density and thermal emission are higher, whereas with distance from the surface, thermal emission as well as current density inside the medium decrease [9]. To obtain uniform distribution of the current one may use a conductor with specific conductance that has the least value at the surface and increases with distance from it. As a result, the current will be distributed more uniformly and, hence, the maximum value of thermal emission will decrease.

The problem of diminution of temperature at the interior surface of pulse solenoids has been numerically solved [10] for a two-layer medium formed by a bronze conductor with stainless steel coating. The field specified on the solenoid surface was given by a unipolar sinusoidal impulse. Heating thus has decreased by 25% in comparison with the homogeneous conductor that was afterwards verified by experimental results. Bimetallic conductors allow reducing the maximum temperature of heating by 30%, as compared with the maximum heating of homogeneous conductors [11].

Estimations of heating modification in conducting media with exponentially decreasing specific resistance show that accessible minimum of maximum thermal emission for them is roughly ~3 times less than this value for the homogeneous conductor.

Thus, the problem of searching of optimum distribution of specific conductance of the medium ensuring maximum reduction of heating in comparison with the homogeneous conductor is of interest.

It has been shown in Section 4.6 that the Lagrange method can be applied for solution of similar problems. It allows the construction of an effective algorithm for searching of distribution of medium parameters, ensuring a minimum for a given functional.

Several statements of the optimization problem represent practical interest. In one of these, it is necessary to obtain a given distribution of magnetic induction and current density in the volume of a conducting body at the instant of time t_0 under the influence of a pulse magnetic field on the body surface.

In the elementary case under the assumption of one-dimensionality of field and current distribution in a conductor of thickness $2d$ the functional can be written down as

$$I = \int_0^{2d} \left[\frac{dB(x,t_0)}{dx} - C \right]^2 dx,$$

where C is a constant, and x is the spatial coordinate.

In another statement it is required to obtain uniform thermal emission $Q(x) = \sigma(x) \int_0^{t_0} E^2(x,t)dt$ at the instant of time t_0 through the whole thickness of the conductor or to ensure minimum of maximal thermal emission $\max Q(x) = \max \sigma(x) \int_0^{t_0} E^2(x,t)dt$ to that instant of time. In the latter case, the functional can be given by:

$$I = max \frac{1}{\sigma(x)} \int_0^{t_0} \left(\frac{dB(x,t)}{dx} \right)^2 dt.$$

Let's consider the solutions of these problems. Let the surfaces x=0 and x=$2d$ of an infinite conducting plate of thickness $2d$ be under the action of pulse magnetic induction $B(x,t)$ equal to $B(0,t) = B(2d,t) = B_m \sin \frac{2\pi}{T} t$ at $0 \le t \le 0.5T$ and $B(0,t) = 0$ at t>0.5T. In a system of rectangular coordinates with axis y and z directed parallel to the plate surfaces we have B=B_y. Taking into account that the plane x=d is a plane of symmetry for the magnetic induction, we have the condition $\partial B / \partial x = 0$ on it.

With this statement of the problem, the equation for the vector of magnetic induction inside the plate is given by

$$\frac{\partial}{\partial x} \left(\mu \sigma(x) \frac{\partial B}{\partial x} \right) = \frac{\partial B}{\partial t}.$$

Magnetic permeability of the material - platinum, is assumed to be constant.

Let's search for such distribution of specific conductance $\sigma(x)$ of the plate material, at which the electric current density $J(x) = J_z = \frac{1}{\mu} \frac{dB}{dx}$ as less as possible differs from the constant $J_w(x) = C$ at $0 \le x < d$ to the instant of time $t_0 = 0.25T$. At that, the desired function $B_w(x)$ is linear on the segment $0 \le x < d$.

We shall write down the minimized functional in the form of

$$I = 0.5 \int_0^d [B(x,t_0) - B_w(x)]^2 dx .$$

Then the augmented functional is given by:

$$L = I + \int_0^d \int_0^{t_0} \lambda(x,t) \left[\frac{\partial}{\partial x} \left(\mu\sigma(x) \frac{\partial B}{\partial x} \right) - \frac{\partial B}{\partial t} \right] dt dx .$$

The equation and boundary conditions for the adjoint variable for the problem under consideration are described in Section 4.6.

This problem has been solved numerically according to the algorithm described in Chapter 4. Values $d = 0.05$ m, $\sigma_{max} = 5.7 \cdot 10^7$ Sm, $B_m = 2$ T, $t_0 = 0.005\pi$ s, $T = 0.02\pi$ s have been assumed.

The number of nodes in the plate was 100, and the time step was 10^{-6}s. The specific conductance was calculated according to the following expression:

$$\sigma_{n+1} = \sigma_n \left(1 + h \frac{\frac{\partial B}{\partial x} \frac{\partial \lambda}{\partial x}}{\left\| \frac{\partial B}{\partial x} \frac{\partial \lambda}{\partial x} \right\|} \right),$$

where $\lambda(x,t)$ is the adjoint variable, n is the number of iteration, and h is a numerical coefficient.

Values of the functional at various values of h for n=1,2,3,4 are shown in the table below.

Table 5.19. Values of the functional at various values of h

Number of iteration	Values of the functional at h=-0.5	Values of the functional at h=-0.75
0 (σ=const)	0.5085	0.5085
1	0.4035	0.3427
2	0.2691	0.1675
3	0.1360	0.0253
4	0.0305	

The first row shows the value of functional at constant specific conductance of the plate material equal to $\sigma = \sigma_{max} = 5.7 \cdot 10^7$ Sm. One can easily see that the functional rapidly decreases as soon as at the first 3-4 iterations. At subsequent iterations, the value of h has been gradually changed down to-0.003 with deceleration of the functional decrease rate.

Figure 5.46 shows the initial distribution of magnetic induction through the plate thickness obtained at $\sigma = \sigma_{max} = 5.7 \cdot 10^7$ Sm (the concave curve), and the desired distribution (the straight line). Solution of the optimization problem results in the distribution of the material's specific conductance, shown in Fig. 5.47.

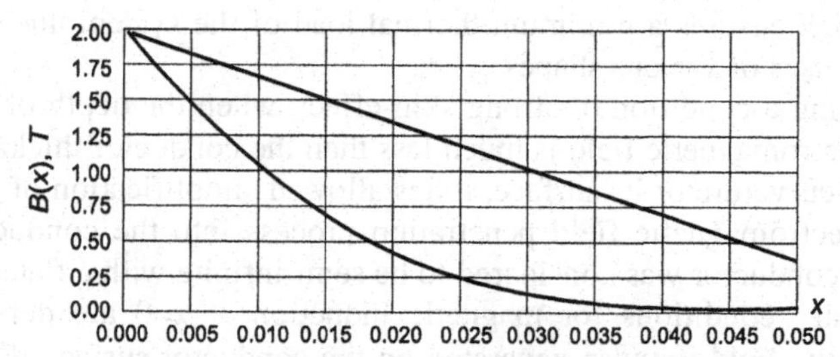

Fig. 5.46. Distribution of magnetic induction through the plate thickness at constant specific conductance and the desired distribution (the straight line)

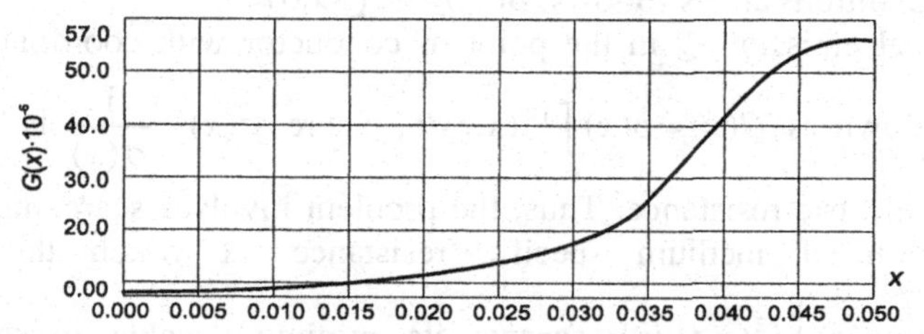

Fig. 5.47. Optimized distribution of specific conductance through the plate thickness

Dependence $B(x)$ at optimum distribution of specific conductance, as well as the straight line of desired distribution $B_w(x)$ are shown in Fig. 5.48.

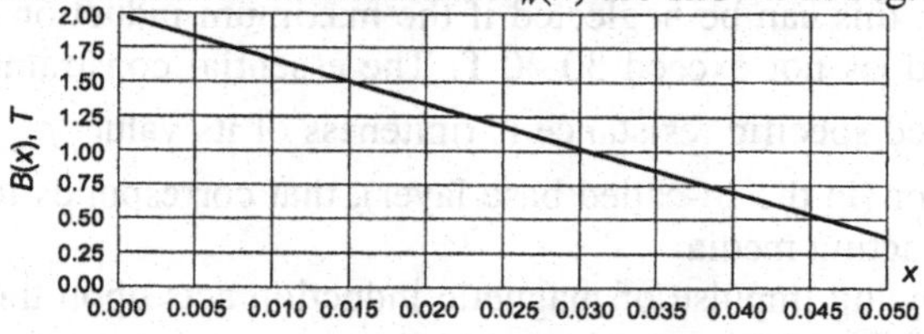

Fig. 5.48. Distribution of magnetic induction through the plate thickness at optimized specific conductance and the desired distribution (straight line)

At the distribution found for specific conductance, the magnetic induction is close to linear, and the current - to the uniform distribution.

Let's consider further the problem of determination of such a distribution of material conductance of a body when the maximum thermal emission $\max Q(x) = \max \sigma(x) \int_0^{t_0} E^2(x,t)dt$ has minimum value at the instant of time t_0, under influence of a magnetic induction impulse on its surface.

We shall give its solution obtained in [12] during a research of pulse solenoids. It has allowed us find a distribution of the medium's electric conductivity which ensures a minimum thermal load of the system under impact of field impulses of various shapes.

Assuming a condition of strong skin-effect, when the depth of penetration of the electromagnetic field is much less than the conductor thickness and the radius of curvature of its surface, it has allowed simplification of the analysis of the electromagnetic field penetration process into the conductor. In this case, the conductor was considered to be semi-infinite with a flat surface.

Boundary conditions for magnetic induction at x=0 are defined by the shape of the field impulse generated on the conductor surface $B(0,t)$. Taking into account the condition of strong skin-effect, we have $B(\infty,t)=0$. Initial conditions are as follows: $B(x,0)$=0, $Q(x,0)$=0.

Thermal emission Q in the point of conductor with coordinate x can be written down as $Q(x)=\rho(x)\int_0^{t_0} J^2(x,t)dt$, where $\rho(x)=\dfrac{1}{\sigma(x)}$ is the medium specific electric resistance. Thus, the problem involves searching for such a distribution of medium specific resistance, at which the functional $I=\max\left(\rho(x)\int_0^{t_0} J^2(x,t)dt\right)$ accepts its minimum value in the segment $0\le x\le d$.

It can be noted that although in general the specific resistance depends on temperature, this can be neglected if the maximum induction on the conductor surface does not exceed 30-40 T. The essential constraint superimposed on the desired specific resistance is finiteness of its value ρ_∞ in the depth of the conductor (in the so-called base layer), that corresponds to the properties of real conducting media.

The following impulse of magnetic induction acts upon the surface of the conductor:

$$B(0,t)=\begin{cases} B_m \sin(2\pi t/T), t<T/2 \\ 0, \quad t\ge T/2 \end{cases}.$$

With the purpose of generalization of the results above, magnitudes have been transformed to dimensionless form. The following values have been assumed to be basal elements: the amplitude of magnetic induction $B_b=B_m$, the specific resistance of the base layer $\rho_b=\rho_\infty$, duration of the external field half-period $t_b=T/2$, depth of the magnetic field penetration into the base layer substance $x_b=\sqrt{\rho_\infty t_b/\mu_0}$, and the specific thermal emission $Q_b=B_b^2/\mu$.

It should be noted that stratified media represent the most practical interest. In this case the conductor is represented in the form of a thick homogeneous base layer exceeding at least ten times the depth of the skin-layer, coated by a large number of layers of identical thickness and various specific resistances. Therefore, optimization is realized by search of specific resistances of the layer material ($\rho_1, \rho_2, \ldots, \rho_N$), corresponding to the minimum possible value of the objective function I. The optimization problem was solved for various number N of coating layers.

Figure 5.49 shows the influence of the number of N layers and the relative thickness of the whole coating $D^* = D/\Delta_\infty$ on the heating of the body (here Δ_∞ is the depth of field penetration into the base layer).

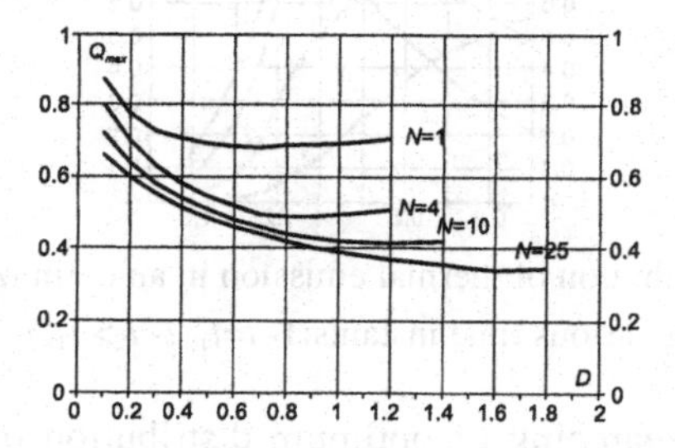

Fig. 5.49. Dependence of thermal emission on full thickness of the coating at various number N of its layers

These dependences are non-symmetrical and have minimum values corresponding to some thickness of coating D^*. Minimums are fuzzy, and, for example, in the case of N=25 it covers a range of values as large as $(1.4\text{-}1.8)\Delta_\infty$. The dependence for a two-layer conductor (N=1), obtained at solution of the optimization problem, corresponds to known results. The optimum thickness of coating in this case is D^*=0.68, and the relative specific resistance of the coating is $\rho^* = 5$. Maximum specific heat thus decreases down to $Q^*_{\max} = 0.67$ (for a homogeneous conductor Q^*=1.09). Use of a coating with a large number of layers (N~25) with optimum distribution of specific resistances allows lowering of the maximum specific heat to $Q^*_{\max} = 0.35$, that is close to the value for substances with a continuous distribution of specific resistance. This value is characteristic for exponential dependence of a medium's specific resistance at ρ_∞ / ρ_0 =0.01 (here ρ_0 is the specific resistance of boundary or near-surface layer of the coating).

Increasing the number of layers results in increase of the optimum thickness of coating corresponding to the largest reduction of heat, as well as in modification of optimum distribution of specific resistance. When increasing the number of layers descending, dependences become sharper which is connected with the increase of specific resistance of the coating boundary layer at optimum conditions. On the basis of this, one can assume that the boundary layer specific resistance is defined by the thickness of coating.

Analysis of the power distribution in the optimized multilayer conductor is of interest. Curves of heat emission power distribution, in an optimized 25-layer conductor at $D_{\Sigma}^{*}=1.2$ in various time instants during a pulse impact ($0<t_1<t_2<t_3<t_4=T/2$), are shown in Fig. 5.50.

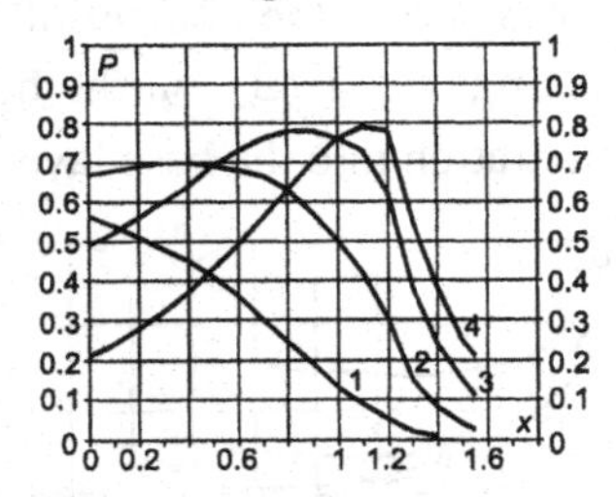

Fig. 5.50. Power distribution of thermal emission in an optimized 25-layer conductor at $D_{\Sigma}^{*}=1.2$, at various time instants: 1- $t=t_1$, 2- $t_2>t_1$, 3- $t_3>t_2$, 4- $t_4>t_3$

The problem of searching of optimum distribution of a conductor's specific resistance has also been solved for the case of action of magnetic induction in the form of damped oscillating impulse:

$$B(0,t)=B_m\sin(2\pi t/T)\exp(-2\delta t/T).$$

Two types of multilayer conductors have been considered with four- and twenty-layer coatings on the base material. Influence of damped oscillating impulse on the coating optimum thickness was found to be insignificant, which allows using results obtained for the case of unipolar impulse action when manufacturing multilayer conductors. In this case decrease of maximum heating was about 54%, as compared with the homogeneous conductor for the four-layer coating, and 68% for the twenty-layer one. This is essential for small damping factors when significant heating is observed. At damping factors $\delta\geq 2$ when the impulse practically becomes unipolar, parameters of the multilayer conductor are close to the corresponding parameters under action of unipolar impulses.

Conductors with inhomogeneous distribution of specific resistance are realized in the form of discrete stacks of conducting layers with different specific resistances. In this connection the following problem arises: how much is the possible limiting number of layers to replace the optimum continuous dependence by stratified conductor in order to access decrease of heating not too differing from the ideal. Replacement of continuous distribution by a stratified one with a large number of layers complicates the design of the conductor, and the use of small numbers of layers does not ensure sufficient decrease of heating.

Based upon the connection of coating optimum thickness with the specific resistance of the boundary layer, it is possible to change the statement of optimization problem by assumption that the specific resistance of the boundary layer is a given parameter. It allows obtaining a series of results describing dependence of the heating decrease degree on the number of layers for each preset value. Increasing the number of layers in the coating up to 20-25 ensures fulfillment of the required condition of closeness of discrete association $\rho(x)$ to the continuous one. We shall consider this problem for the case of unipolar sinusoidal pulse magnetic field as results for damped sinusoidal impulse do not lead to any basic differences.

Dependences of relative maximum heating of the optimized stratified conductor on the number of layers N for various specific resistances of the boundary layer, approximated by continuous curves (1-ρ_0/ρ_∞=5, 2-ρ_0/ρ_∞=10, 3-ρ_0/ρ_∞=16) are shown in Fig. 5.51.

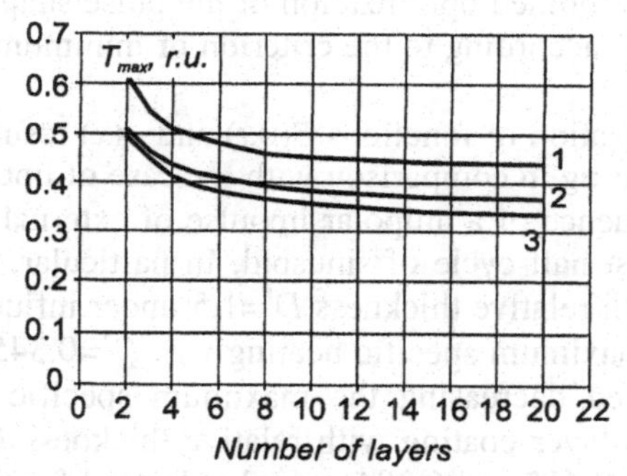

Fig. 5.51. Dependences of relative maximum heating of the optimized stratified conductor on the number of layers N for various specific resistances of the boundary layer (1-ρ_0/ρ_∞=5, 2-ρ_0/ρ_∞=10, 3-ρ_0/ρ_∞=16)

Heating change at maximum rate occurs in the range of the first 3-4 layers. Largest diminution of maximum heating, that makes 52-62 % from the heating of the homogeneous medium depending on the value of the boundary layer, specific resistance occurs within this same area. When increasing the number of layers decreasing of the heating gradually decelerates and becomes insignificant in the range of 10-20 layers when the maximum level of its drop, depending on the boundary layer, specific resistance is reached. Thus, coatings with 4-8 layers are most expedient.

Figure 5.52 shows the optimized ideal continuous dependence of the coating specific resistance (1) and its 5-layer approximation (2) at the given relative resistance of the boundary layer ρ_0/ρ_∞=10.

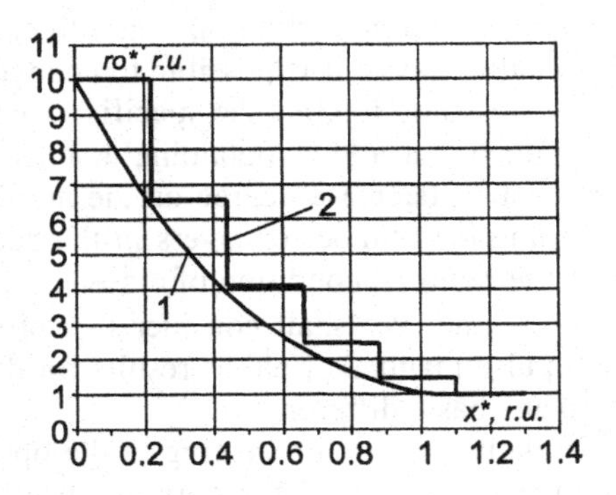

Fig. 5.52 Optimized ideal continuous dependence of the coating specific resistance (1) and its 5-layer approximation (2)

The problem of combined optimization of the pulse shape and dependence of specific resistance according to the criterion of minimum heating has been solved.

Combined optimization of functions $B(0,t)$ and $\rho(x)$ results in an insignificant decrease of heating in comparison with the case of optimization of function $\rho(x)$ under influence of a unipolar impulse of external magnetic field in the shape of the first half-cycle of sinusoid. In particular, for the optimized 25-layer coating with relative thickness $D^*=1.5$ under influence of a unipolar impulse field, the maximum specific heating was $Q^*=0.345$. Combined optimization has allowed decreasing the maximum specific heating down to $Q^*=0.318$. For a 10-layer coating with relative thickness $D^*=1.2$, the corresponding values are 0.415 and 0.384 accordingly, and for the one-layer coating they are equal to 0.68 and 0.63. For all considered cases the additional decrease of heating does not exceed 8 % as compared to corresponding values obtained under the action of a unipolar impulse.

Thus, replacement of the optimum continuous distribution of a specific resistance with a large number (up to 20) of layers allows reaching a triple drop of specific heat in comparison with the homogeneous conductor. As effectiveness of heating decrease, due to adding of layers, sharply drops beginning from the case of 8-layer coating, this allows considering conductors with such number of layers as a good approximation for optimum continuous dependences of specific resistances.

Both for unipolar, and for oscillating damped impulse external field, the four-layer conducting coatings that are more practical from the point of view of their manufacturing allow more than twice the reduction of the maximum heating.

References

1. Johnk, C.T.A. (1976). *Engineering Electromagnetic Field and Waves*. New York: Wiley&Sons.
2. Demirchian, K.S. et al. (2003). *Theoretical Fundamentals of Electrical Engineering* (in Russian). vol 2, St.Petersburg: Piter.
3. Korovkin, N.V., and E.E. Selina (1992). *Simulation of wave processes in the distributed electromagnetic systems* (in Russian). St.Petersburg: SPbGTU.
4. Korovkin, N.V. et al. (2000). New transmission line equivalent circuits of the increased accuracy for the EMC problems solution. *Proc 4th Europ Symp on EMC*, Brugge, Belgium: 243-248.
5. Korovkin, N.V. et al. (2001). Synthesis of forming lines with the help of genetic algorithm (in Russian). *Proc 4th Intern Symp on EMC and Electrom Ecol*, St. Petersburg, Russia, : 273-278.
6. Korovkin, N.V., and E.E. Selina (1998). An efficient method of wave processes in transmission line simulation using discrete models. *Proc of IEEE Intern Symp Electrom Compat*, Denver, USA.
7. Kuchinskiy, G.S., L.T. Vehoreva, and O.V. Shilin (1997). Concepts for designing of powerful high-voltage lines for nano- and microsecond pulses generation (in Russian). *Electrical Technology*, n9.
8. Kuchinskiy, G.S. et al. (1999). The powerful non-uniform high-voltage lines for nano- and microsecond pulses generation (in Russian). *Electrical Technology*, n 8.
9. Shneerson, G.A. (1992). *Fields and transients in devices of extra high currents* (in Russian). Moscow: Energoatomizdat.
10. Farynski, A, L.Karpinski, and Nowak (1979). A Layer conductor of cylindrical symmetry in non-stationary magnetic field. Part 1, *J of Techn Phys*, 20, 2, 265-280, Warzawa, Polish Academy of Sc, Institute of Fundam Technol Res.
11. Karpova, I.M., and V.V.Titkov (1988). Features of two layer conductors heating by pulse current (in Russian). *Izv AN USSR, Energetica i transport*, no5, 83-90.
12. Karpova, I.M., and V.V.Titkov (1999). Optimization of multilayer conductors to minimize its maximum heating in pulse electromagnetic field (in Russian). *Electrical Technology*, n12,20-26.

Appendix A. A Method of Reduction of an Eddy Magnetic Field to a Potential One

This method of reduction of eddy magnetic field into potential one is based on resolution of the magnetic field vector $\boldsymbol{H}$ into eddy $\boldsymbol{H}_{eddy}$ and potential $\boldsymbol{H}_{pot}$ components. Let's represent the strength of magnetic field in the form of the sum $\boldsymbol{H} = \boldsymbol{H}_{pot} + \boldsymbol{H}_{eddy}$. At that, we assume that the current with density $\boldsymbol{J}$ creates only the unknown field $\boldsymbol{H}_{eddy}$, which defines the eddy part of the magnetic field, i.e rot $\boldsymbol{H}_{eddy} = \boldsymbol{J}$. As shown below, the component $\boldsymbol{H}_{pot}$ of the desired field can be expressed by the scalar magnetic potential φ_m.

The method under consideration proves useful to calculate both stationary and quasi-stationary fields. Here we shall limit ourselves to its statement for the case of magnetic fields generated by direct currents. In literature, the eddy component $\boldsymbol{H}_{eddy}$ is often designated as vector $\boldsymbol{T}$ and the method is named $\boldsymbol{T}$-Ω method. This method is based on the possibility of finding the eddy component $\boldsymbol{H}_{eddy}$ by the given spatial distribution of current density $\boldsymbol{J}$ in the conductor by means of simple calculations.

As $\operatorname{div}\boldsymbol{H}_{eddy}$ is not a given quantity, then equation rot $\boldsymbol{H}_{eddy} = \boldsymbol{J}$ has a set of solutions not depending on magnetic properties of the medium. Therefore, the equation rot $\boldsymbol{H}_{eddy} = \boldsymbol{J}$ can be solved for the case of homogeneous medium.

As $\operatorname{div}\boldsymbol{J} = 0$ and $\boldsymbol{J} = \operatorname{rot}\boldsymbol{H}_{eddy}$, then the equation

$$\operatorname{rot}\left(\boldsymbol{H} - \boldsymbol{H}_{eddy}\right) = \operatorname{rot}\boldsymbol{H}_{pot} = 0,$$

$$\boldsymbol{H}_{pot} = -\operatorname{grad}\varphi_m$$

is valid for the eddy component equal to $\boldsymbol{H}_{pot} = \boldsymbol{H} - \boldsymbol{H}_{eddy}$.

Thus, the component $\boldsymbol{H}_{pot}$ can be presented in the form of the gradient of scalar magnetic potential φ_m. Then the magnetic field strength is given by $\boldsymbol{H} = \boldsymbol{H}_{eddy} - \operatorname{grad}\varphi_m$. In this case, the equation $\operatorname{div}\mathbf{B} = 0$ can be written down as $\operatorname{div}\mu\left(\boldsymbol{H}_{eddy} - \operatorname{grad}\varphi_m\right) = 0$. Then we have:

$$\operatorname{div}\mu\operatorname{grad}\varphi_m = \operatorname{div}\mu\boldsymbol{H}_{eddy}, \tag{A.1}$$

where φ_m and $\boldsymbol{H}_{eddy}$ are unknown quantities.

In the general case of three-dimensional magnetic field it is necessary to find four scalar functions, i.e. three components of $\boldsymbol{H}_{eddy}$ and the scalar magnetic potential φ_m. In the equation (A.1), sources of the potential φ_m are defined by $\boldsymbol{H}_{eddy}$. Therefore, at first we shall find the solution of equation rot $\boldsymbol{H}_{eddy} = \boldsymbol{J}$. Since only the curl of the vector $\boldsymbol{H}_{eddy}$ is given, then the div $\boldsymbol{H}_{eddy}$ can be an arbitrary quantity. The possibility of arbitrary choice of div $\boldsymbol{H}_{eddy}$ distribution is an essential advantage of the method of transformation of the eddy magnetic field into potential one. Then there is a possibility to choose a solution from the set of solutions that allows calculation of the desired field with strength vector $\boldsymbol{H}$ at minimal costs. One of the possible methods of calculation of $\boldsymbol{H}_{eddy}$ is based on use of the Biot-Savart law:

$$\boldsymbol{H}_{eddy} = \frac{1}{4\pi} \int_V \frac{[\boldsymbol{J}\boldsymbol{r}]}{r^3} dV, \tag{A.2}$$

where volume V is a simply connected domain.

The expression (A.2) allows finding the magnitude of $\boldsymbol{H}_{eddy}$ in all points through the domain of calculations, including conductors with currents. The magnitude of $\boldsymbol{H}_{eddy}$ can be equal to zero everywhere where there is no current. When $\boldsymbol{J} \neq 0$, then we have $\boldsymbol{H}_{eddy} \neq 0$.

As noted above, the domain V of integration in (A.2) should be simply connected. If it is multiply connected, it is possible to specify a current enveloping loop l on which $\oint_l \mathbf{H}_{eddy} d\mathbf{l} = 0$. As the integral $-\oint_l \mathrm{grad}\varphi_m d\mathbf{l}$ is equal to zero, then their sum $\oint_l \boldsymbol{H}_{eddy} d\mathbf{l} - \oint_l grad\, \varphi_m d\mathbf{l} = \oint_l \boldsymbol{H}\, d\mathbf{l}$ also becomes zero. However, the integral $\oint_l \mathbf{H} d\mathbf{l}$ should be equal to the current. This inconsistency is eliminated if the integration domain is simply connected.

To provide simple connectedness of the integration domain we shall introduce cuts that are forbidden for intersection by the contour of integration. In case of contour l with current i (Fig. A.1) this cut can be any surface S stretched by the contour l.

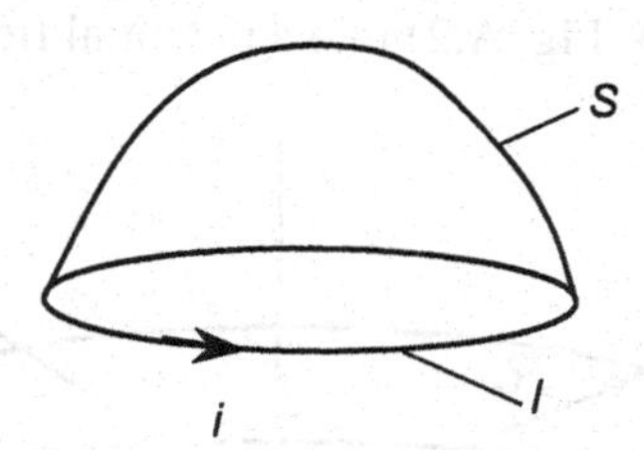

Fig. A.1. A contour l with current i and surface S of the cut, transforming the domain into a simply connected one

From the analogy of equations for the potential of electrostatic field $\operatorname{div}\varepsilon \operatorname{grad}\varphi_e = -\rho$ and for the scalar magnetic potential $\operatorname{div}\mu \operatorname{grad}\varphi_m = \operatorname{div}\mu\mathbf{H}_{eddy}$, follows that the quantity $-\operatorname{div}\mu\mathbf{H}_{eddy}$ can be considered as the volume density $\rho_m = -\operatorname{div}\mu\mathbf{H}_{eddy}$ of fictitious magnetic charge. This magnetic charge is a formally introduced quantity for calculations. From this analogy also follows that for the relation $-\oint_S \varepsilon \operatorname{grad}\varphi_e dS = q$ there is a similar relation

$$-\oint_S \mu \operatorname{grad}\varphi_m dS = m,$$

where m is the magnetic charge.

In a homogeneous medium with magnetic permeability μ the scalar magnetic potential satisfies the Poisson equation $\operatorname{div}\operatorname{grad}\varphi_m = \operatorname{div}\mathbf{H}_{eddy} = -\rho_m/\mu$. In view of the aforesaid, its solution can be written down as

$$\mathbf{H} - \mathbf{H}_{eddy} = \frac{1}{4\pi\mu}\int_V \frac{\rho_m \mathbf{r}}{r^3} dV.$$

Then, the expression for the magnetic field strength becomes

$$\mathbf{H} = \frac{1}{4\pi\mu}\int_V \frac{\rho_m \mathbf{r}}{r^3} dV + \mathbf{H}_{eddy}.$$

Thus, determination of vector $\boldsymbol{H}$ in the whole space, including the domain with currents, requires solving a single equation concerning the scalar magnetic potential inasmuch calculation of $\boldsymbol{H}_{eddy}$ by (A.2) does not require solution of any equations.

As an example, let's consider the procedure of reduction of the magnetic field of current in a toroidal conductor of rectangular cross-section and given dimensions h and (R_e-R_i) (Fig. A.2) into a potential field.

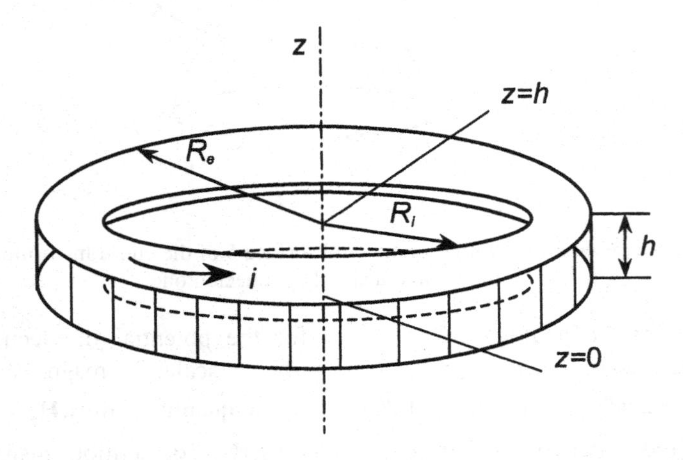

Fig. A.2. A toroidal conductor of cross-section h*(R_e-R_i) with current i and simple layers of magnetic charges of radiuses R_e (at z=0 and z=h) that are equivalent to the current

Here h is the conductor height, and R_e, R_i are its external and internal radiuses, accordingly. The current density J in the conductor is constant.

Let the magnetic permeability be constant everywhere. We shall limit the simply connected domain with the current by the lower (with coordinate z=0), upper (with coordinate z=h) and the lateral surfaces of the cylinder of radius R_e. At that, we assume $\boldsymbol{H}_{eddy} = 0$ everywhere outside of the volume formed by these surfaces. In the cylindrical system of coordinates r, α, z the vector $\boldsymbol{J}$ has a single component J_α.

Since rot $\boldsymbol{H}_{eddy} = \boldsymbol{J}$, then at $\boldsymbol{J}$=$\boldsymbol{j}$ J_α (here $\boldsymbol{j}$ is the unit vector of axis z) we can use the single component $\boldsymbol{H}_{eddy} = \boldsymbol{H}_{eddyz}$. Then, for $\boldsymbol{H}_{eddy}$ we have

$$-\frac{\partial H_{eddy}}{\partial r} = J_\alpha,\ H_{eddy}(r) = -\int_{R_e}^{r} J_\alpha dr = (R_e - r)J_\alpha \ .$$

Therefore, $\boldsymbol{H}_{eddy}(r) = (R_e - r)J_\alpha$ when R_i<r<R_e and $\boldsymbol{H}_{eddy}(r) = (R_e - R_i)J_\alpha$ when 0<r<R_i.

The dependence $\boldsymbol{H}_{eddy}(r)$ is shown in Fig. A.3.

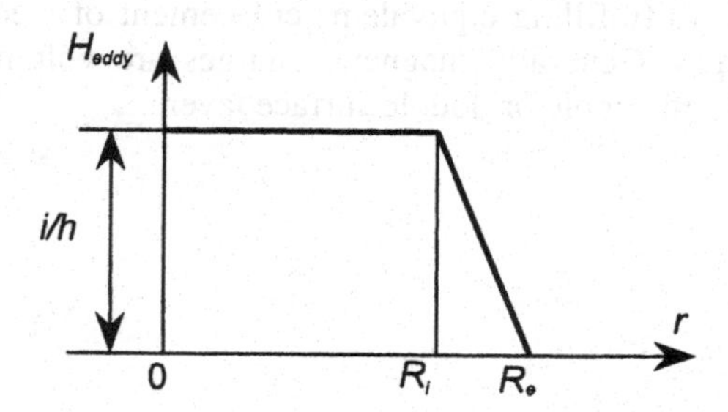

Fig. A.3. Dependence of eddy component of the magnetic field $\boldsymbol{H}_{eddy}$ on radius

Let the current be directed in such a way that the positive direction of vector $\boldsymbol{H}_{eddy}(r)$ coincides with the direction of axis *z*. At that, $\boldsymbol{H}_{eddy}(r)$ varies stepwise by a magnitude $\boldsymbol{H}_{eddy}$ on the surface *z*=0 and by a magnitude $-\boldsymbol{H}_{eddy}$ on the surface *z*=*h*. According to the condition

$$\mathrm{div}\mu\mathbf{H}_{eddy} = \mu\frac{\partial H_{eddyz}}{\partial z} = -\rho_m,$$

there will be magnetic charges on these surfaces, i.e. charge –*m* on the lower surface, and charge +*m* on the upper surface. There are no magnetic charges in all remaining points of the domain.

The volume density of magnetic charges on these surfaces is infinite, but their surface density σ_m is finite: $\sigma_m(r) = -\mu\boldsymbol{H}_{eddy}(r) < 0$ on the lower surface and $\sigma_m(r) = +\mu\boldsymbol{H}_{eddy}(r) > 0$ on the upper surface.

Thus, found magnetic charges completely define the magnetic field in all points of the domain, where $\boldsymbol{H}_{eddy} = 0$, i.e. in all points where it can be described by means of the scalar magnetic potential. In those parts of space where the eddy component $\boldsymbol{H}_{eddy}$ is present, we have the following expressions for components of the magnetic field strength:

$$H_r = -\frac{\partial\varphi_m}{\partial r},$$

$$H_\alpha = -\frac{\partial\varphi_m}{r\partial\alpha},$$

$$H_z = H_{eddy} - \frac{\partial\varphi_m}{\partial z}$$

Thus, application of the method of reduction of eddy magnetic field to potential one allows fulfilling equivalent replacement of electric currents with magnetic charges. Generally magnetic charges are volume distributed, but they can also form simple or double surface layers.

Appendix B. The Variation of a Functional

Under the variation δy of a function $y(x)$ set on an interval (a,b), one understands such function of the argument x which is defined as a difference $\delta y=Y(x)-y(x)$ of a new function $Y(x)$ and the function $y(x)$. The new function $Y(x)$ can be defined as $Y(x)=y(x)+\alpha\eta(x)$, where α is a number tending to zero, and $\eta(x)$ is an arbitrary smooth function. The function $Y(x)$ coincides with $y(x)$ at boundary values $x=a$ and $x=b$. Therefore, $\eta(a)=\eta(b)=0$. Thus, the variation of function $y(x)$ is equal to $\delta y=\alpha\eta(x)$ when $\alpha\to 0$.

Increment of functional $I(y)=\int\limits_a^b f(x,y,y')dx$ at variation $\delta y=\alpha\eta(x)$ of function $y(x)$ is equal to $\Delta I=\int\limits_a^b f(x,y+\alpha\eta,y'+\alpha\eta')dx-\int\limits_a^b f(x,y,y')dx$. Let's expand it in Taylor's series in powers of α:

$$\Delta I=\alpha\frac{dI}{d\alpha}+\frac{\alpha^2}{1\cdot 2}\frac{d^2I}{d\alpha^2}+\frac{\alpha^3}{1\cdot 2\cdot 3}\frac{d^3I}{d\alpha^3}+\dots$$

The leading, linear part of the functional increment is designated as $\delta I=\alpha\frac{dI}{d\alpha}$ and is referred to as its first variation. Accordingly, the expression $\delta^2 I=\frac{\alpha^2}{2}\frac{d^2I}{d\alpha^2}$ is referred to as second variation of the functional I, etc.
In order that function $y(x)$ provides an extremum to the functional I, it is necessary that equality $\frac{dI}{d\alpha}=0$ or $\delta I=0$ was valid for small α. Thus, setting a variation $\delta y=\alpha\eta(x)$ for functions $y(x)$, one can search for an extremal of the functional from the necessary condition $\delta I=0$ or $\frac{dI}{d\alpha}=0$. The extremal is understood as such a function $y(x)$ at which it becomes its extreme value.

Examples of calculation of some functionals' variation are given below.
Let's find the variation of functional $I=\frac{1}{2}\int\limits_V \varphi^2(x,y,z)dV$, in which $\varphi(x,y,z)$ is the field scalar potential. Assuming $\delta\varphi=\alpha\eta(x,y,z)$, we find

$$\delta I = \alpha \frac{dI}{d\alpha}\bigg|_{\alpha=0} = \alpha \frac{d}{d\alpha}\left[\frac{1}{2}\int_V (\varphi + \alpha\eta)^2 dV\right] =$$

$$\alpha \int_V (\varphi + \alpha\eta)\eta dV\bigg|_{\alpha=0} = \alpha \int_V \varphi\eta dV = \int_V \varphi\delta\varphi dV.$$

Similarly, variation of the functional $I = 0.5 \int_V \left(\frac{\partial\varphi}{\partial x}\right)^2 dV$ can be calculated:

$$\delta I = \alpha \frac{dI}{d\alpha}\bigg|_{\alpha=0} = \alpha \frac{d}{d\alpha}\left[\frac{1}{2}\int_V \left(\frac{\partial(\varphi+\alpha\eta)}{\partial x}\right)^2 dV\right] =$$

$$= \alpha \int_V \frac{\partial\varphi}{\partial x}\frac{\partial\eta}{\partial n} dV = \int_V \frac{\partial\varphi}{\partial x}\frac{\partial(\delta\varphi)}{\partial x} dV.$$

Using this expression we can write down variation of the functional

$$I = W_e = \frac{1}{2}\int_V DEdV = \frac{1}{2}\int_V \varepsilon E^2 dV = \frac{1}{2}\int_V \varepsilon\left(\left(\frac{\partial\varphi}{\partial x}\right)^2 + \left(\frac{\partial\varphi}{\partial y}\right)^2 + \left(\frac{\partial\varphi}{\partial z}\right)^2\right) dV,$$

which defines the energy of electric field in the volume V:

$$\delta I = \alpha \frac{dI}{d\alpha}\bigg|_{\alpha=0} = \int_V \left(\frac{\partial\varphi}{\partial x}\frac{\partial(\delta\varphi)}{\partial x} + \frac{\partial\varphi}{\partial y}\frac{\partial(\delta\varphi)}{\partial y} + \frac{\partial\varphi}{\partial z}\frac{\partial(\delta\varphi)}{\partial z}\right) dV =$$

$$= \int_V (grad\varphi)(grad\delta\varphi)\, dV$$

When $I = \frac{1}{2}\int_S \left(\frac{\partial\varphi}{\partial n}\right)^2 dS$ where n is the normal to the surface S, we have

$$\delta I = \alpha \frac{dI}{d\alpha}\bigg|_{\alpha=0} = \alpha \frac{d}{d\alpha}\left[\frac{1}{2}\int_S \left(\frac{\partial(\varphi+\alpha\eta)}{\partial n}\right)^2 dS\right] =$$

$$= \alpha \int_S \frac{\partial\varphi}{\partial n}\frac{\partial\eta}{\partial n} dS = \int_S \frac{\partial\varphi}{\partial x}\frac{\partial(\delta\varphi)}{\partial n} dS.$$

It should be noted that if the functional includes the function $y(x)$, as well as its derivative $y' = \frac{dy}{dx}$, then variation $\delta y'$ caused by variation δy is equal to $\delta y' = \delta \frac{dy}{dx} = \frac{d}{dx}(\delta y)$.

Index

accuracy of linear connections,137
active *RC*-filters diagnostic,97
adjoint variables,197,202,217
admittance matrix,163,164
algorithms,
-genetic,25,109,111
-search for shape and structure,201
-for searching the node number,177
analogy of,
-electric current and magnetic charges,319
-equations for forward and conjugate problems,199
-forward and conjugate potentials,199
anisotropy axes,216
antigradient,62
antigradient vector,62,63
a priori information,107,202
area
-of effective solution,49
-with magnetic flux concentrations,222

Biot-Savart law,8,318
boundary condition for
-forward variable,195
-conjugate variable,197
boundary layer
-duration,128
boundary-value problem,193

chromosome,112,113
circuit
-diagnostic,156
-modeling,275
-nonreciprocal,171
-observable,158
-reduced,176,178
-reciprocal,174
circuit branches
-admittances,157
-resistances,3,157
circuit with
-a single special cut,165
-independent special cuts,164
-two special cuts,165
classification
-of the fault character,93
-problems,93
closeness of neurons,96
clusters,97
cluster algorithm,107
concentration of substances,216
concept of smallness,94
conditionality number,10,135
condition number,161
conditional extremum,71
constraints
-equality,69
-inequality,60,69
convergence
-acceleration,63
-rate,91
correlation,94
criteria of closeness,94
criterion
-scalar,55
-vector,51,52

crossover,113

defectoscopy,4,43,44
delta function,203,204
degeneration,114,116,241,304
derivatives
-"material",201
-numerical,62,76
-of objective function,91,194
-partial,85,251,278
descendants,112
directed crossbreeding,
-graphic interpretation,116
discontinuous function,246
discrete analogues,11,14
discreteness,11,18,118
distance
-definition,94,148,239,288
-Euclidean,17,95,133,147
distributed parameter systems,42
domain of
-admissible position of optimized body,202,217
-external sources,239,240
duality of forward and conjugate problems,199

eddy current,5,42,44
electrical circuit
-diagnostic,4,13,43,97,121,145,156
-graph links,30
-graph tree,30,158
-synthesis,4,7,12,257
-topology,1,21,33,188
eigenvalues,9,65,126
-small,148,150,164,171
-large,172
eigenvectors,67,146
-right,146
-left,147
electromagnetic force,206,229
electromagnetic transient analysis,122
elite,115
equation
-of electric circuit,42,79
-descrete,11
-integral,36,239,240
-Kirchhoff,3,278
-operator,44
-state,83,102
-stiff,10,125,136
equivalent magnetic charges,319
evolutionary
-inheritance,113
-natural selection,112,113
-principles,113
-variability,112,113
error
-approximation,101,255
-construction of linear connections,180
-identification,161,164,178
-matrix elements,148
-modeled process,135
-measurement,13,16,170,173
-numerical integration of equations,129,281
-numerical solution,11
relative,134,141
-root-mean-square,59,101,
-experimental data,145,155,152

fast and slow components,124
field
-electrostatic,3,39
-homogeneous,296
-quasistatic,42,306
-steady magnetic,6,40,48,296,301
finite-difference derivative,62,76
fitness-function,112
floating-point arithmetic

standards,172
forecasting of power systems
performance,97
Fourier-image of the kernel,100
frequency dependences
of the condition numbers,179
function
-bipolar,87
-identifying,185
-membership,57,59
-neuron activation,87
-objective,2,60
-sigmoid unipolar,87
-transfer,93
-unit step,87
functional,2,9,38
-augmented,194,196
-derivatives,61,109,194
-expanded,69,71,73,74,80,84
-ravine,63,67,133
-variation,73,196,323
fundamental tree
with branches,158
fuzzy
-logic,56
-method,59
-sets,56
gemmating process,114
gene,112
genetic algorithm,111
-disadvantage,118
-multiple processor,117
-practical accuracy,118
global
-membership,59
-minimum,105
gradient methods,62
gradient of functional,62
Green's theorem,196

Hemming interval,94
heuristic rules,97
homogeneous magnetic
fields,234,296
half-fall time,122

identifications of AC circuits,29
individual,112
induction heating,42
inductively connected coils,74
isolated subcircuit,173
isoperimetric variational
problems,73
iterative calculation,241
iterative process convergence,
inverse problem,1,2,37
-criteria,3,4
-defectoscopy,44
-diagnostics,5
-electrostatics,39
-identification,5,32,43
-incorrect,11
- incorrectness,11
-in frequency domain,275
-in operator domain,24
-in time domain,81
-macromodeling
(macromodel),5,26
-magnetostatics,40
-multiobjective,10
-multicriterion,10,47
-non-uniqueness,6
-optimization,33
-quasistatics,42
-restrictions,6
-rigid,9,10
-*r*-ravine,67
-stability,7
-stiffness,8
-synthesis,4

kernel of k-th order,28,100

kernel of the class,95
Kirchhoff equations,3
Kohonen's rule,96

Lagrange multipliers,
-continious,196
-discrete,71
layer of magnetic charges
-double,205
-single,210
level lines,60
linear connections,137,143,
150,151
-independent,172,177,178
Lipschitz constant,108
local minimum,60
loop
-admittances,160,162
-resistances,160,163

macromodel of,5,26
-nonlinear electric circuit,
79,102
-non-unimodal,105
-operational amplifier circuit,
macromodeling,5,26,102
-object,27
-problem,5,26,98
-synthesis,27
magnetic charge,319
magnetic flux,202
-focusing,221
-redistribution,223
magnetomotive force,36,206
magnetostatics,40
main criterion method,52
matrix
-degenerate,128
-generalized,16
-Hesse,9,10,64,66,133
-Jacobi,126
-magnetic permeability,1,216
-nodal,159
admittances,162,165,176
-of contours,3
-of cutsets,3
-pseudo-inverse,16,17
media
-anisotropic,35,216
-homogeneous,214
-isotropic,214
medium
-classes,200
-composite,216
-homogeneous,216
membership function,57
method
-configurations,134
-convolution,50,54,55
-coordinate rotation,134
-criterion,55
-evolutionary,112
-finite difference,220
-finite element,235
-global optimization,105
-gradient,62
-linear convolution,50
-loop resistances,160,163
-loop admittances,158
-minimax,54
-minor alterations,121
-multistart,107
-Nelder-Mead simplex,130
-nodal impedances,158,159
-penalty,68
-Rosenbroke,63
-simulated annealing,110
minimization
-constrained,68
-unconstrained,59,68
-weighting coefficients,50
minimum of approximation
error,101
modification of a diagnosed
circuit,168
multicriterionity,10

multistart,107
multiterminal network,5,157
mutation,117

nervous pulse,87
neural networks,29
- -classification without a teacher,95
- -coefficients,87
- -error of inverse transmission,91
- -error on the network output,92
- -layers,89
- -learning,94
- -learning factor,91
- -learning steps,91
- -learning without a teacher,95
- -learning with a teacher,91
- -matrices of neural network,89
- -self-organizing,95
- -single layer,88
- -three-layer,89
- -testing,5,27

neuron,87
- -activation function,87
- -model,88
- -properties,88

non-uniform,193,200
norm of matrix,147
norms of discrepancies,90
numerical gradient approximation,62,76

objective function,2,60
observability of electric circuit,161
Ohm's law,3
optimal exponent,172
optimal pole shape,234
optimality conditions,248
optimality criteria,285
optimization,
- -criteria,1
- -one-parameter,61
- -parametric,18,22
- -problem,24,41
  - -multicriteria,24
- -topology,200
- -shape,200
- -unconstrained,59

optimization method
- -Broyden-Fletcher-Goldfarb-Shanno,66,237
- -conjugate gradients,67
- -Davidon-Fletcher-Powell,66
- -gradient,62
- -Hook and Jeeves,61
- -multi-criterion,49
- -Newton,65
- -quasi-Newtonian,66
- -quickest descent,62
- -random search,62

parametric
- -identification,166
- -optimization,18,22
- -synthesis,22

parents' generation,112
Pareto,49
- -optimal set,49
- -optimal solutions,49

penalty functions method,68
pole optimum shape,234
population,112
potential,3
- -vector of current,42
- -electric,206
- -scalar magnetic,205
- -vector magnetic,202,205

principle of
- -quasi-stationarity of derivatives,121,136

-repeated
measurements,121,168
problem
-identification,4,5,29,33,43
-minimax,53
-multiextremal,49,52
-two-criterion,54,55
-stiff,122
-synthesis,8,127

quasi-stationarity of
-derivatives “in reverse”,141
-first derivative of
variables vector,139

random search algorithms,61,109
ravine,9,60,132
ravine functional,9,63
ravine functional definition,133
regularization
-parameters,14
-procedure,13
-Tikhonov,14
removing a topological
singularity,173
Roth-Hurwitz conditions,21
roots of the characteristic
equation,123
roulette rule,113,115
saddle point,106
scalarization,50
self-organizing network,97
sensitivity
-of coefficients,147
-of eigenvalues,151,154
-of equation solution,10
-of objective functional,108
separated spectrum,148
sets of
-tree currents
and voltages,158
-cotree currents
and voltages,158
shape of
-ferromagnetic pole,300
-polar tip,47,90,234,236
simply connected region,318
simulated annealing method,110
singularly
perturbed problem,127
small parameters,121,127,147
soft optimization methods,109
solution
-analytical,12,24,71,125,253
-effective,18,49,136
-generalized,16
-normal,14
-non-unique,6,13,17,25,
68,105,177
-pseudo,16
-system
of equations,14,76,141
-weakly effective,49
-of identification
problem,168
sources
-of adjoint variable,202
-external,239
special
-cuts,161,178
-loops,161,179
-split signal,28
splitting stiff problems,125
stabilizer,14
stiffness,8
-of mathematical model,121
-of initial problem,125,154
-system of
differential equations,126
-of state equations,127
style of evolution,113
subcircuits,163,173
-poorly connected,163
synthesis
-parametrical,22,36
-structural,20,35

system of equation
-ill-conditioned,128
-overdetermined,15
-poorly conditioned,168
-stiff,124,126,136
-underdetermined,15

Taylor expansion,
theorem
-Bauer-Fike,147
-Frechet,98
-Green,196
-Vieta,124
Tikhonov's functional,14
time constant,123
Todd's number,161
topological singularities
-of the circuit,161,172,177
-of embedded type,163
trajectory of
quickest descent,63,133
TTL-logic element,92
transfer function,93
transient
-characteristic,1,122,257,284
-process,139,149
-response,128,251
transient conductivity,122
transient process,122
-boundary layer part,128,154
-outside of the boundary
layer part,128,154
transmission line,251
type of repeated experiments,169
types of identification
problems,157

variation
-calculus,323
-of function,323
-of augmented functional,197

Volterra
-amplitude and phase of
polynomials,101
-convergence of series,98
-functional polynomial,98
-functional series,27,28
-kernel,28
-series,28
Volterra - Picard series,27

weak actions,121
winner
-neuron,96
-gets all strategy,96